Polymers and polymer composites in construction

Edited by L. Hollaway

Thomas Telford Ltd, London

Published by Thomas Telford Ltd, Thomas Telford House, 1 Heron Quay, London E14 9XF

First published 1990

British Library Cataloguing in Publication Data
Hollaway, L. (Ed)
Polymers and polymer composites in construction
1. Construction. Use of polymers
I. Hollaway, L.
624.1′892

ISBN: 0 7277 1521 6

Typeset in Great Britain by The Alden Press
Printed and bound in Great Britain by Butler and Tanner Ltd, Frome and London

To Patricia and Suzanne

Preface

This book is intended for engineers, architects, designers and students of engineering who need to acquaint themselves with various aspects of the use of polymers and polymer composites in the construction industry. The engineer/designer must be able to synthesize the information available when planning on paper the construction that is to be built; this includes assessing physical and chemical composition, stress analysis, cost, manufacturing, erection, appearance and performance in use. There must be a greater understanding between the needs of the architect and the requirements of the fabricator to enable the most efficient structural components to be placed in service. The hesitation on the part of the architect and engineer to use these polymer composites and polymers stems from the lack of basic tuition at undergraduate level and the insufficient experience in the design techniques generally and the utilization of the material in practice.

The aim of this book, therefore, is to bring together various aspects of polymers and polymer composites to enable the forward-looking designer to use these unique materials effectively, and, by considering a number of specialist areas, to develop a greater confidence in them. The intention is to present the various structural properties of polymers and polymer composites in a quantitative manner and to show how these materials can be fabricated, designed and used in various areas within the construction industry.

The areas which the book covers include technical details of the mechanical properties of the material, the method of manufacture of composites for construction, and the design analysis of polymeric composites in their various forms, including bonded and bolted joints. Also included is technical information on the inservice properties of polymeric composites, the uses of bearings and resilient seatings, information on geosynthetics and polymers for ropes and for cables, the latter for prestressing concrete, and on polymers for pavement construction.

The various chapters of the book have been written by experts in their field who are responsible for the contents and viewpoints expressed. The authors of Chapter 3 acknowledge the permission of NPL to publish the material in this Chapter and also acknowledge the assistance of Mr K. T. Stevens CEng MRAeS for undertaking the computation in the design case study of Section 3.5.2. The authors of Chapter 11 acknowledge the assistance of ICI Linear Composites Ltd, Dr Gowripalan of the University of Leeds, Dr Hannant of the University of Surrey, Bayer AG and Fibremesh Ltd. The editor is indebted to Dr B. Grieveson, Visiting Reader in the Department of Civil Engineering University of Surrey and formerly of Shell International Chemical Company, London, for his help and advice.

The information in this book has clearly been derived from many sources including individuals, firms and organizations and the various authors have attempted to acknowledge, in the appropriate parts of the text, the assistance that they have been given. It would be quite impossible for them to express their thanks to everyone concerned for their co-operation in producing this book but on their behalf the editor would like to extend due gratitude.

Leonard Hollaway
University of Surrey
January 1989

List of contributors

D. B. S. Berry	Fulmer Yarsley, Redhill, Surrey
C. J. Burgoyne	Engineering Department, University of Cambridge, Cambridge
J. R. Crowder	Building Research Station, Garston, Watford
P. Grootenhuis	Department of Mechanical Engineering, Imperial College of Science and Technology, London
P. R. Head	G. Maunsell and Partners, Consulting Engineers, London
L. Hollaway	Department of Civil Engineering, University of Surrey
C. M. Howard	Department of Civil Engineering, University of Surrey
A. F. Johnson	National Physical Laboratory, Teddington, Middlesex
C. R. Lawson	Exxon Chemical Geopolymers Ltd, Gloucestershire
A. J. Leggatt	NCL Stewart Scott, London
A. Marchant	Associated and Marine Technology, Romsey, Hampshire
R. Templeman	G. Maunsell and Partners, Consulting Engineers, London

Contents

1 Introduction

L. Hollaway

The construction industry has always been associated with heavy bulk material and it is therefore difficult to realize that about 30% of all polymers produced each year are used in the civil engineering and building industries. Polymers offer many advantages over conventional materials, including lightness, resilience, resistance to corrosion and ease of processing. They can also be combined with fibres to form composites which have enhanced properties, enabling them to be used as structural members and units.

Polymers and polymer composites are used in many different forms, ranging from geosynthetics through to structural composites in the construction industry and to the high technology composites of the aerospace and space satellite industries.

The strength of materials which have to support in-plane loads in whatever form should be within required limits. If, in addition, the unit has to support bending and shear forces, then the material must possess adequate stiffness. In the civil engineering and building industries it is generally essential that the shape of the unit or structure under the design load should not deform greatly, and as the stiffness of the material and the shape of the unit or structure are generally proportional to the deformation, the engineer requires materials with relatively high strength.

From an engineering point of view it is desirable, although not entirely essential, that the stress–strain relationship of the material should be similar whether it is under an increase or a decrease of load. If it is, then an elastic relationship exists; also, if the relationship is linear, then the material is said to be linear elastic. The term "linear elastic" would normally imply that deformation would cease when the load has been applied to the material. If, however, deformation continues to increase after the application of the load, the material is said to be undergoing creep. This could be detrimental to the material or to the structure using it if the design is incorrect.

Creep does occur in most materials to a certain degree, and it becomes greater the higher the stress level in the structural component or the higher the temperature of the environment in which the material is placed.

The shape and the structural requirement of the unit are invariably related to the physical properties demanded of the component which, in turn, are associated with the properties of the material used. For instance, a long span bridge would not be manufactured from a low modulus material with poor shear strength, and likewise a high modulus material would not be used in the manufacture of a tension cable roof. A further observation is that most materials of construction are manufactured by man; two notable exceptions are stone and timber. The manufactured materials may be placed broadly under two headings

(*a*) the homogeneous materials which on a macroscopic scale may be considered isotropic

(*b*) the composite materials which consist of two or more components, each one being used to its maximum benefit.

Both groups of materials are of significance in the context of this book; specifically, the polymeric and elastomeric materials come under the first heading and the composite materials, which are described as fibre-reinforced polymer matrix materials, come under the second one.

Plastics, which are described in detail in chapter 2, are generally considered to be a relatively recent development. In fact, they are members of the much larger family of polymers, which is the basic component of animal life.

Until the beginning of the 20th century, the materials used in the construction industry were basically inorganic. During the 20th century, organic materials have been developed. Phenol-formaldehyde (Bakelite) was the first completely synthetic plastic produced and it was named after the American chemist Leo Baekeland. In the 1930s materials such as nylon, polyethylene and acrylic were produced but during the latter part of the 1940s plastics acquired the reputation of being a cheap substitute for many other materials. However, nowadays their reputation for being high technology materials with special mechanical and inservice properties has been redeemed.

During the Second World War, radomes were constructed from glass fibre/polyester composites and since then considerable advances have been made in the use of this material. Load-bearing and infill panels have been manufactured using composites and Fig. 1.1 shows an example of this system. Also, complete structures have been fabricated where units manufactured from glass-reinforced polyester are connected together to form the complete system in which the shape provides the rigidity. An example of this form is shown in Fig. 1.2. The icosahedron geometric shape which was chosen to form the prototype classroom was constructed from a series of identical pyramids manufactured from the same mould. Glass-reinforced plastics have been used in many other structural forms including the material of construction for pressure pipes, and in their unreinforced polymeric state have been used as drainage and ducting etc.

Fibre-reinforced polymeric materials are now being considered for use in more sophisticated systems, such as double-layer skeletal structures, reinforcement for concrete beams and mooring cables.

Since the beginning of the 1980s, geotextiles technology has made great

Fig. 1.1. Load bearing infill panels at London Brixton railway station; architects B.R. Croydon (courtesy Brensal GRP Products)

Fig. 1.2. Prototype classroom in GRP, developed by Lancashire County Council, analysed by Composite Structures Research Unit, University of Surrey

Fig. 1.3. The use of geotextiles in construction

advances into the civil engineering industry and it is difficult now to find a construction site which does not use geotextiles in some form to provide, for example, drainage channels or soil and embankment stabilization, or geomembrane for situations requiring an impervious membrane. Fig. 1.3 shows a construction site on which geotextiles are being used.

In contrast to the use of the conventional design procedures of the more traditional materials of construction, that of polymer and polymer composite materials for structural applications requires greater development effort and wider understanding of the material. It will become evident, after reading the following chapters, that this is due to the fact that the material properties of the final component are the result of a design process which must consider many topics. These latter are characterized by the anisotropic behaviour of the material and cover the micro-mechanical, elasticity, strength and stability properties; these will be influenced by manufacturing techniques, environmental exposure and loading histories. It will become apparent that designing with composites is an interactive process between the design of the constituent materials used, the design of the composite material and the understanding of the manufacturing technique for the composite component; these three elements of design must be considered simultaneously.

The following chapters deal specifically with various aspects of polymers and of polymer composites from the point of view of the properties of the materials and the uses to which these are put, the analyses and design of composites, the inservice properties of the material and the construction and erection of the components.

The contents of this book are directed mainly towards the construction industry but the principles used in the analysis and the design procedures are relevant to all composite engineering designs.

It is hoped that the subsequent chapters will provide the architect and engineer with an understanding of the basic principles of polymers and polymer composites and will demonstrate to them how the material can be utilized in engineering.

2 Polymer, fibre and composite material properties and manufacturing techniques

L. Hollaway

2.1. Introduction

Composite fibre/matrix materials are not new to the aircraft and allied industries and they are now becoming important in the construction industry with great potential in certain areas.

The component parts of composite materials consist of two or more distinct physical phases, one of which, the fibrous, is dispersed in a continuous matrix phase. Only the polymer composites will be considered here and those to be discussed will range from the elastomeric to the rigid plastics and from the extruded and drawn polymer fibres to those made by sheet fibrillation.

Composites and polymers offer the designer a combination of properties not available in other materials. It is possible to introduce the fibres in position, direction and volume in the matrix material at highly stressed regions in a certain position, direction and volume in order to obtain the maximum efficiency from the reinforcement and then, within the same member, to reduce the reinforcement to a minimal amount at a region of low stress value. Other advantages offered by the material are lightness, resistance to corrosion, resilience, translucency and greater efficiency in construction compared with the more conventional materials.

translucence: imperfect transmission of light

2.2. Polymeric materials

Polymers are the products of combining a large number of small molecular units called monomers by the chemical process known as polymerization (which is the process by which molecules or groups of atoms are joined together) to form long-chain molecules. Natural materials such as bitumen, rubber and cellulose have this type of structure. There are two main types of polymerization. In the first type, a substance consisting of a series of long-chain polymerized molecules, called *thermoplastics*, is produced. All the chains of the molecules are separate and can slide over one another. In the second type, the chains become cross-linked so that a solid material is produced which cannot be softened and which will not flow. Such solids are called *thermosetting polymers*. These two groups classify polymer materials.

Polymers are usually made in one of two polymerization processes. In condensation polymerization the linking of molecules creates by-products, usually water, nitrogen or hydrogen gas. In addition polymerization no by-products are created.

The final property of the plastics material will be determined by the type of additive mixed with the pure polymer. Additives may be simply fillers or extenders designed to reduce the quantity of polymers used or to toughen the product. Pigments or stabilizers can be added to provide colouring for the material or to reduce degradation, and plasticizers are incorporated to alter the characteristics of the polymers. The glossary outlines the purpose of the main additives used with polymers.

2.2.1. Thermoplastic polymers

Thermoplastic polymers consist of linear molecules which are not interconnected. The chemical valency bond along the chain is extremely strong, but the forces of attraction between the adjacent chains are weak. Because of their unconnected chain structure, thermoplastics may be repeatedly softened and hardened by heating and cooling respectively; with each repeated cycle, however, the material tends to become more brittle.

The thermoplastic group of materials can be further subdivided into the groups: crystalline ordered structure; amorphous random structure. It is not possible in practice to mould a polymer which has a pure crystalline structure due to the complex physical nature of the molecular chains; consequently the system would be described as semi-crystalline; nylon is an example of this group. An example of an amorphous thermoplastic polymer is polymethylmethacrylate (acrylic sheet). Further examples and use of various types of thermoplastic polymer are given in chapters 10 and 11.

Recently a large number of high performance, high technology polymers (for example polyethersulphone, PES, and polyetheretherketone, PEEK) have been developed. The inservice properties of these polymers are vastly superior to those of the normal thermoplastics. In addition, the softening temperature of the former group is much higher than that of the latter. For instance the softening temperatures of PES and PEEK are 280°C and 350°C respectively while that of acrylic is 100°C.

2.2.2. Thermosetting polymers

The thermosetting polymer is formed by a chemical reaction. In the first stage, a substance consisting of a series of long-chain polymerized molecules, similar to those present in thermoplastics, is produced. In the second stage of the process, the chains become cross-linked; this reaction can take place either at room temperature or under the application of heat and pressure. The resultant material will not flow and cannot be softened by heating. The thermosetting polymers which are used in the construction industry are the polyesters and the epoxides; these are discussed in depth in sections 7.3.1 and 7.3.2, respectively.

2.2.3. Foamed polymers (cellular rigid foam)

Rigid foams are a two-phase system of a gas dispersed in a solid polymer. The system is produced by dispersing an inert gas throughout the molten resin directly before moulding. This is achieved by adding to the resin formulations a chemical blowing agent which, during the exothermic reaction of polymerization, releases a gas which causes the polymer to expand, increasing its original volume many times by the formation of small gas cells.

Thermoplastic and thermosetting polymers may be foamed and the foamed polymers will share the advantages and limitations of the solid phase. Foam polymer systems may also include fibres in the material to improve their mechanical properties.

Applications of foamed polymers are given in sections 2.6.2 and 11.5.

2.2.4. Elastomers

Elastomers, the commonest of which is rubber, consist of long-chain molecules and are members of the polymer family. The long chains are coiled and twisted in a random manner and the flexibility is such that the material is able to undergo very large deformations. In the uncured state the elastomers are unable to recover completely from large deformations due to the sliding of the molecules over each other during load. After a curing process (known as vulcanization) the molecules are cross-linked and are thus held firmly together. This process is similar to the cross-linking of thermosetting polymers. As the vulcanization prevents sliding of the molecules relative to each other but does not change the form of the coiled molecules, the elastomeric material will completely recover its original shape after the removal of a force.

The properties of vulcanized rubber are wide and include resilience, flexibility at low temperatures, and resistance to oils, greases, ozone and many acids and bases. A new family of thermoplastic rubbers has been developed. The advantages of these over the conventional rubbers are

(*a*) the material can be melted and moulded more easily
(*b*) the chemical composition can be varied in order to produce a novel range of properties thus making them adaptable to a wider variety of uses than conventional rubbers; thermoplastics rubbers (TPR) (elastomers) are produced by block polymerization of a mixture of substances such as butadiene or isoprene and styrene.

2.3. Fibres

If polymers are to be used as engineering materials, their stiffness and strength must frequently be enhanced. This is achieved by the addition of reinforcing fillers. The resulting combination of materials is known as a composite. A composite material is made from two or more components, generally a low modulus, low strength material (e.g. a thermosetting resin, a thermoplastic resin, an elastomer) and a high strength, high stiffness reinforcing-fibre material (e.g. glass, carbon, Kevlar).

A wide range of amorphous and crystalline materials can be used as the fibre. In the construction industry the most common fibre used is glass fibre. Carbon fibre can be used separately or in conjuction with the glass fibre as a hybrid to increase the stiffness of the structural member or the area within the structure, so that the stiffness exceeds the value which would be possible using the glass fibre only. Kevlar fibres can be used instead of glass fibres to give increased stiffness to the composite.

2.3.1. Glass fibres

Glass fibres are manufactured by the continuous rapid drawing of molten glass from an electrically heated furnace through platinum bushings at high speed. Their manufacture is shown in Fig. 2.1. The filaments cool from the liquid state, at a temperature of about 1200°C, to room temperature in approximately 10^{-5} s. On emerging from the bushings, 204 filaments are bundled together and bonded to one another by a lubricant or size. This process is known as sizing and has three main functions

- to reduce the abrasive effect of filaments rubbing against one another
- to reduce the damage to fibres during mechanical handling
- to facilitate the moulding process.

There are four types of glass fibre.

(*a*) E-glass, of low alkali content, is the commonest glass in the market and is used in the construction industry. It was not until the advent

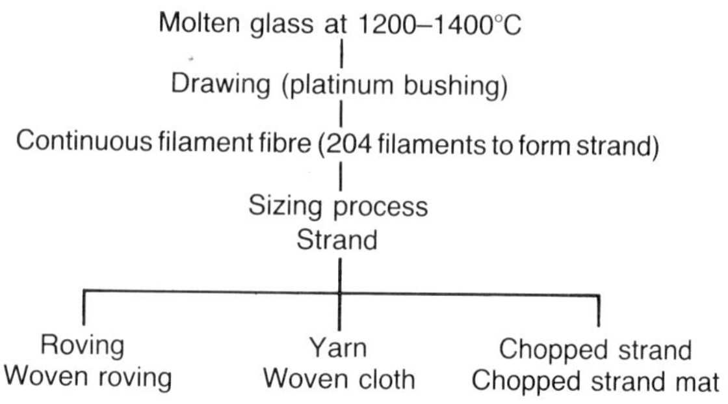

Fig. 2.1. Schematic diagram for the manufacture of glass fibre

of polyester resin in 1942 that this glass was used successfully. Now it is employed widely, especially with polyester and epoxy resins.

(*b*) Z-glass (zirconia glass) has been developed in fibre form to be used as a reinforcing fibre for cements mortars and concretes. This glass has a high resistance to alkali attack.

(*c*) A-glass of high alkali content was formerly used in the aircraft industry but is now gradually going out of production.

(*d*) S-glass fibre is produced for extra high strength and high modulus applications in aerospace technology and space research.

Bundles of 204 filaments or multiples of these are called strands and these are usually combined to form thicker parallel bundles called rovings. Strands may also be twisted to form several types of yarn; rovings or yarns may be used either individually or in the form of woven fabric. A great variety of fabrics is available, most of them being plain or square weave, having identical characteristics in the weft and in the warp direction.

Glass strands for reinforcing thermosetting resins may be used in a number of different forms.

(*a*) *Chopped fibres* may be subdivided into milled fibres, short chopped fibres and long chopped fibres. Milled fibres are finely ground fibres which have lengths of 30 to 3000 microns. Their aspect ratio is generally about 30. They are used mainly as the reinforcement for the injection moulding process. Short chopped fibres have lengths up to about 6 mm and their aspect ratio is generally about 800. Long chopped fibres have lengths up to about 50 mm and are used in the manufacture of composites by the sheet moulding compound (SMC), dough moulding compound (DMC) (see section 2.8.2.2.) and the spray-up technique. The fibre lengths for the DMC and the SMC processes would be about 3 mm and 12 mm, respectively, and for the spray-up technique would be about 50 mm length.

(*b*) *Chopped strand mats* are manufactured from chopped strands which are randomly orientated in the form of a mat. The glass strands are loosely bonded together with a resinous binder which must be compatible with the laminating resin. The composites which utilize chopped strand mats would be manufactured by the hand lay-up technique or by resin injection (see sections 2.8.1.1. and 2.8.2.4.).

(*c*) *Woven rovings* are used in composites to produce high directional strength characteristics. Unidirectional roving composites have great strength in one direction. This is achieved by using a high percentage of roving for the warp direction and a small percentage for the weft direction. Bidirectional roving composites have high strength properties in two directions at right angles to each other.

(*d*) *Surface tissue* is a thin glass-fibre mat bonded with a readily wetted medium. It is used when a resin-rich surface is required or when the coarse fibre pattern of a chopped strand mat is to be concealed.

2.3.2. Carbon fibre

There are two general sources for the commercial production of carbon fibres: synthetic fibres, similar to those used for making textiles, and pitch, which is obtained by the destructive distillation of coal.

The textile fibre polyacrylonitrile (generally known as PAN) is a synthetic which is spun as a bundle of continuous filaments which, in the process, are stretched so that their molecular chains are aligned parallel to the fibre axis. It is essential that this parallel orientation is preserved throughout the carbonization process.

The preliminary step in the manufacture of carbon fibre is to heat the PAN fibre under tension in air at a temperature of 250°C; this process prevents shrinkage. During this heating, the fibre will absorb oxygen and gain in strength. It will change its colour from white to yellow and then to brown and it will eventually form a black polymer containing nitrogen. At this stage the fibre, unstressed, is strongly heated in an inert atmosphere. This step is known as carbonization and leaves the fibre free of hydrogen and nitrogen atoms. The filament now consists of a relatively crystalline carbon embedded in amorphous carbon. To increase the stiffness of the carbon filament, it can be further heated up to 2200°C, which enlarges the crystalline region at the expense of the amorphous one. By varying the process conditions, mechanical property modifications can be achieved.

There are three grades of carbon fibre

(*a*) type I which is the stiffest fibre and will have the highest modulus of elasticity value of the three
(*b*) type II which is the strongest fibre
(*c*) type III which is the least stiff of the three fibres and has a strength mid-way between type I and type II fibres.

2.3.3. Kevlar fibre

Kevlar is an aromatic polyamide and its generic name "aramid" is coined from these two words. In addition to the Kevlar used in the rubber industry, two variations of the fibre are manufactured for special application; these are Kevlar 29 and Kevlar 49.

Kevlar 29 is mainly used in woven and twisted form for personnel protection, friction materials and woven fabrics and for ropes and cables. This fabric is not suitable for reinforcement purposes as the surface treatments applied are not compatible with most resin systems.

Kevlar 49 has a higher modulus and is designed for reinforcement applications. When used with polymers to form a composite, it has up to 35% and 10% advantage in laminate weight compared with glass fibre and carbon fibre respectively. Composites of Kevlar 49 and polymer matrix display good resistance to fatigue, weathering and chemical attack. The fibre is flame resistant but will char and decompose at temperatures above 425°C.

To enhance the properties of the composite over single reinforcement systems, Kevlar 49 can be combined with glass and/or carbon fibres in hybrid form. To form composites in single reinforcement systems, woven or unidirectional reinforcements can be combined with epoxy vinyl-ester, isophthalic polyester or phenolic resins by using the prepreg systems, the hand lay-up, the resin-injection and the filament-winding techniques (these are described in sections 2.8.1 and 2.8.2).

2.4. Composites

The mechanical properties of polymers can be greatly improved by using techniques from nature. Few natural materials consist of one substance only: most are a mixture of different components which, when combined, produce a material with enhanced properties. Bone, for example, achieves its lightness and strength by combining crystals of apatite with fibres of the protein collagen. This combination of two or more materials is known as a composite.

Polymers are often combined with fillers and/or fibres to improve their mechanical or physical properties. The fillers usually consist of wood flour, china clay, quartz powder or other powdered minerals. The filler is incorporated not only to improve the physical property of the composite but sometimes to reduce the polymer content and hence the cost.

The use of fibres in combination with polymers enhances the latter's mechanical properties for structural use. Most fibres can be used, provided there is chemical compatibility between the two parts.

Both thermoplastics and thermosetting polymers can be combined with fibre reinforcement. In the past, thermoplastics have been combined mainly with short fibres but these composites were unable to take full advantage of the fibre strength (section 2.10.1.3). However, users have taken advantage of this system's toughness property by manufacturing a range of injection mouldings. Conversely, the thermosetting composite system tends to be brittle, because of the cross-linking of the matrix. Long fibres have been utilized in the composite (section 2.10.1.2), to take full advantage of the strength of the fibre. This system has found application in large panels and pressure pipes and containers.

Currently, however, the traditional market areas of the thermoplastics and thermosetting polymer composites are less well defined and both long and short fibres are being used with thermoplastics and thermosetting matrices.

2.5. Mechanism of reinforcement in composite materials

The reinforcement of a low modulus matrix with a high modulus, high strength fibre uses the plastic flow of the matrix material under stress to transfer the load to the fibre; this results in a high strength, high modulus composite. The aim of the combination is to produce a two-phase material in which the primary phase (i.e. the fibres, which determine stiffness and strength, are in the form of particles of high aspect ratio) is well dispersed and bonded by a weak secondary phase (i.e. the matrix). The principal constituents which influence the strength and stiffness of composites are the reinforcing fibres, the matrix and the interface between the fibres and the matrix.

2.5.1. Fibres

The functional requirements of fibres in a fibre/matrix composite are that they should have

(*a*) a high modulus of elasticity to give stiffness to the composite
(*b*) a high ultimate strength
(*c*) a low variation of strength between individual fibres
(*c*) stability during handling
(*e*) a uniform diameter.

2.5.2. Matrix

The matrix should fulfil certain functions. These are

(*a*) to bind the fibres together and to protect their surface from damage during the service life of the composite
(*b*) to transfer stresses to the fibres efficiently by adhesion and/or friction
(*c*) to disperse the fibres and to separate them
(*d*) to be chemically and thermally compatible with the fibres.

2.5.3. Interface

The interface between the fibre and the matrix is an anisotropic transition region exhibiting a graduation of properties. It is the component through which the transfer of stresses between fibre and matrix occurs. It must provide adequate chemically and physically stable bonding between the fibres and the matrix. Its functional requirements vary considerably according to the performance requirements of the composite during its various stages under service conditions.

There are various techniques for the manufacture of polymer components and composites. The processes can be broadly divided into those suitable for the thermoplastic polymers, the thermoplastic composites and the thermosetting composites.

2.6. Processing methods for the manufacture of unreinforced thermoplastic polymers

The principal processing methods for thermoplastic polymers fall under four main headings: extrusion process, injection moulding, thermoforming process and calendering process.

2.6.1. Extrusion process

The polymer is fed from a hopper into a screw which is inside a heated barrel. At the end of the extruder the polymer passes through a die, and is extruded into the desired shape. The extrusion process may be used to

(*a*) convert polymer formulations and additives into granules
(*b*) produce continuous lengths of plastic mouldings of uniform cross-section.

The extrusion technique is very versatile and can be adapted to produce a wide range of products, merely by using the appropriate dies. Some of the production processes which are based upon the extrusion process are

- profile production
- film-blowing plastic sheet
- blow-moulding hollow plastic articles
- extrusion blow-moulding hollow plastic articles
- co-extrusion moulding in which two or more polymers are combined in a single process to produce a multi-layer film of the required mechanical properties
- extrusion of sheets for highly orientated grids.

For the civil engineer, by far the most important extrusion process is the extrusion of sheets for highly orientated grids. This technique is used for the

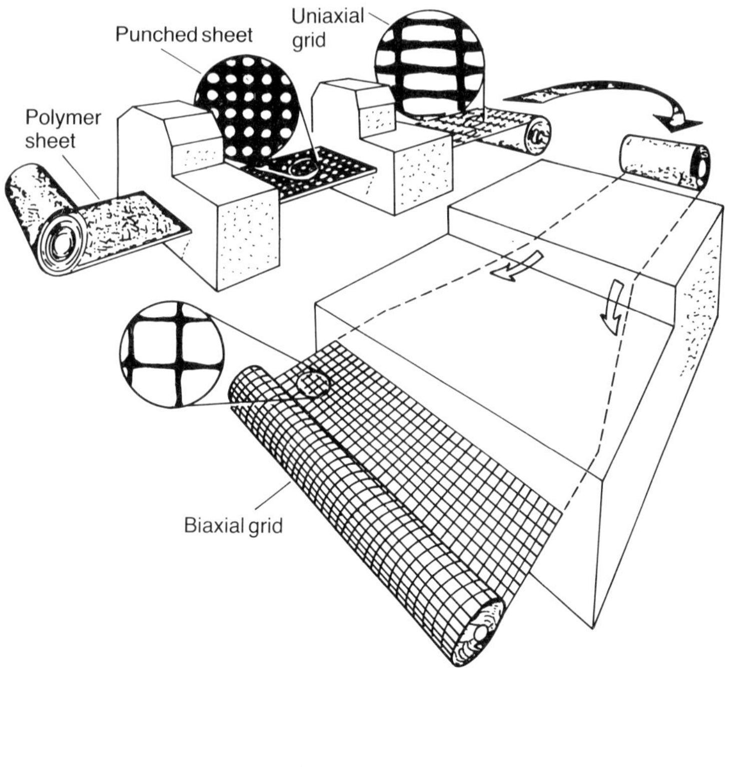

Fig. 2.2. Manufacturing process for sheet-formed grids[2.17]

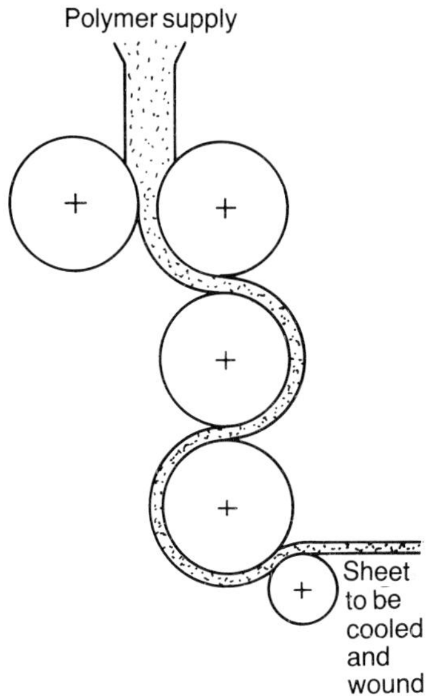

Fig. 2.3. Typical arrangement for calendering

manufacture of high strength polymer geogrids as a non-corroding soil-reinforcement material in which the open grid structure permits load transfer by positive interlocking of the soil-fill into grid apertures.

The polymer grid derives its high strength from the orientation of the polymer molecules during manufacture. The manufacturing process is shown in Fig. 2.2. In this particular process, a continuous sheet of polymer, usually high density polyethylene or polypropylene, has a pattern of holes stamped into it. The perforated sheet is then stretched in one direction to align the long chain molecules, giving a high tensile strength in this direction. The sheet then undergoes a second stretching operation in the transverse direction. The resulting product is a highly orientated bidirectional grid sheet that is strong and stiff in the two orthogonal directions. The technique is discussed from a practical point of view in sections 10.3.2.2 and 10.4.3.

2.6.2. Injection moulding

The principle of injection moulding is simple. Softened thermoplastic polymer is forced through a nozzle into a clamped cold mould. When the plastic has become cold, the mould is opened and the article is ejected; the operation is then repeated. The process may be subdivided into

(*a*) sandwich moulding, which provides mouldings with a rigid and stiff face material and a low density core material
(*b*) reaction injection moulding, where the liquid reactants are combined prior to being injected into the mould; polymerization then takes place inside the mould
(*c*) injection blow-moulding, in which molecular orientation of the polymer is introduced to the moulding; in this case the properties of the material are significantly improved.

2.6.3. Thermoforming process

Thermoforming techniques shape heat-softened but not molten thermoplastic polymers. There are basically two subdivisions

(*a*) vacuum forming, where the thermoplastic sheet material is heated and then shaped by reducing the pressure between it and a mould
(*b*) pressure forming, where the heat-softened thermoplastic sheet is shaped by an applied pressure above the sheet.

2.6.4. Calendering process

In the calendering process the thermoplastic material is squeezed through counter-rotating cylinders to produce thin sheet or film of plastic material. It is important that the accuracy of the roll temperature is maintained: this varies between 100° and 300°C depending on the polymer type. The facility to adjust the rotational speed of each roller independently is also important. Calendering can achieve a tolerance of 0·005 mm in the thickness of the sheet. Fig. 2.3 shows a typical arrangement of calender rolls. In the civil engineering industry the calendering process is used to enhance some of the properties of the non-woven geotextiles as discussed in section 10.3.1.2.

2.7. Processing methods for the manufacture of reinforced thermoplastic polymers

2.7.1. Injection moulding technique

Fibre-reinforced thermoplastics can be moulded by many of the processes stated above, but the injection moulding technique is the most important. The length of the fibres used in the various processes is between about 0·2 and 0·4 mm; consequently the full loading capacity of the fibre is not developed. Various improvements have taken place over recent years to overcome the short fibre problem by producing continuous fibre tapes or mats which can be embedded into the thermoplastic matrix (e.g. the aromatic polymer composite, APC, and the glass mat reinforced thermoplastic, GMT). A composite of carbon fibres in a matrix of polyethersulphone (PES) can be made by arranging wide tapes of carbon/PES in aligned unidirectional layers in a matched split mould. The mould is then closed and pressure applied.

2.7.2. Film-stacking technique

The film-stacking process may be used to produce composites in which the fibres are manually aligned in the desired direction.

This process consists of interleaving lightly impregnated fibres of fabric with thin films of pure thermoplastic polymer (polyethersulphone or polyetheretherketone) in a split mould. When the required stacking sequence has been achieved, the two split moulds are brought together and heat and pressure are applied. The result is a high quality laminated thermoplastic composite of superb appearance.

This process is utilized in the manufacture of high technology composites for the aerospace and space industries. The materials are advanced thermoplastics polymers such as polyetheresulphone (PES) and polyetheretherketone (PEEK). Further discussion on these systems is given in chapter 11.

2.8. Processing methods for the manufacture of fibre-reinforced thermosetting polymers

There are various techniques for the manufacture of fibre-reinforced thermosetting polymers and these may be considered under two main headings

(*a*) the open mould process in which, during, the moulding operation, the material is in contact with the mould on one surface only
(*b*) the closed mould technique in which the composite is shaped between the male and female moulds.

Both the open and closed mould process can also be divided into three categories: manual, semi-manual and automatic. The manual techniques include the hand lay-up, spray-up and pressure bag. The semi-manual techniques cover the cold press, the hot press and the resin-injection methods. The automatic techniques include pultrusion, filament winding and injection moulding.

2.8.1. Open mould systems

These production methods take full advantage of an important characteristic of the polyester resin: it does not require heat or pressure for complete polymerization to occur. The resins are known as cold cure resins. There are many variations of the open mould process, but only principal ones will be given here.

2.8.1.1. Hand lay-up technique

In this technique only one mould is used and this may be either male or female. Most materials are suitable for mould making, but probably the most

Fig. 2.4. Hand lay-up moulding method

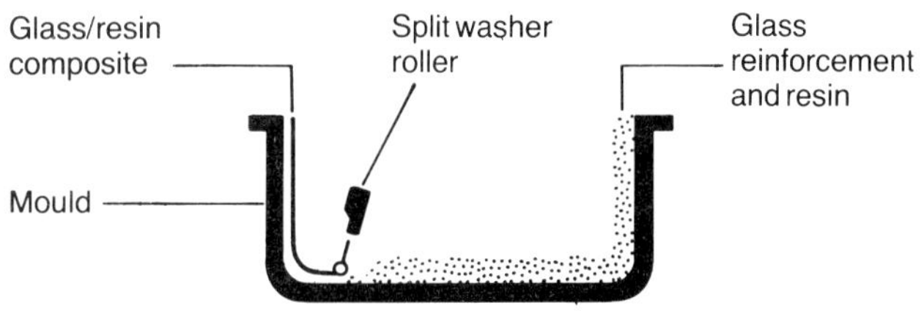

common one is glass-reinforced polymer (GRP). A suitable masterpattern is prepared and from this GRP moulds may readily be made. Fig. 2.4 shows the hand lay-up operation. To prevent bonding of the GRP components, a release agent is applied to the mould and is then allowed to dry before any lay-up is undertaken.

The durability of GRP composite in the construction industry is dependent upon the quality of the surface which is exposed to the atmosphere and it is necessary to protect the fibres on the exposed surface of the composite with a resin-rich coating, known as a gel coat. The function of the gel coat is

(*a*) to protect the glass fibre from external influences, the main one being moisture penetration to the interface of the fibre and matrix, with consequent breakdown of the interface bond
(*b*) to provide a smooth finish and reproduce precisely the surface texture of the mould.

The thickness of the gel coat, which is sprayed or painted on to the mould surface, should be controlled to about 0·25 mm and should be uniform throughout. If uneven thicknesses do occur, different cure rates over its surface will result, with induced stresses in the resin coat causing crazing. On occasions, a surface tissue mat is used to reinforce the gel coat.

After the gel coat has become tacky but firm, a liberal coat of resin is brushed over it and the first layer of glass reinforcement is placed in position and consolidated with a roller. The glass fibre may be in the form of chopped strand mat or woven fabric, which is precut to the correct size. Subsequent layers of resin and reinforcement are then applied until the required thickness of the composite is reached. Curing of the composite can take place at room temperature in the fabrication shop; this would normally take about 28 days at 20°C. It is bad practice to place units in a yard at the mercy of the elements immediately after demoulding. Instead, it is recommended that at the latter stage units should be placed in an oven at say 60°C for 8 hours. An oven could take the form of heaters placed in a foil enclosure.

To facilitate a greater compaction of the polymer into the fibre, one of two improvements can be made to the basic hand lay-up process. These are: vacuum bag moulding; pressure bag moulding. In the first process, a flexible bag is placed over the lay-up in the mould and clamped around the latter's edge. A negative pressure is then applied which forces the bag onto the surface of the wet composite, thereby improving the latter's consolidation as well as expelling excess air and resin. The second process achieves the same result as the first method, but here a positive pressure is applied above the bag, forcing it on to the wet lay-up.

2.8.1.2. Spray-up technique

The spray-up technique is shown diagramatically in Fig. 2.5. The preparatory stages of this method are similar to those of the previous one, but the actual technique of applying the resin and fibre to the mould is less labour-intensive. During the spray-up operation, glass-fibre roving is fed continuously

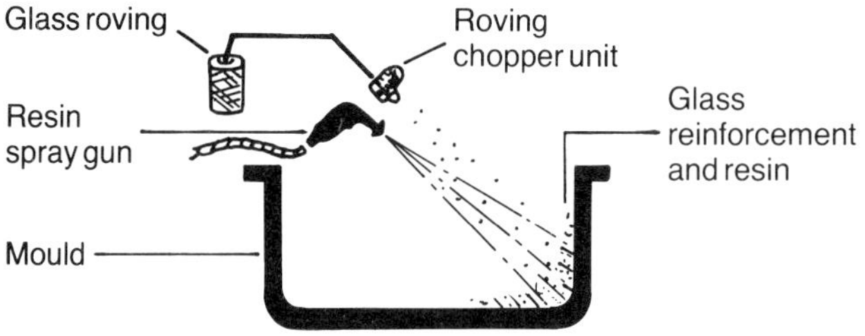

Fig. 2.5. Spray-up moulding technique

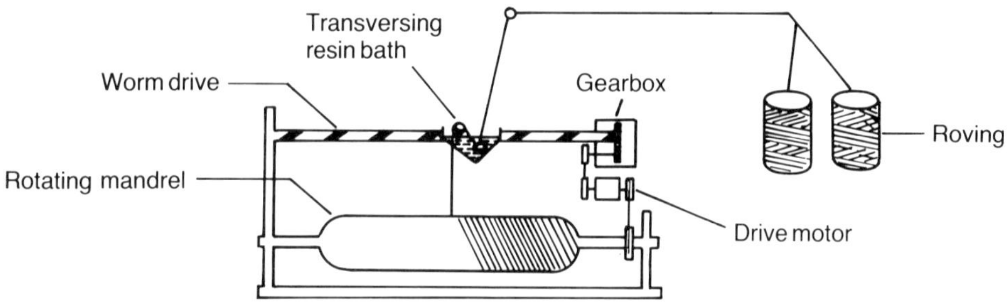

Fig. 2.6. Filament-winding technique

through a chopping unit and the resulting chopped strands are projected on to the mould in conjunction with a resin jet. The glass-fibre resin matrix is then consolidated with rollers. The technique requires considerable skill on the part of the operator to control the thickness of the composite and to maintain a consistent glass/resin ratio.

The comments made regarding the curing of the finished composite manufactured by the hand lay-up process are also relevant to this process.

The hand lay-up and the spray-up techniques are relatively simple and low tooling costs can result in considerable versatility from the designer's point of view. Very large mouldings have been produced by these methods, but particularly by the second one.

In the construction industry these two methods are used to form load-bearing and infill panels for buildings and face materials for sandwich panels.

2.8.1.3. Filament-winding technique

In this technique, continuous rovings or single strands are passed through a bath of activated resin and then, via a comb fitted with a swivelling head, are wound on to a rotating mandrel. Fig. 2.6 illustrates the process. It can be seen that the angle of helix is determined by the relative speeds of the traversing bath and mandrel, thus enabling the maximum benefit of the fibre strength to be utilized. After the winding process, the resin is hot cured, either by means of heating lamps fitted to the machine or by passing the mandrels along a curing oven. The process has now become highly automated through the development of computer-controlled winding equipment. This method of manufacture can be used to produce pressure pipes, pressure vessels and load-bearing tubes.

2.8.2. Closed mould systems

For many years fibre/matrix composites have been produced by moulding techniques using matched dies. These techniques produce mouldings with a good finish on both surfaces of the composite. Compared with the open mould process, this system enables a higher glass content to be used in the composite, with improved mechanical properties.

2.8.2.1. Cold-press moulding technique

In the cold-press moulding technique, unheated matched tools under pressure are used to impregnate the resin throughout the fibre reinforcement which has been placed in the mould. A release agent and gel coat are applied to the surface of the moulds before the fibre and resin are placed in position.

2.8.2.2. Hot-press moulding technique

In the hot-press moulding technique, the fibre-filled polymer is confined between heated, matched, polished metal dies (temperature approximately

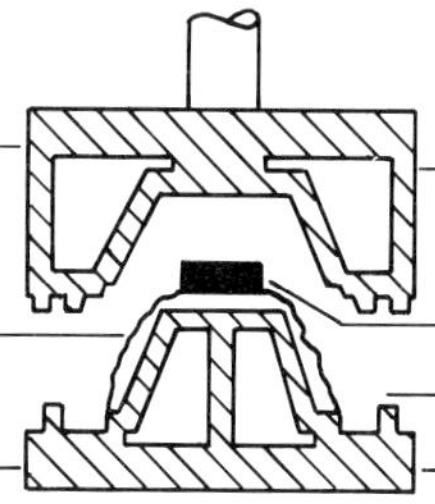

Fig. 2.7. Hot-press moulding technique

180°C) brought together under pressure (pressure approximately 15 MN/m^2). Fig. 2.7. shows a typical hot-press mould.

Both pre-form moulding and compression moulding are used in this technique.

(*a*) *Pre-form moulding.* In this system, chopped rovings are projected on to a rotating fine-mesh screen which is approximately shaped to the profile of the article being moulded. The strands are bound together by spraying the pre-form with a resinous binder in the form of a powder or an emulsion and the whole is transferred to an oven at about 150°C for 2 or 3 min, after which time the pre-form is ready for the press.

(*b*) *Sheet moulding compound (SMC).* Sheet moulding compound (or prepreg) is a resin-based polymer (usually polyester) moulding material with chopped strands which vary in length between 12 to 55 mm and in content between 20% and 35% by weight. They are produced and supplied in the form of a continuous sheet wound into a roll and protected on both sides by sheets of polyethylene film. The latter are removed before loading into the press. The desired composite shape and rapid cure are obtained by the application of heat and pressure in suitable tools.

As the material flows uniformly to produce a homogeneous composite, complex and deep draw mouldings can be produced. It is necessary to comply with the correct conditions for moulding. These include using a suitable press and a mould designed specifically for the material, charging the mould in the correct manner and using the optimum temperature, pressure and curing time. Moulding pressures of between 3·5 and 5 MN/m^2 and moulding temperatures of between 125° and 155°C are generally required; the lower pressures and temperatures are for moulding simple flat shapes and the higher values for the more complex mouldings.

(*c*) *Dough moulding compound (DMC).* Dough moulding compound (or pre-mix) contains an unsaturated polyester resin, an unsaturated cross-linking monomer such as styrene, suitable mineral fillers and fibrous reinforcement which is usually composed of chopped strands. The fibres vary in length between 3 and 12 mm and in content between 5% and 20%. Because DMC flows readily, it may be moulded by compression, transfer or injection. The pressure required to produce a component is relatively low, so that large mouldings can be produced easily. These compounds do not give such high composite mechanical properties as the SMC due to the geometry of the fibres (see section 2.10.1.3), but they can be used for intricate mouldings.

The hot-press moulding technique can be used to produce certain forms of street furniture etc. The requirement, from an economic point of view, is that the high grade of products produced are for a volume market.

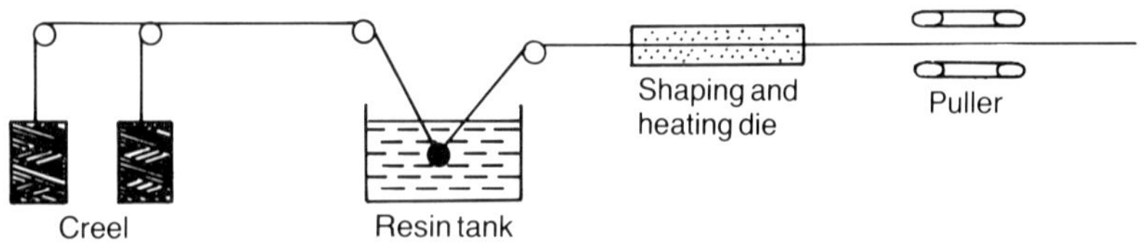

Fig. 2.8. Pultrusion technique

2.8.2.3. Pultrusion

A pultrusion machine is designed to produce a continuous length of composite material with a uniform cross-section. Both solid and hollow profiles can be made.

The pultrusion technique is one in which continuous strands of a reinforcing material, together with any additional required layers of fabric, are drawn from creels and are passed through a resin bath fitted with a constant level device. After picking up an excess of matrix, the saturated fibre is passed through a number of wiper rings and into the mouth of a heated die.

The formed composite is gripped firmly between rubber blocks and is pulled through the die at a predetermined speed to enable curing of the composite to take place. It is then cut into the desired length by means of a cut-off wheel. Fig. 2.8 shows a diagramatic representation of the technique.

Most forms of geometric sections can be produced by this method. They can be stored and used as structural units when required.

Thermosetting resins are used almost exclusively, with polyester resin comprising the majority of the annual volume of pultruded stock.

Thermoplastic matrix pultrusion sections can be made, using high technology matrices which have high softening temperatures and high physical properties with relatively low melt viscosities. Compared to the thermosetting polymers, thermoplastics matrices will greatly improve the toughness of the composites. Much of the initial work has been undertaken using standard prepregs, such as the solvent-applied polysulphone and polyethersulphone polymers in combination with glass, Kevlar or carbon roving tow, tape or cloth reinforcements. These component parts are fed through short heat and chill dies to be pultruded into profiles.

It is becoming difficult for manufacturers to utilize the solvent-application methods because of growing government legislation against work and environment exposure to these chemicals. Adequate solvent has to be retained in the material to enable it to be sufficiently flexible for machine and manual handling, but the excess has to be drawn off and this operation considerably increases the processing time.

2.8.2.4. Resin injection

In this process, glass-fibre reinforcement is placed in position in the bottom mould and allowed to extend beyond the sides of the mould. The top mould is then placed in position over the bottom one and secured. Thus, the

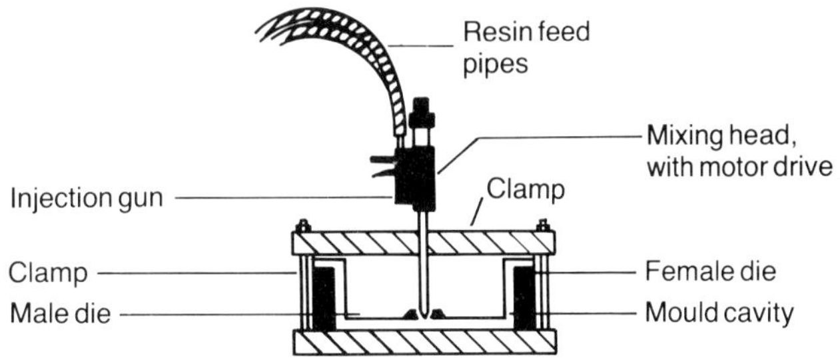

Fig. 2.9. Resin-injection technique

Table 2.1(a). Typical mechanical properties of polymers (thermosets)

Properties	Polyester resin	Epoxy resin	Phenolic resin with fillers
Density: Mg/m^2	1·2–1·4	1·1–1·35	1·35–1·75
Tensile strength MN/m^2	45–90	40–100	45–60
Compression strength: MN/m^2	100–250	100–200	
Modulus of elasticity in tension: GN/m^2	2·5–4·0	3·0–5·5	5·5–8·0
Poisson's ratio	0·37–0·40	0·38–0·40	0·37–0·40
Coefficient of thermal expansion: $10^{-6}/°C$	100–120	45–65	30–45
Shrinkage on curing: %	5–8	1–2	
Water absorption, 24 hours at 20°C: %	0·15–0·20	0·10–0·35	

Table 2.1(b). Typical mechanical properties of polymers (thermoplastics)

Properties	Polyvinylchloride PVC rigid	Nylon 66	High density polyethylene HD PE
Density: Mg/m^2	1·40	1·14	0·96
Ultimate tensile strength: MN/m^2	58	70	32
Modulus of elasticity: GN/m^2	2·8	2·8	1·2
Coefficient of thermal expansion: $10^{-6}/°C$	50	100	120
Elongation at break: %	80	60	120

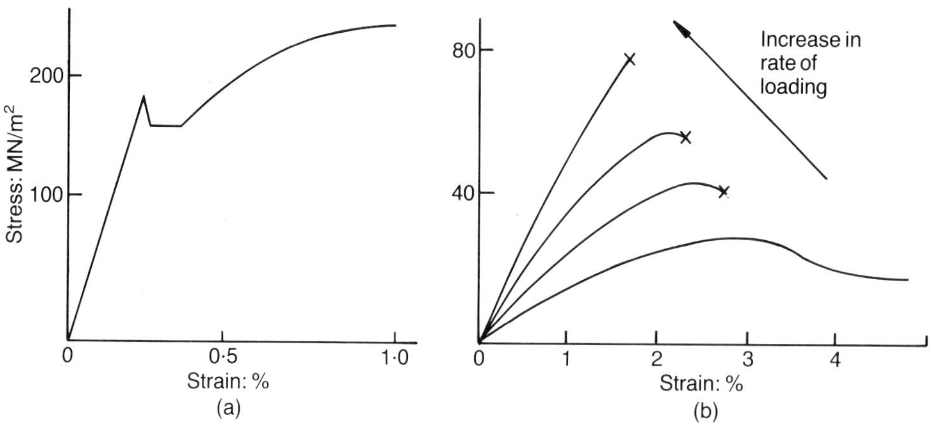

Fig. 2.10. Typical set of curves for: (a) metals; (b) polymers

reinforcement is held firmly in position whilst the resin is injected. A pressure of about 400 kN/m^2 is sufficient for most mouldings and when resin seeps through holes specially placed at the highest point of the mould, the resin inlet is then sealed. Fig. 2.9 illustrates the resin-injection process.

2.9. Mechanical properties

2.9.1. Polymer properties

From a structural point of view, polymers exhibit poor mechanical behaviour when under load; Tables 2.1(a) and 2.1(b) give the most important mechanical properties of some unreinforced polymers. In addition Table 10.1 gives some representative mechanical properties of available geosynthetics. Continuous load on polymers (particularly thermoplastic polymers) causes a straining of the molecular bonds which leads to the phenomenon known as creep or viscoelastic behaviour (see chapter 7). This molecular response is slow to reach an equilibrium state and consequently creep continues for a considerable time after application of the load. On removal of the load, some non-elastic behaviour of the material will normally be evident. Therefore, as the mechanical behaviour of polymers is highly dependent upon the magnitude, the time, the rate and the frequency of load application and upon the temperatures throughout, it is not possible to obtain a single stress–strain relationship for polymer materials. Fig. 2.10 shows a typical set of curves. At high rates of loading the curves will tend to be linear with negligible yielding or plastic flow, whereas at low rates considerable plastic flow is evident. If polymers are exposed to high temperatures, a characteristic behaviour similar to that for low rates of loading is encountered. Lockett[2.1] has discussed the significance of viscoelastic behaviour in engineering applications and has highlighted the difficulty for designers caused by the range of material behaviour and the small number of recognized design procedures for polymers. Likewise the test procedure, derivation and presentation of test data have not, as yet, been standardized; in the past, it has been the responsibility of the individual research body or company to develop test methods for all new polymers. However, it is now the expressed intention of governments and industry to establish international standardization for all test methods.

2.9.2. Fibre properties

When fibres are added to the polymer matrix, the above effects are reduced and indeed the main function of the fibre is to carry the majority of the load applied to the composite and to improve the stiffness characteristics of the resin. The final properties of the composite, therefore, will depend upon the properties of its component materials; the nature of the fibre and the matrix must be defined.

The most widely used reinforcement for polymers employed in construction is the E-glass fibre; to a much lesser extent carbon fibre is used as unidirectional reinforcement placed in the high stress region of the composite. Glass fibre combines good mechanical behaviour with relatively low cost and carbon fibre provides an extremely stiff laminate.

Synthetic and natural fibres such as nylon, rayon, polyester, acetate and Kevlar 49 have been combined with resins to form fibre/matrix composites, but none of them has had the same impact on the construction industry as composites manufactured from glass fibre/matrix materials. Table 2.2 gives typical mechanical properties of some of the more common fibres, and Fig. 2.11 gives typical tensile stress–strain behaviour of fibres. Table 10.2 gives typical values of the mechanical properties for the most important synthetic fibres used in geotextiles.

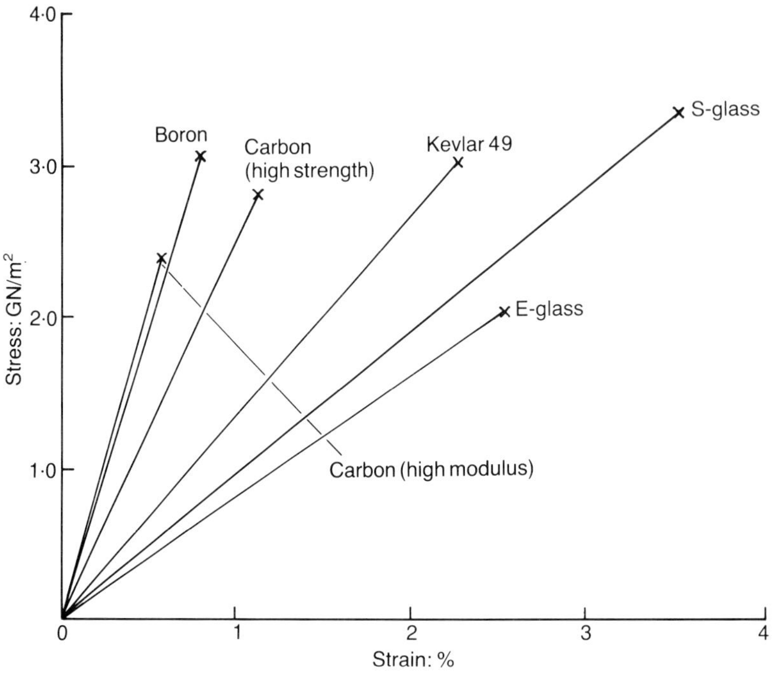

Fig. 2.11. Typical tensile stress–strain behaviour of some fibres

Glass fibres may be considered linear elastic up to failure, whilst the resin, although exhibiting essentially linear behaviour in the working range of stress, becomes nonlinear in the ultimate load region. However, as the ultimate strain in the brittle glass fibre is less than that for the more ductile matrix, when the line of action of the load is parallel to the direction of the fibres, the matrix stress at composite failure could be low enough to justify the assumption of linear elasticity.

2.9.3. Mechanical properties of composites

The mechanical properties of the composites are highly dependent on

(*a*) the relative proportions of fibre to matrix and the fibre orientation within the matrix material
(*b*) the components used to form the composite
(*c*) the method of manufacture of the composite.

Table 2.2. Mechanical properties of E-glass, carbon and Kevlar 49 fibres

Material	E-glass	Carbon high modulus type I	Carbon high strength type II	Aromatic polyamide Kevlar 49
Diameter : m	7–14	7–9·8	7·5–8·8	11·8
Density : 10^3 Kg/m^3	2·56	1·95	1·75	1·45
Modulus of elasticity : GN/m^2	70	380	240	120
Tensile strength: GN/m^2	1·5–2·5	2·0	2·8	2·7–3·5
Elongation to fracture: %	1·8–3·0	0·5	1·0	2·0–2·7
Coefficient of thermal expansion: 10^{-6}/°C	5·0	−0·6 to −1·3	−0·2 to −0·6	−2·0

Composites which have been formed under pressure will generally have a higher fibre volume fraction than those which have been manufactured by hand. Consequently, the ratio of fibre to matrix is highly dependent on the method of manufacture and the fibre orientation within the composite.

The mechanical properties of the composite are mainly dependent upon the value of the fibres' modulus of elasticity and ultimate strength. The inservice properties of the composite would be mainly dependent on the physical properties of the matrix material.

The degree of impregnation of the matrix into the fibre is dependent on the manufacturing process of the composite. The units or panels that are often required in the construction industry are large in area, thus requiring large moulds. In addition, the number of identical panels (e.g. the infill panels for the facade of a building) is generally small. Therefore, it is not economic to produce them using a mechanical production method. In these cases, the manual or semi-manual manufacturing processes would be used with a consequent reduction in fibre volume fraction over the automated production methods.

Table 2.3 gives typical mechanical properties for glass and carbon fibre reinforced plastics composites. Clearly these composites would be used for the manufacture of structural units and chapters 3 and 4 have presented analyses and design techniques relevant to these materials. Geotextiles, which are discussed in chapter 10, consist of highly orientated polymers and fibre/matrix composites. Typical mechanical properties of some of these materials are given in Table 10.3.

2.10. Micro-mechanical analysis—elastic properties

2.10.1. Unidirectional aligned fibre laminae

When unidirectional fibres are dispersed within the matrix material, as shown in Fig 2.12, it is assumed that they are uniform and continuous and that there is complete bond at the interface between the fibres and matrix. In this condition the composite will possess orthotropic properties with the greatest stiffness and strength in the direction of the fibres. The properties of the composite will depend upon the fibre volume fraction and the mechanical

Table 2.3. Typical mechanical properties for glass and carbon fibre reinforced plastics composites

Material	Glass content: % by weight	Density: Mg/m^2	Tensile modulus: GN/m^2	Tensile strength: MN/m^2
Undirectional laminate glass fibre/polyester	50–80	1·6–20	20–50	400–1250
Glass fibre/polyester randomly orientated hand lay-up	25–45	1·4–1·6	6–11	60–180
Glass fibre/polyester matched metal dye	25–50	1·4–1·6	6–12	60–200
Glass fibre/polyester woven rovings hand lay-up	45–62	1·5–1·8	1·2–2·4	300–350
SMC hot-press moulding undirectional laminate	20–25	1·75–1·9	9–13	60–100
Carbon fibre/epoxy	70	1·93	120	800

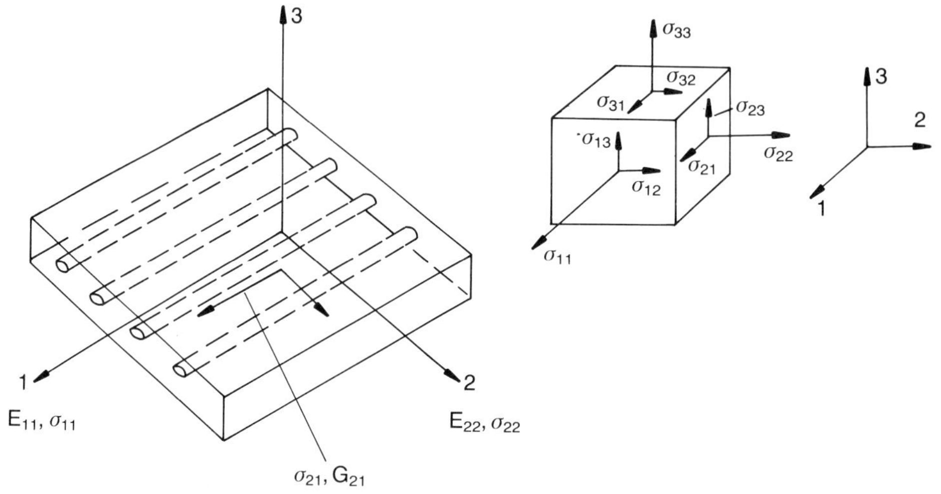

Fig. 2.12. Uniform, continuous and unidirectional composite

properties of the component parts. The elastic properties of the composite may be expressed as:

$$E = f(E_f, V_f, \nu_f, E_m, V_m, \nu_m)$$

where E, ν and V are the elastic modulus, Poisson's ratio and volume fraction respectively. The subscripts f and m refer to fibre and matrix materials respectively.

2.10.1.1. Elastic properties of unidirectional fibre laminae

(*a*) *Modulus of elasticity* E_{11}. When a tensile or compression load is applied parallel to the fibre in a unidirectional laminae, it is assumed that the strains on the fibre, matrix and composite in direction 1 are equal. The resultant axial force P_{11} of the composite is shared between the fibre load P_f and the matrix load P_m in the following way

$$P_{11} = P_m + P_f$$

$$\sigma_{11} A_{11} = \sigma_m A_m + \sigma_f A_f$$

or

$$\sigma_{11} = \sigma_m V_m + \sigma_f V_f \qquad (2.1)$$

where A is the area of the phase, V is the volume fraction of phase $V_c = 1$ and σ is the stress in the phase.

$$\sigma_{11}/\varepsilon_{11} = E_m V_m + E_f V_f$$

where E is the modulus of elasticity of phase and $\sigma_{11}/\varepsilon_{11}$ is the modulus of elasticity of composite in direction 1. Therefore

$$E_{11} = E_m V_m + E_f V_f \qquad (2.2)$$

or

$$E_{11} = (E_f - E_m)V_f + E_m \qquad (2.3)$$

Equation (2.2) is the "law of mixtures" equation and it has been found experimentally that it predicts E_{11} with a fair degree of accuracy.

(*b*) *Poisson's ratio*. By considering the element ABCD in Fig. 2.13, which is under an average normal stress σ_{11}, an expression for ν_{21} may be derived. The Poisson's ratio ν_{12} refers to the strain developed in

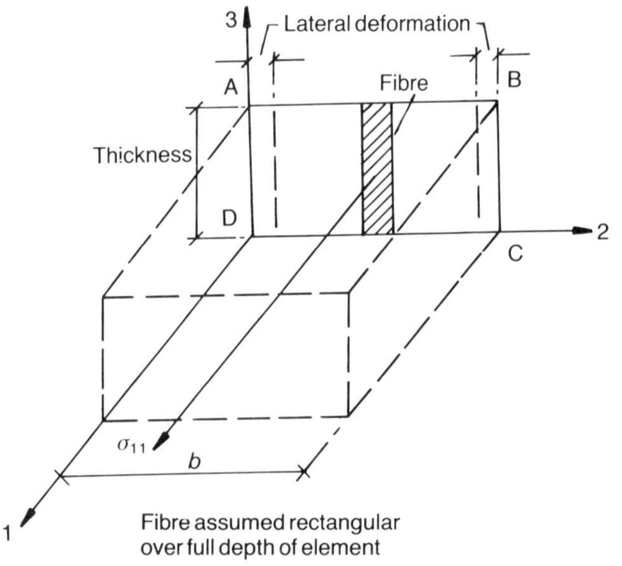

Fig. 2.13. Lateral strain in element under mean normal stress σ_{11}

direction 2 when the laminate is stressed in direction 1. Therefore, ν_{12} is equal to $-\varepsilon_{22}/\varepsilon_{11}$. The strain ε_{22} is a function of two components: the fibre component, $-\nu_f\varepsilon_{11}$, and the matrix component, $-\nu_m\varepsilon_{11}$.

These two component strains will produce lateral displacements which are dependent on the geometry and upon the cross-sectional shape of the fibre. In the following discussion it will be assumed that the fibre is rectangular and extends over the whole thickness of the element, as shown in Fig. 2.13, and that the laminae strain effects in direction 3 are negligible.

The lateral displacement in the fibre is $-\nu_f\varepsilon_{11}(V_f b)$.

The lateral displacement in the matrix is $-\nu_m\varepsilon_{11}(V_m b)$.

The total lateral displacement in direction 2 is $[-\nu_f\varepsilon_{11}V_f - \nu_m\varepsilon_{11}V_m]b$ where b is the breadth of the element.

The lateral strain in direction 2, ε_{22}, is $-(\nu_f V_f + \nu_m V_m)\varepsilon_{11}$.

Therefore

$$\nu_{12} = [\nu_f V_f + \nu_m V_m] \qquad (2.4)$$

Equation (2.4) predicts the value of ν_{12} with a fair degree of accuracy.

(*c*) *The modulus of elasticity E_{22}*. Fig. 2.14 shows an element of the composite in which the load is applied in the direction 2 perpendicular to the fibre direction; consequently the composite is under a stress σ_{22}.

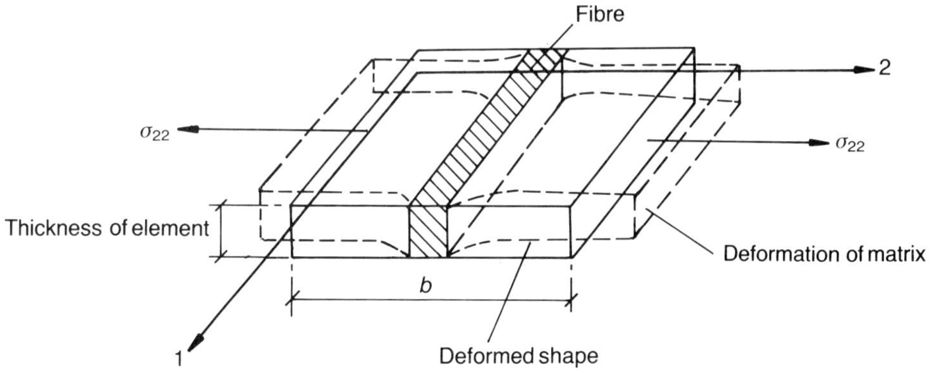

Fig. 2.14. Element under normal stress σ_{22}

Again it is assumed that the cross-section of the fibre is rectangular and extends over the whole thickness of the element. As the stiffness of the fibre is large compared with that of the matrix, the displacement differential in the two components introduces varying values of shearing stresses in the plane 1–2 which are especially noticeable near the fibre/matrix interface.

For this discussion, the Poisson's ratio is neglected and therefore σ_{22} will be constant throughout the element.

The transverse strains induced in the fibre and matrix are respectively

$$\left.\begin{aligned}\varepsilon_f &= \sigma_{22}/E_f \\ \varepsilon_m &= \sigma_{22}/E_m\end{aligned}\right\} \qquad (2.5)$$

The transverse displacements in the fibre and matrix are $\varepsilon_f(V_f b)$ and $\varepsilon_m(V_m b)$ respectively. Therefore the total transverse displacement is

$$(\varepsilon_f V_f b) + (\varepsilon_m V_m b) \qquad (2.6)$$

Combining equations (2.5) and (2.6) gives

$$\varepsilon_{22} = \sigma_{22}(V_f/E_f + V_m/E_m)$$

$$E_{22} = \sigma_{22}/\varepsilon_{22}$$

Therefore

$$E_{22} = E_f E_m/(V_f E_m + V_m E_f) \qquad (2.7)$$

(*d*) *The modulus of rigidity*, G_{12}. Fig. 2.15 shows a fibre/matrix element under a uniform shearing stress σ_{12} and the complementary shearing stress σ_{21} acting along its boundary. The full derivation G_{12} is analogous to that for E_{22} and it can be shown that

$$G_{12} = G_f G_m/(G_m V_f + G_f V_m) \qquad (2.8)$$

2.10.1.2. Elastic properties of long, randomly orientated fibre laminae

Randomly orientated discontinuous fibres of length greater than 40 mm, which have been impregnated with polymer, have elastic properties which are macroscopically isotropic in the plane of the laminae. The length to diameter ratio, l/d, known as the aspect ratio, of the fibre can vary between 100 and 5000. Providing that its length is much greater than the critical length required for shear strength transfer, the short length fibre will approach the efficiency of the continuous fibre.

Figure 2.16(a) shows an isolated high modulus fibre embedded in a low modulus matrix. The latter is subjected to a tensile load parallel to the fibre.

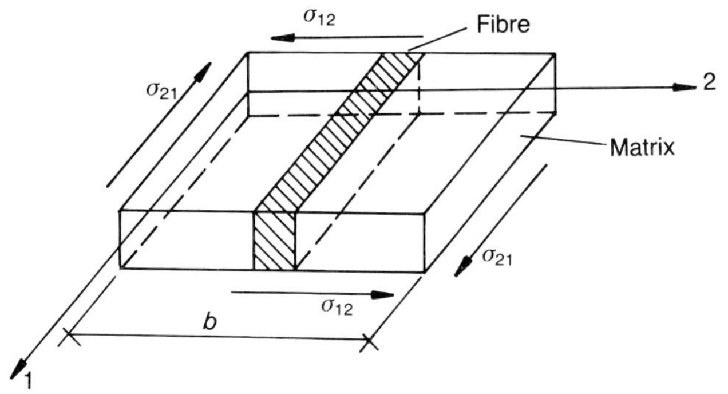

Fig. 2.15. Element under uniform shearing stress σ_{12}

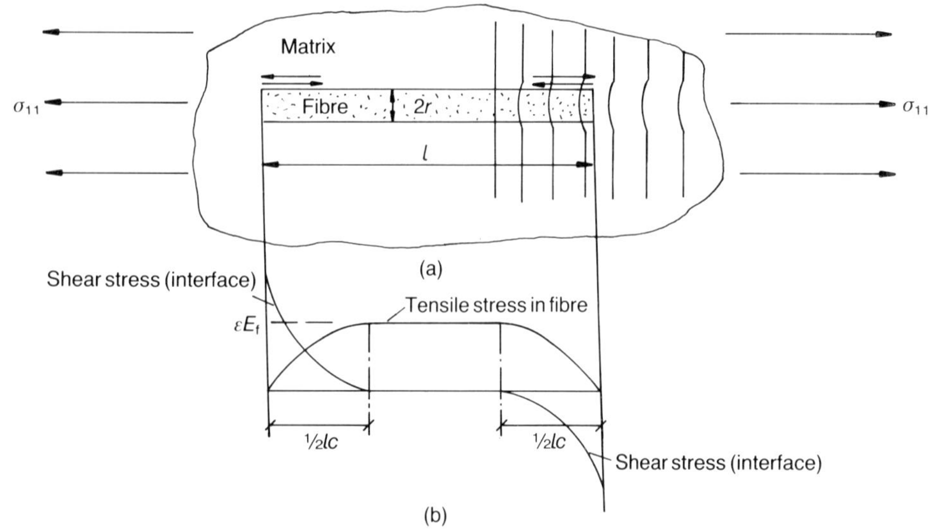

Fig. 2.16. Diagrammatic representation of discontinuous fibre in a uniaxially aligned state embedded in a matrix: (a) isolated fibre in matrix; (b) stress distribution

Provided the fibre is completely bonded to the matrix, the applied stress on the latter will be transferred to the fibre by interface shear stresses. Fig. 2.16(b) shows the variation of the tensile stress on the fibre along its length and the shear stress at the fibre/matrix interface. The term l_c is the critical length of the fibre for maximum stress to be developed in it when the composite is under a particular external load. Consequently, with all short length fibres there is a finite length at the end of each fibre in a randomly orientated laminae which is stressed to less than the maximum stress of a continuous fibre.

Figure 2.17 shows an ideal situation in which discontinuous fibres are arranged in a uniaxially aligned state. The transfer length depends upon the stress σ_f in the fibre and the effective shear strength of the fibre/matrix interface σ_{xy} and is given by

$$l_1 = \sigma_f d_f / 4\sigma_{xy} \qquad (2.9)$$

where d_f is the diameter of the fibre or

$$l_c = \sigma_{fmax} d_f / 4\sigma_{xy} \qquad (2.10)$$

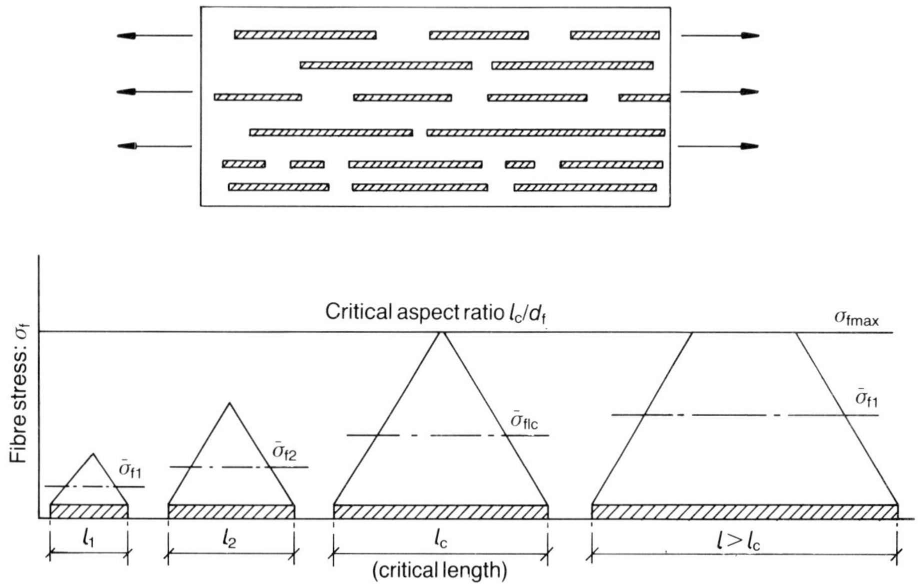

Fig. 2.17. Schematic representation of discontinuous unidirectional fibre/matrix composite. Average fibre tensile stress σ_{fav} (σ_{f1} etc.) along the fibre length

where σ_{fmax} is the maximum value of fibre stress and l_c is the critical length of the fibre.

The critical transfer length over which the fibre is decreased from σ_{fmax} to zero is referred to as $\frac{1}{2}l_c$ and by considering Fig. 2.17 it will be seen that $\sigma_{fmax}(l - l_c) + \sigma_{fmax}l_c/2 = \sigma_{fav}l$, or

$$\sigma_{fav} = \sigma_{fmax}(1 - l_c/2l) \tag{2.11}$$

where σ_{fav} is average axial stress in fibres with lengths not less than l_c.

$$\sigma'_{fav} = \sigma_{fmax}\, x/(2l_c) \tag{2.12}$$

where σ'_{fav} is the average axial stress in fibres in which the length is less than the critical length.

2.10.1.3. Elastic properties of short discontinuous fibre composites

The elastic theory given above is relevant to long discontinuous fibres where the length is greater than 40 mm. If the length of fibre is between say 0·1 mm and 10 mm, which would be the case for resin-injection and SMC processes respectively, then the above theory must be modified slightly.

The reinforcing efficiency of the short fibre is less than that for the long fibre and consequently the effective modulus of elasticity will also be less.

The relationship between the modulus of elasticity for the composite, the fibre, the matrix and the fibre volume fraction can be defined by a modification of the law of mixtures equation, so that

$$E_c = \eta_l E_f V_f + E_m(1 - V_f) \tag{2.13}$$

The modification factor η_l has been expressed by Cox[2.2] as

$$\left.\begin{aligned} \eta_l &= 1 - (\tanh \tfrac{1}{2}\beta l)/\tfrac{1}{2}\beta l \\ \beta &= [2G_m/E_f r^2 l_n(R/r)]^{1/2} \end{aligned}\right\} \tag{2.14}$$

where $2R$ is the interfibre spacing, $2r$ is diameter of fibre and l is length of fibre.

Equation (2.13) is further modified, if the fibres are randomly orientated, to

$$E_c = \eta_o \eta_l E_f V_f + E_m(1 - V_f) \tag{2.15}$$

where η_o = orientation factor.

Krenchel[2.3] has obtained values for η_o at different fibre orientations by calculation. The two extreme values are 1 and 0 when the load application is parallel and perpendicular to the fibres respectively. For in-plane laminae with randomly orientated fibres η_o is given the value of 3/8.

2.10.2. Strength characteristics of unidirectional laminae

(*a*) *Uniaxial tension.* To a first approximation, glass, carbon and Kevlar 49 fibres behave elastically in tension up to the fracture stress σ_f^*. Polyester and epoxy resins have nonlinear stress–strain relationships and may undergo considerable viscoelastic deformation before fracture. Figs 2.18(a) and 2.18(b) show two possible stress–strain curves of fibre and matrix; one when $\varepsilon_f^* > \varepsilon_m^*$ and the other when $\varepsilon_m^* > \varepsilon_f^*$ where ε^* represents the failure strains.

The law of mixture equation (2.2) may be modified to approximate the composite stress σ_{11} in terms of the fibre volume fraction V_f and the stresses induced in the matrix σ_m and fibre σ_f. The equation becomes

$$\sigma_{11} = \sigma_f V_f + \sigma_m(1 - V_f) \tag{2.16}$$

where σ_f and σ_m are the stresses in the fibre and matrix respectively.

The response of the composite is dependent upon the relative strains to failure of the matrix and the fibre. If $\varepsilon_f^* > \varepsilon_m^*$ and V_f is small then the strength of the composite will depend primarily upon σ_m^*. When the matrix fails, the fibres will be unable to carry the load and the composite will fail. This situation is shown in Fig. 2.18(c) and is denoted by region 1. Then

$$\sigma_{11}^* = \sigma_f' V_f + \sigma_m^*(1 - V_f) \tag{2.17}$$

where σ_f' is the stress in the fibre at failure of the matrix.

When V_f is large, the matrix will carry only a small fraction of the load so that when the matrix fails the load is transferred to the fibres. If the system is able to continue to transfer load, then the composite stress will increase until the fibres fail. This situation is shown in Fig. 2.18(c) and is denoted by region 2. Then

$$\sigma_{11}^* = \sigma_f^* V_f \tag{2.18}$$

The above two equations would be used for the analysis of a glass fibre/polyester resin composite.

If $\varepsilon_m^* > \varepsilon_f^*$ and V_f is small, when the fibres fracture the extra load on the matrix does not cause the latter to fail. As the load is increased the composite will fail when σ_m^* is reached. Thus

$$\sigma_{11}^* = \sigma_m^*(1 - V_f) \tag{2.19}$$

This condition is shown in Fig. 2.18(d) and is denoted by region 1.

When V_f is large, the transfer of load to the matrix at fibre fracture is large and the matrix fails, being unable to support it. Thus

$$\sigma_{11}^* = \sigma_f^* V_f + \sigma_m'(1 - V_f) \tag{2.20}$$

Fig. 2.18. Stress–strain curves of fibre and matrix: (a) when $\varepsilon_f^* > \varepsilon_m^*$; (b) when $\varepsilon_m^* > \varepsilon_f^*$; (c) variation of fracture strength of unidirectional laminate with $V_f - E_f^* > E_m^*$; (d) variation of fracture strength of unidirectional laminate with $V_f - E_m^* > E_f^*$

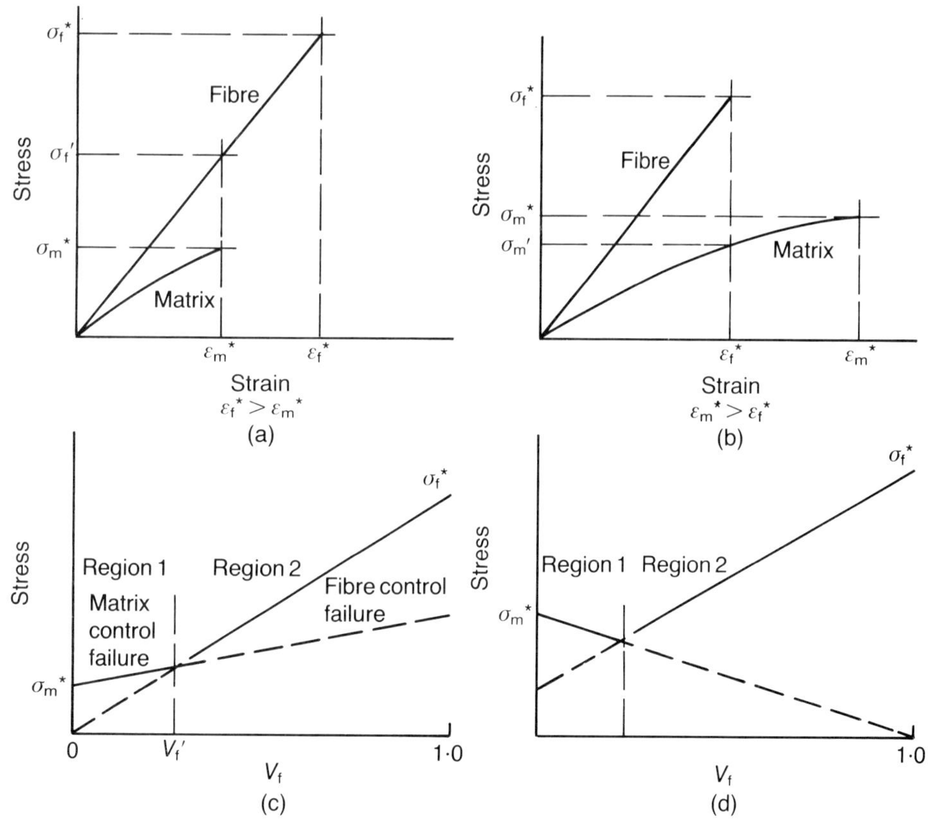

where σ'_m is stress in the matrix at failure of the fibre. This condition is shown in Fig. 2.18(d) and is denoted by region 2.

Equations (2.19) and (2.20) would be used for the analysis of a carbon fibre/epoxy resin composite.

It can be seen from the above that if the composite is to be reinforced by the fibres, there must be a minimum fibre volume fraction $V_{f\,min}$. Thus, the strength of the composite above that of the matrix alone is

$$\sigma_c^* = \sigma_m^* = \sigma_f^* V_{f\,min} + \sigma'_m(1 - V_{f\,min})$$

$$V_{f\,min} = (\sigma_m^* - \sigma'_m)/(\sigma_f^* - \sigma'_m) \qquad (2.21)$$

However, in the civil/structural engineering areas the fibre volume fraction is much greater than this value: typically 20% for manual production and 40% for mechanical production.

The stress relationship given in equations (2.17) and (2.18) is an oversimplification. Although it provides reasonable estimated values, no account is taken during the computations of statistical scatter in fibre strengths or of the influence of the micro-structural defects. However, equation (2.16) may be modified to include the effect of these quantities.

$$\sigma_{11} = \beta \sigma_\beta V_f + \sigma_m(1 - V_f) \qquad (2.22)$$

where σ_β is fibre bundle strength and β is the matrix efficiency factor, usually between 1 and 2. Consequently, the fibre strength is the dominant component in the strength of the composite; the contribution from the matrix is small.

(*b*) *Transverse tensile strength.* The transverse strength of a unidirectional fibre array is influenced by many parameters. These include

(*i*) the properties of the fibre and the matrix
(*ii*) the interface strength
(*iii*) the internal stresses and strains set up by the interaction between the above parameters.

In an analysis of composites under transverse tensile stress, it is assumed that either there is no bond between the fibre and matrix or there is complete bond. In either case, stress concentrations will occur and these will cause a stress reduction in the matrix material. In addition, it is well known that the tensile strength of brittle polymers is affected by flaws in the material. Consequently, it is difficult to interpret strength in the transverse direction of unidirectional fibre composites.

(*c*) *Unidirectional compressive strengths.* The longitudinal compressive strength of a continuous unidirectional fibre is dependent upon the fibre and the resin properties and upon the interface bond strength. The failure mechanism is sensitive to the fibre volume fraction and the compressive failure strength of unidirectional composites appears to be largely dependent upon the buckling of the fibres and it is generally assumed that adjacent fibres will act independently of each other. Fig. 2.19 shows two possible basic modes of buckling and it can be seen that the adjacent fibres will buckle with equal wave lengths at the ultimate load either out-of-phase with each other (an extensional mode) or in-phase with each other (a shear mode). The out-of-phase and the in-phase modes of buckling are associated with low and high fibre volume fraction respectively.

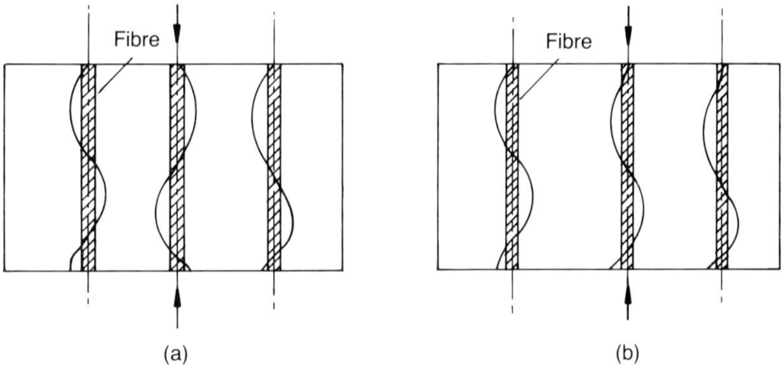

Fig. 2.19. Schematic representation of the two basic modes of buckling in unidirectional laminate: (a) out-of-phase buckling (extension mode); and (b) in-phase buckling (shear mode)

The critical compressive stress associated with the extensional mode is

$$\sigma_c^* = 2V_f[E_f E_m V_f/3(1 - V_f)]^{1/2} \tag{2.23}$$

and the critical compressive stress associated with the shear mode is

$$\sigma_c^* = G_m/(1 - V_f) \tag{2.24}$$

In the practical range of V_f for civil engineering applications, the shear mode tends to dominate and the compressive strength is highly dependent upon the shear modulus of the matrix.

(*d*) *Shear strength*. Shear stresses which are generated in the laminae may also be sufficient to cause failure by a shear mode. Ewins and Ham[2.4] have suggested that the ply shear stresses developed are given by

$$\sigma = \sigma_{11} \sin\theta \cos\theta \tag{2.25}$$

which has a maximum when the plane $\theta = 45°$. Thus $\sigma_{max} = \frac{1}{2}\sigma_{11}$. If the shear strength is less than the buckling strength, shear failure will be the criterion. If shear failure occurs in the individual fibres due, say, to misalignment of molecular structure, the strength will follow a rule of mixture equation thus

$$\sigma_{11}^* = 2[V_f\sigma_{sf}^* + (1 - V_f)\sigma_{sm}^*] \tag{2.26}$$

where σ_{sf}^* is shear strength of the fibres and σ_{sm}^* is shear strength of the matrix.

If equation (2.11) is substituted into the law of mixtures equation (2.16) and the ultimate strength values of the fibre are used, then

$$\begin{aligned} \sigma_{11}^* &= \sigma_{fav}^* V_f + \sigma'_m V_m \\ &= \sigma_f^*(1 - l_c/2l)V_f + \sigma'_m V_m \end{aligned} \tag{2.27}$$

where σ_f^* is ultimate strength of the fibres and σ'_m is the stress in the matrix at failure of the fibre.

2.11. Current position of test methods for composite material properties

The test methods for determining the mechanical properties of long fibre composite materials are much more complicated than those for the more conventional civil engineering materials. This is because the former material, which consists of stiff fibres embedded in a relatively flexible matrix, is inherently orthotropic and also because of the nature and the number of failure modes. The latter could be fibre fracture, matrix cracking, delamination, interface shear failure between the fibre and the matrix, etc.

The need for developing test methods for fibre composite materials was first recognized by the Aircraft Industry and in 1980 CRAG (Composites Research Advising Group) was set up and charged with the task of defining test methods for the measurement of the engineering properties of fibre-reinforced composites. Its first report was published in 1984.[2.5] In 1985 an extended report was published[2.6] and the third and final report was published in 1988.[2.7] The final report describes the test methods for determining the characteristics of long fibre composite materials such as carbon-, glass- and Kevlar-fibre reinforced polymers. The properties which may be determined from these tests are shear, flexural, tensile, compressive, fatigue, toughness, physical, and environmental.

The BSI Committee PLM 39-3 has been charged with the task of preparing a General Engineering Standard on the test methods for composite materials. The European Standards Commission (CEN) is developing European Standards for composite materials and the representatives of AECMA (Association Européenne des Constructeurs de Matériel Aérospatial) are providing much of the input to the development of these standards. When the CEN has approved the document as a European Standard, the national standards organizations of the member countries (BSI in the case of the UK) are required to include it in their lists of recognized standards. This European Standard will, however, appear in the Aerospace Series rather than the General Engineering Series.

In the USA, the ASTM (American Society for Testing and Materials) is the body which deals with the development of test methods for long-fibre composite materials. Its approach has been to develop a standard for each individual test method which has resulted in a large number of documents being produced.[2.8–2.16] This approach is not convenient for the practising structural engineer, who would prefer a single document describing all tests.

References

2.1. Lockett, F.J. *Engineering design basis for plastics products*. HMSO, London, 1982.

2.2. Cox, H.L. The elasticity and strength of paper and other fibrous materials. *Brit. Jnl. Appl. Phys.* **3**, 1952, 72–79.

2.3. Krenchel, H. *Fibre reinforcement*. Akademisk Forlag, Copenhagen, 1964.

2.4. Ewins, P.D. and Ham, A.C. The nature of compressive failure in unidirectional carbon fibre reinforced plastics. *Technical report 73057*, Royal Aerospace Establishment, Farnborough, 1973.

2.5. Curtis, P.T. (Ed.) CRAG test methods for the measurement of the engineering properties of fibre reinforced plastics. Technical report 84102. Royal Aerospace Establishment, Farnborough, 1984.

2.6. Curtis, P.T. (Ed.) CRAG test methods for the measurement of the engineering properties of fibre reinforced plastics. Technical report 85099, Royal Aerospace Establishment, Farnborough, 1985.

2.7. Curtis, P.T. (Ed.) CRAG test methods for the measurement of the engineering properties of fibre reinforced plastics. Technical report 88012, Royal Aerospace Establishment, Farnborough, 1988.

2.8 American Society for Testing and Materials. Test method for apparent tensile strength of ring or turbular plastics and reinforced plastics by split disk method. *Standard D2290-87*. ASTM, Philadelphia, 1987.

2.9 American Society for Testing and Materials. Test method for apparent interlaminar shear strength of parallel fibre composites by short beam method. *Standard D2344-88*. ASTM, Philadelphia, 1988.

2.10 American Society for Testing and Materials. Method for preparation and tension testing of filament-wound pressure vessels. *Standard D2585-68*. ASTM, Philadelphia, 1985.

2.11. American Society for Testing and Materials. Test method for hydrostatic

compressive strength of glass-reinforced plastic cylinders. *Standard D2586-68*. ASTM, Philadelphia, 1985.
2.12. American Society for Testing and Materials. Test method for tensile properties of fibre-resin composites. *Standard D3039-76*. ASTM, Philadelphia, 1982.
2.13 American Society for Testing and Materials. Test method for compressive properties of unidirectional or crossply fibre-resin composites. *Standard D3410-87*. ASTM, Philadelphia, 1987.
2.14. American Society for Testing and Materials. Test methods for tension-tension fatigure of orientated fibre, resin matrix composites. *Standard D3479-76*. ASTM, Philadelphia, 1982.
2.15. American Society for Testing and Materials. Practice for inplane shear stress-strain responses of unidirectional reinforced plastics. *Standard D3418-76*. ASTM, Philadelphia, 1982.
2.16. American Society for Testing and Materials. Guide for testing inplane shear properties of composite laminates. *Standard D4255-83*. ASTM, Philadelphia, 1983.
2.17. van Zanten R.V. *Geotextiles and geomembranes in civil engineering*. Balkema, Rotterdam, 1986.

Further reading

Benjamin, B.S. *Structural design with plastics*. Van Nostrand Reinhold, New York, 1982, 2nd edn.
Broutman, L.J. and Krock, R.H. *Modern composite materials*. Addison-Wesley, Reading, Mass., 1967.
Hollaway, L. *Glass reinforced plastics in construction engineering aspects*. Surrey University Press in association with International Textbook Company, Glasgow, 1978.
Hollaway, L. Pultrusion. In: *Developments in plastics technology* (3 Edn), A. Whelan and J.L. Craft (Eds), chapter 1. Elsevier Applied Science, London, 1986.
Holmes, M. and Just, D.J. *GRP in structural engineering*. Applied Science, London, 1983.
Morgan, P. (Ed.) *Glass reinforced plastics*. Iliffe, London, 1955.

3 Design and analysis of composite structures

A. F. Johnson and A. Marchant

3.1. The design process

Design is an interactive process, embracing several disciplines, talents and skills including science, art, engineering, technology, and manufacturing. The designer can be likened to the conductor of an orchestra who brings together a number of talents to create a harmonious product.

Designing in composite materials is a similar interactive process with an emphasis on the synergy between the designer's drawing board and the production engineer's manufacturing processes. It is essential that a design methodology is selected and rigorously used because of the many different materials to hand which can be affected by quality, the environment and the manufacturing process. It is also important that the design methodology recognizes the product cost because the constituents of composite materials, the fibre and the polymer, can vary significantly in price and the manufacturing process can range from the simplistic contact moulded room cured method to the sophistication of high temperature and pressure cure. The need, therefore, for interaction between the various input skills is paramount when designing in fibre reinforced polymers.

The design process can be divided into four main phases (Fig. 3.1). The first phase is typical to any design situation, that is writing the design brief and estimating the cost of design. The importance of this phase must not however be understated as without a correct brief and properly assessed design costs there is little point in commencing.

The second phase is concerned with schematics and the identification of manufacturing methods and estimates for final product cost. At the commencement of the second phase, and with the brief to hand, the designer must address the question of the most suitable material or materials to be considered for the product. If composite materials are selected, then he must understand and be able to quantify the benefits and disadvantages of the selected materials. Composite materials are by no means a panacea for all the designer's problems: they too have their advantages and disadvantages, which can be summarized as follows.

Advantages

- lightweight
- good specific mechanical properties
- good environmental behaviour
- readily formed to complex shapes
- low thermal conductivity
- the ability to tailor the mechanical properties by fibre choice and direction

Disadvantages

- high cost of material compared to other civil engineering materials
- the need for tooling for manufacture
- low stiffness of some composites and the high cost of improved stiffness composites
- reduced robustness

Clearly there will be other factors to consider and the experienced designer will be able to reduce the investigatory work of material selection to a minimum. However, the considerable range of fibres and polymer matrices could in some respect be considered disadvantageous; certainly this would be so for the inexperienced designer.

The next important decision, and one which must be made very early in this second phase, is that of the choice of the manufacturing method to be used.

Fig. 3.1. Design process for composite structures

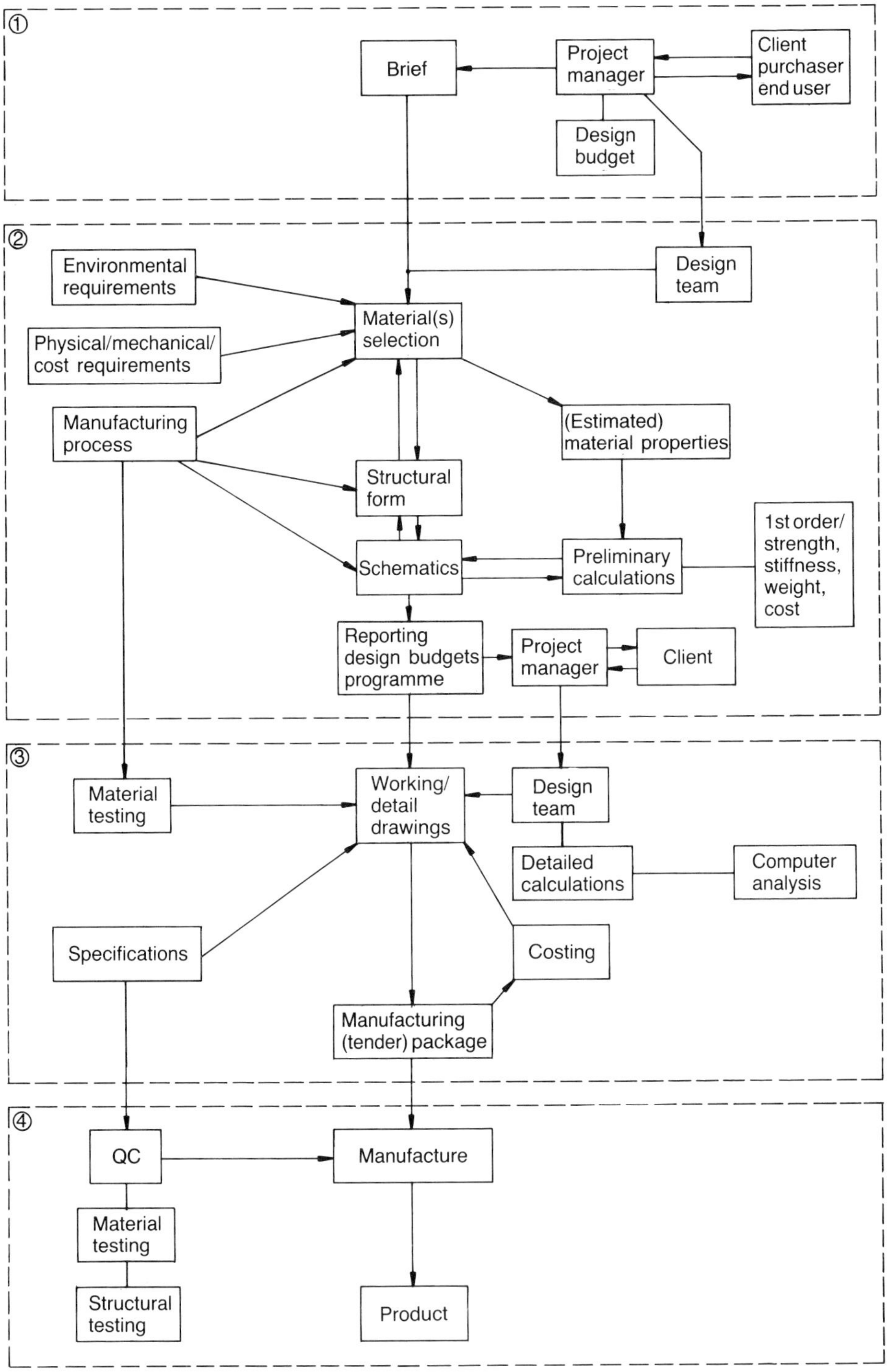

This is because the manufacturing method will determine both the mechanical properties of the materials and the production numbers, this having a great influence on product cost. For instance, pipes are generally made by a filament-winding process which gives a high fibre content and therefore good mechanical properties. The process requires expensive and sophisticated machinery. However, high volume production is possible. Pipes could also be made by rotational moulding, using less expensive machinery, but giving a lower fibre content and reduced mechanical properties, with a more labour-intensive production process. Water tank panels could be manufactured by contact moulding, pressing or injection, with each process requiring machinery or tooling of differing cost and resulting in materials of different mechanical properties with different production volumes.

Material choice and manufacturing process are fundamental to the cost-effectiveness of the design solution and can only be made by a continuous interaction of skills. This second phase will identify the most suitable materials and manufacturing methods to be employed, leading to a first-order product cost. The design work undertaken will be based on calculated or assessed material properties which will be quantified in the detail design phase. Again, the experienced designer will be able to reduce the preliminary investigating work by making a selection based on accumulated data, but even the experienced designer may have to make two or three preliminary designs to explore the optimum solution.

The third phase will be concerned with the detail work and the production of manufacturing drawings and specifications. To undertake factual calculations, the engineer must have reliable materials property data and appropriate design-analysis methods. A preliminary design analysis, which for some products is all that is required, may be carried out using some of the methods and materials data described in the following sections. Standards and codes of practice now exist for certain polymer composite components and these may also provide valuable design information. For more detailed design studies, or under critical loading conditions, the provision of suitable materials data can often only be achieved by testing the selected laminate, made by the chosen manufacturing process and in the anticipated environment. Should the application involve long-term loading conditions, then previous data on the effect of this on the mechanical properties must be sought. If this is not adequate, then tests have to be put in hand to provide the data. It is possible that both material and structural testing would be required to validate fully a polymer composite component development, particularly if the product, say a water tank panel, was to be made in quantity.

An important part of the design analysis is the selection of design factors of safety. Should a code of practice be used, then the selection of design factors is likely to be covered by the code. Where, however, the design is conducted by first principle analysis, the designer must select his design factors of safety, bearing in mind the extent and exactness of his calculations, the manufacturing process, the working environment, the quality of manufacture, the product life, the exactness and nature of the loading etc. Generally speaking, the selection of such design factors would follow the pattern for any other material but with perhaps more emphasis on considering the quality reflected by the manufacturing process and with the anisotropic nature of composites leading to the likely use of more approximate calculations.

The final phase is manufacture. The designer's job is not finished here as the quality control he instigates will determine the performance of the product. The interaction between the designer and those responsible for manufacture and quality control must be a continuous process until production is finished.

3.2. Design-analysis principles

Design-analysis procedures for polymer composites and for traditional structural materials such as metals are essentially the same and consist of two main steps. First, stress and/or strain levels in the component are determined under relevant design loads. Then, critical stress, strain or deformation values are compared with appropriate design criteria to ensure that the component satisfies product requirements and material limitations. However, the implementation of this design-analysis process for polymer composites is often more complex than traditional procedures established for metals. At ambient temperatures, metals such as steel are isotropic elastic at stresses below yield, and the required engineering design properties are the modulus of elasticity, Poisson's ratio, and a yield or proof failure stress.

In contrast, polymer composites are usually macroscopically inhomogeneous and anisotropic due to the presence of reinforcing fibres, and viscoelastic from the properties of the polymer matrix. Because of these differences in material structure, further materials property data are required to characterize polymer composites mechanical behaviour. More complex stress-analysis procedures are required to determine stress and deformation levels in a component. The modifications to traditional design procedures are briefly described in this section. There are three main aspects to consider. The effects of anisotropy and inhomogeneity in polymer composites are handled by materials design analysis, which determines materials design properties in terms of fibre and matrix properties, ply orientation angles, layer thicknesses and other aspects of the composite structure. For design under short-term loads elastic stress analysis methods may be used provided anisotropy is taken into account. Viscoelastic effects may be significant under long-term loads when it is necessary to modify the short-term elastic design procedures.

3.2.1. The need for materials design

Most polymer composite materials consist of several thin plies of resin-impregnated fibres laminated together as shown in Fig. 3.2. Each ply may contain one of several types of fibre reinforcement such as unidirectional (UD) fibres, woven fibre rovings (WR), fibre cloth, and chopped strand mat (CSM) which consists of chopped fibres randomly oriented in a mat. During the component fabrication process it is possible to change ply thicknesses and orientations to meet different load requirements. Whilst this freedom to tailor materials properties is an important factor in the success of polymer composites, it is also the main reason for the additional complexity in design analysis with these materials.

One approach to designing with polymer composites is to use a finite element analysis (FEA) for the structural design, in which elements are defined in each material ply in the structure and the material properties are varied as required from element to element as fibre orientations change. Such an approach to structural design, which combines both structural and materials analysis, is time consuming and expensive in computer time and is only appropriate for critical analysis in high technology applications such as in the aerospace industry. Even then FEA is often only suited to detailed checking of a construction, which has already been defined by a simpler hand calculation. For most other medium technology applications a simpler approach may be used in which the materials design is considered independently from the structural design. In materials design the properties of a chosen fibre lay-up are calculated or measured and these composite element design properties are subsequently used in a structural design analysis. This approach is appropriate for sandwich structures and for many polymer composite materials with WR and CSM reinforcement, or where UD reinforcement is used in standard lay-ups, as in UD cross-ply and angle-ply laminates. It is then possible to use simpler methods for the structural analysis, based on hand calculation using design formulae, charts and tables.

Since most polymer composite structures are composed of relatively thin plate or shell laminates, a plate element (Fig. 3.2) can be taken as the basic materials element in the structure. The properties of a thin laminated plate element may be calculated in terms of laminate structure and ply thicknesses and properties using laminated plate theory, see for example references 3.1–3.3, or by commercially available microcomputer programs. Usually the materials element will have orthotropic symmetry, that is there are two perpendicular symmetry directions in the laminate plane. The most general loading of the plate element, as shown in Fig. 3.2, consists of in-plane direct and shear loads N_{xx}, N_{yy}, N_{xy}, bending moments M_{xx}, M_{yy} and twisting moment M_{xy}. The element properties required in response to these loads are two principal tensile stiffnesses, the shear stiffness, two principal flexural rigidities and the torsional rigidity, along with corresponding strength values in tension, flexure and shear. It is assumed that, as a result of laminate analysis, or by mechanical tests of a plate element, the main element stiffnesses and strengths are known. The structural design analysis is then based on relating these few material element properties, rather than ply properties and detailed ply lay-up, to component geometry and applied loads in order to satisfy the appropriate design criteria.

The materials design problem is discussed in more detail in section 3.3 where design formulae are given for the properties of commonly used polymer composite laminate constructions and sandwich materials.

3.2.2. Short-term properties for design

Loading tests on polymer composites indicate that stress–strain curves are approximately linear elastic, with a brittle failure usually associated with fracture of the fibres. This is illustrated in the typical tensile stress–strain curve for a CSM/polyester laminate shown in Fig. 3.3. Deviations from linearity in the stress–strain curve are apparent[3.2] and indicate the onset of microdamage, such as fibre/resin debonding, resin cracking or transverse ply failures. For design analysis, polymer composites may be characterized as linear eleastic materials, and suitable elastic moduli and failure stresses are required. Depending on the application, ultimate strengths or a lower microdamage failure stress may be used for design. Because of the strong influence of fibre orien-

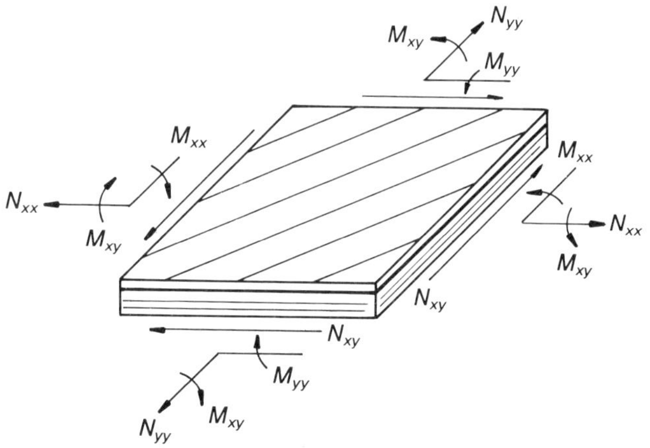

Fig. 3.2. Schematic diagram of a composite laminate element

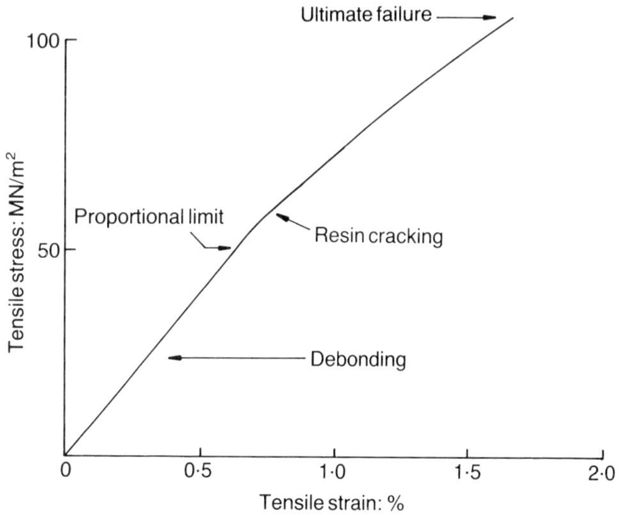

Fig. 3.3. Tensile stress–strain curve for typical CSM/polyester laminate under short-term loads

tation on properties, the required moduli and failure stresses are anisotropic and depend on load direction relative to the fibre direction in the composite.

For design analysis it is convenient to classify composites into two main types, depending on whether they contain chopped or continuous fibre reinforcement. The most widely used polymer composite in civil engineering applications is CSM/polyester. This material is isotropic in its elastic properties since the chopped glass fibres are randomly oriented in the plane of the mat. The required mechanical properties for design are thus a modulus of elasticity E, a Poisson's ratio v and a failure stress σ^*. The shear modulus G satisfies the isotropic elastic relation

$$G = E/2(1 + v)$$

Higher performance composites contain continuous fibres in UD plies or woven fibre fabrics, and are anisotropic. Most materials in current use have orthotropic symmetry, that is they possess two orthogonal symmetry directions in the plane of the laminate. For thin orthotropic materials under plane stress conditions, four orthotropic elastic constants are required to characterize the material stiffness properties. The 1 and 2 axes are defined in the material symmetry directions, i.e. in a UD composite the 1 axis is parallel and the 2 axis is transverse to the fibres. The four orthotropic elastic constants are then the tensile moduli E_{11}, E_{22} in the 1 and 2 directions, the in-plane shear modulus G_{12} and Poisson's ratio v_{12}. The standard convention is that v_{12} is the strain in the 2 direction, per unit strain in the 1 direction, for a material loaded in the 1 direction. The second Poisson's ratio v_{21} is connected to v_{12} by the relation

$$v_{21} = v_{12}E_{22}/E_{11} \tag{3.1}$$

To characterize strength properties, corresponding tensile failure stresses σ^*_{11}, σ^*_{22} for loads applied in the 1 and 2 directions, along with the in-plane shear strength σ^*_{12} are required.

A typical set of modulus and strength design data for a number of polymer composite materials is given in Table 3.1. Since composite properties depend critically on fibre-volume fraction v_f, the data quoted refer to a typically used fibre-volume fraction for each type of reinforcement. The table gives data on glass-reinforced plastics (GRP) with CSM, WR and UD fibre reinforcement. For a more detailed discussion of the engineering design properties of GRP see reference 3.4. Comparative data on UD composites with aramid and carbon fibre are also given in the table. Note that the data given were obtained under tensile loads. Because of the inhomogeneity in polymer composites and the different failure modes which may occur, properties under flexural and

Table 3.1. Design data for typical polymer composite materials (ult—ultimate failure, dam—damage stress)

Property: units	CSM/ polyester	WR/ polyester	UD glass/ epoxy	UD aramid/ epoxy	UD carbon/ epoxy
V_f	0·2	0·35	0·6	0·7	0·7
E_{11}: GN/m^2	8	16	42	76	180
E_{22}: GN/m^2	8	16	12	8	10
G_{12}: GN/m^2	3	4	5	3	7
v_{12}	0·3	0·15	0·3	0·34	0·28
σ^*_{11} ult MN/m^2	95	250	700	1400	1500
σ^*_{11} dam MN/m^2	50	100	–	–	–
σ^*_{11} ult MN/m^2	95	250	30	12	40
σ^*_{11} dam MN/m^2	50	100	–	–	–
σ^*_{12}: MN/m^2	80	95	72	34	68

compression loads will differ from the tensile data quoted. The failure stress data refer mainly to ultimate fracture. Lower failure stresses such as resin cracking stresses may also be required for design and some limited data are given where these are available.

The polymer composites data in the table show the significance for the designer of anisotropic effects. Thus mat-reinforced materials are seen to be isotropic, whereas UD composites have moduli and strengths parallel to the fibres significantly higher than the transverse values. Although WR materials have equal properties in the principal fibre directions, off-fibre-axis properties are much lower, and the relationship between tensile and shear properties is far from isotropic. Transverse failure stresses in UD composites are particularly low, so that the designer must be aware that failure could occur in UD composites as a result of low secondary stresses transverse to the main load-bearing fibre direction. It is this low transverse strength in UD composites which has led to the development for many applications of composite laminates with cross-ply and angle-ply reinforcement, as discussed in section 3.3.2.

3.2.3. Design under long-term loads

Under long-term loads polymer composites no longer exhibit linear elastic behaviour. The polymer matrix is in general viscoelastic with time-, frequency- and rate-dependent properties, see sections 7.6–7.10. It follows that creep effects may be significant in polymer composites under long-term continuous loads, especially in materials with low fibre volume fractions. Under dynamic fatigue loads, complex interactions take place between the fibres and matrix, leading to a degradation in composite mechanical properties. Both these aspects of polymer composites' long-term behaviour require consideration by the designer.

Stress and deformation analysis can be carried out for viscoelastic materials at small strains, see for example reference 3.5. However such analyses are not usually carried out in practice for polymer composites, since the analysis is mathematically involved and there is a lack of suitable viscoelastic materials property data. Under many constant-load situations, it has been found that a simpler pseudo-elastic analysis may be used[3.6] which is in good agreement with test results under creep loads. The basis of the pseudo-elastic method is that for many loading conditions of practical interest, the viscoelastic stress–strain response may be approximated by a set of pseudo-elastic stress–strain curves in which material properties such elastic modulus and failure stress are no longer constants, but may be time-, rate- or frequency-dependent. These pseudo-elastic material properties are then used in conventional elasticity-based design procedures to predict component behaviour at the appropriate time, rate or frequency. In this way orthotropic elastic design formulae or finite element analysis (FEA) may be used for design under long-term loads, provided that appropriate materials property data are used.

For design analysis of a polymer composite component under load for time t, a set of long time moduli E_{11}, E_{22}, G_{12}, ν_{12} are required and failure stresses σ^*_{11}, σ^*_{22}, σ^*_{12} at time t. Suitable values may be obtained from material creep tests as discussed in reference 3.4. For example, in a uniaxial tensile creep test at constant stress σ applied in the 1 direction in the material, the measured strain–time function $\varepsilon(t)$ may be used to define the time-dependent modulus $E_{11}(t) = \sigma/\varepsilon(t)$. The measured creep rupture time at this stress level gives one point on the failure stress-time function $\sigma^*_{11}(t)$. Since the creep strain $\varepsilon(t)$ increases with time under load, it follows that $E_{11}(t)$ is lower than the short term elastic modulus E_{11}. Similarly the creep rupture stress $\sigma^*_{11}(t)$ is lower than the short term strength σ^*_{11}.

Table 3.2. Typical long-term load-reduction factors for GRP materials—data refer to long-term strength/short-term strength ratios at ambient test conditions (see reference 3.4)

Long-term strength-reduction factor	CSM/ polyester	WR/ polyester	UD glass/ epoxy
Fatigue stress ratio			
10^3 cycles	0·6–0·7	0·35–0·45	0·4–0·6
10^6 cycles	0·25–0·35	0·2–0·3	0·2–0·3
Creep rupture stress ratio			
10^2 hours	0·65–0·75	0·55–0·65	0·75–0·9

In dynamic fatigue loading of polymer composites, microdamage such as resin cracking or fibre/resin debonding is initiated and with further cycling leads to ultimate failure at stresses below the short-term strength.[3.4] In a fatigue test programme the number of cycles N required to fail a test specimen at different stress levels is measured. The results are presented on S–N curves showing failure stresses $\sigma_{11}^{*}(N)$, $\sigma_{22}^{*}(N)$ as a function of the number of load cycles, under a range of load conditions. There is also some degradation in the modulus of polymer composites during cyclic fatigue loading, although this is not usually considered to be significant enough to modify the stiffness aspects of the design analysis.

A complete set of time-dependent moduli, failure stresses, and fatigue strengths as in Table 3.1 is not generally available for the main polymer composites. Some typical data, taken mainly from reference 3.4, showing the orders of magnitude of long-term load effects on selected material properties are given in Table 3.2. Presentation of long-term data is simplified, as in Table 3.2, by expressing long-term properties as a proportion of the short-term values. The typical long-term strength and stiffness-reduction factors shown in the table, which for tensile strength are in the range 0.2–0.4 after 10^4 hours constant load or 10^6 cycles fatigue load, indicate the importance of allowing for long-term load effects where relevant in design analysis.

Knowledge of these long-term strength- and stiffness-reduction factors may be used as the basis for design analysis of polymer composites under long-term creep and fatigue using the pseudo-elastic method. These reduced modulus and strength values are then used in conventional short-term elastic analysis to analyse component behaviour at a fixed time t, or after fixed load cycles N. This process may be formalized as in the limit state design method (see chapter 4), by using the modulus- and strength-reduction factors as partial material coefficients. A similar approach is also used in design codes, such as BS 4994[3.7] for GRP tanks, by using the long-term property-reduction factors as design factors for creep and cyclic loads, with analogous factors to allow for environmental damage from high temperatures or water immersion.

3.3. Materials design

Materials design is the determination of the properties of a composite laminate in terms of laminate structure and individual ply materials properties. The laminate properties required for component design analysis must be defined first. Attention is restricted here to specially orthotropic laminates, which are commonly used in practice and are simpler to analyse than the general orthotropic laminate discussed in reference 3.2. Most laminates which are symmetric about their midplane with balanced reinforcement (i.e. each ply oriented at $+\theta$ is balanced by another ply $-\theta$) are specially orthotropic. Just as four elastic constants were required to define the elastic properties of an orthotropic material, four in-plane stiffnesses and four flexural stiffnesses are needed to define the properties of a specially orthotropic laminate element under the general load system shown in Fig. 3.2.

Laminate behaviour can be defined with respect to an x–y co-ordinate system in the plane of the laminate, with x and y axes in the laminate symmetry directions. When subjected to the in-plane direct and shear loads N_{xx}, N_{yy}, N_{xy} shown in Fig. 3.2, the corresponding in-plane strains are ε_{xx}, ε_{yy}, ε_{xy}. The in-plane load–strain relation for a specially orthotropic laminate is then given in matrix notation as

$$\begin{pmatrix} N_{xx} \\ N_{yy} \\ N_{xy} \end{pmatrix} = \begin{pmatrix} A_{11} & A_{12} & 0 \\ A_{12} & A_{22} & 0 \\ 0 & 0 & A_{66} \end{pmatrix} \begin{pmatrix} \varepsilon_{xx} \\ \varepsilon_{yy} \\ \varepsilon_{xy} \end{pmatrix} \qquad (3.2)$$

where, in conventional composites notation, A_{11}, A_{22}, A_{12}, A_{66} are the four in-plane laminate stiffnesses.[3.1,3.2]

Under the bending and twisting moments M_{xx}, M_{yy}, M_{xy} shown in Fig. 3.2, the laminate has curvatures k_{xx}, k_{yy}, k_{xy} and the corresponding moment–curvature relations for a specially orthotropic laminate are defined as

$$\begin{pmatrix} M_{xx} \\ M_{yy} \\ M_{xy} \end{pmatrix} = \begin{pmatrix} D_{11} & D_{12} & 0 \\ D_{12} & D_{22} & 0 \\ 0 & 0 & D_{66} \end{pmatrix} \begin{pmatrix} k_{xx} \\ k_{yy} \\ k_{xy} \end{pmatrix} \qquad (3.3)$$

where D_{11}, D_{22}, D_{12}, D_{66} are the four laminate flexural rigidities. In the laminate materials design analysis the in-plane stiffnesses [**A**] and the flexural rigidities [**D**] are defined in terms of the laminate structure and ply properties. Note that in a generally orthotropic or anisotropic laminate some of the zero coefficients in equations (3.2) and (3.3) may be non-zero, which would imply a coupling for example between bending and twisting. Other quantities which define a coupling between in-plane and bending properties may also be required in anisotropic laminates and these are discussed in more detail in references 3.1–3.3.

3.3.1. Mat, woven and UD fibre reinforcement

Many polymer composite materials are composed of a single reinforcement type, with each ply oriented in the same direction. Examples are GRP laminates composed of all CSM and all WR reinforcement, or UD composites fabricated from prepregs in which all the plies have parallel fibres. These composites will usually consist of alternate layers of fibre separated by resin-rich layers. These local variations in material structure have a small effect on mechanical properties; for example they cause differences in tensile and flexural properties. However, for design analysis they are usually neglected and these materials with a single reinforcement type are assumed to be macroscopically homogeneous. It follows that ply moduli and strengths, as given in Table 3.1, will also be the materials' properties for design analysis and a separate laminate analysis is not required.

In the case when there are no basic ply data, materials design analysis to determine ply properties in terms of fibre and resin properties may be required. Several theoretical formulae are available for predicting composite properties, ranging from simple rule of mixtures approximations to more rigorous analyses of the complete set of elastic constants in a UD composite (see section 2.10 and references 3.1 and 3.2 for a more detailed discussion). These materials calculations for UD composites are included in commercially available design software.[3.8–3.10] Use of these microcomputer programs therefore gives a rapid means of estimating the effect of new reinforcing fibres or fibre-volume fraction on UD ply properties, and could be used as a substitute for a complete set of measured mechanical property data.

In the case considered here of a single reinforcement type the laminate properties A_{11}, D_{11} are readily determined from the orthotropic material properties and the laminate thickness h. It is also assumed that the (1,2) material symmetry axes are the same as the (x, y) plate element symmetry axes, that is in a UD composite the fibres are aligned in the x direction. It then follows from integrating the laminate stress–strain relations[3.2] that the in-plane stiffness components are

$$\begin{aligned} A_{11} &= E_{11}h/\mu \quad & A_{22} &= E_{22}h/\mu \\ A_{12} &= \nu_{12}E_{22}h/\mu \quad & A_{66} &= G_{12}h \end{aligned} \tag{3.4}$$

where

$$\mu = 1 - \nu_{12}^2 E_{22}/E_{11} \tag{3.5}$$

and the flexural rigidities are

$$\left.\begin{aligned} D_{11} &= E_{11}h^3/12\mu \quad & D_{22} &= E_{22}h^3/12\mu \\ D_{12} &= \nu_{12}E_{22}h^3/12\mu \quad & D_{66} &= G_{12}h^3/12 \end{aligned}\right\} \tag{3.6}$$

The laminate failure loads N_{xx}^*, N_{yy}^*, N_{xy}^* and bending moments at failure M_{xx}^*, M_{yy}^*, M_{xy}^* under uniaxial conditions are given in terms of the material failure stresses by the relations

$$\begin{aligned} N_{xx}^* &= h\sigma_{11}^*,\ N_{yy}^* = h\sigma_{22}^*,\ N_{xy}^* = h\sigma_{12}^* \\ M_{xx}^* &= h^2\sigma_{11}^*/6,\ M_{yy}^* = h^2\sigma_{22}^*/6,\ M_{xy}^* = h^2\sigma_{12}^*/6 \end{aligned} \tag{3.7}$$

The composite in-plane and bending properties given above define stiffness and strength properties when loads are applied along the material principal directions, the 1 and 2 axes. In certain applications it is necessary to know the orthotropic material properties for loads applied at an angle to the principal material directions. In these cases the plate element x–y axes are inclined at an angle θ to the material 1–2 axes, as shown schematically in Fig. 3.4. Using equations for the transformation of stresses from the 1–2 to the x–y co-ordinate system, it is possible to derive formulae for the rotated components of the [**A**] and [**D**] stiffness matrices. Details of these transformed components, which are functions of E_{11}, E_{22}, G_{12}, ν_{12}, h and θ, can be found in references 3.1 to 3.3.

Of specific interest for design analysis are the modulus of elasticity E_θ and tensile strength σ_θ for a uniaxial load applied in the x direction of Fig. 3.4, at an angle θ to the fibre or 1 direction in the composite. These quantities may be determined from the more detailed analysis of the transformed stiffness matrix referred to above. The modulus E_θ is given in references 3.1 and 3.4 as

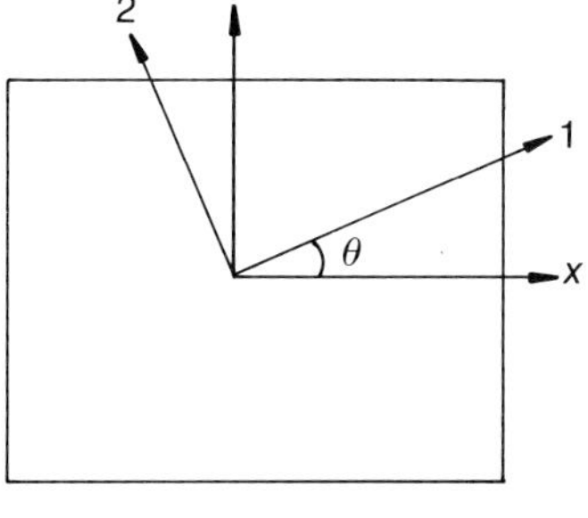

Fig. 3.4. Material (1–2) and plate (x–y) co-ordinate systems

$$\frac{1}{E_\theta} = \frac{\cos^4\theta}{E_{11}} + \left(\frac{1}{G_{12}} - \frac{2\nu_{12}}{E_{11}}\right)\sin^2\theta\cos^2\theta + \frac{\sin^4\theta}{E_{22}} \tag{3.8}$$

Similar expressions may be derived for the modulus in the y direction, the shear modulus and the Poisson's ratio.[3.1]

For an orthotropic material, application of a uniaxial load in the x direction will lead to both tensile and shear stress components in the 1–2 axes. The tensile strength σ_θ will then depend on the multiaxial failure criterion adopted for the material. At the present time there is no universal criterion valid for the main polymer composite materials. The two criteria most widely used are the maximum stress and quadratic stress criteria.[3.3] In the maximum stress criterion, failure occurs when one of the material stress components σ_{11}, σ_{22}, σ_{12} first reaches the corresponding failure stress value σ^*_{11}, σ^*_{22}, σ^*_{12}. In one form of the quadratic stress criterion, the Tsai-Hill criterion which is based on the strain energy in the composite, failure occurs when the following quadratic condition amongst the stress components is satisfied

$$\left(\frac{\sigma_{11}}{\sigma^*_{11}}\right)^2 - \frac{\sigma_{11}\sigma_{22}}{\sigma^{*2}_{11}} + \left(\frac{\sigma_{22}}{\sigma^*_{22}}\right)^2 + \left(\frac{\sigma_{12}}{\sigma^*_{12}}\right)^2 = 1 \qquad (3.9)$$

Using the maximum stress criterion the off-axis tensile strength σ_θ is given by reference 3.1

$$\sigma_\theta = \min\{\sigma^*_{11}/\cos^2\theta,\ \sigma^*_{22}/\sin^2\theta,\ \sigma^*_{12}/\sin\theta\cos\theta\} \qquad (3.10)$$

that is, by the lowest of the three bracketed stress values. Use of the Tsai-Hill condition from equation (3.9) gives the alternative design formula

$$\sigma_\theta = \left[\frac{\cos^4\theta}{\sigma^{*2}_{11}} + \left(\frac{1}{\sigma^{*2}_{12}} - \frac{1}{\sigma^{*2}_{11}}\right)\sin^2\theta\cos^2\theta + \frac{\sin^4\theta}{\sigma^{*2}_{22}}\right]^{-\frac{1}{2}} \qquad (3.11)$$

Both equations (3.10) and (3.11) give reasonably good predictions of the failure stress in off-axis UD composites.

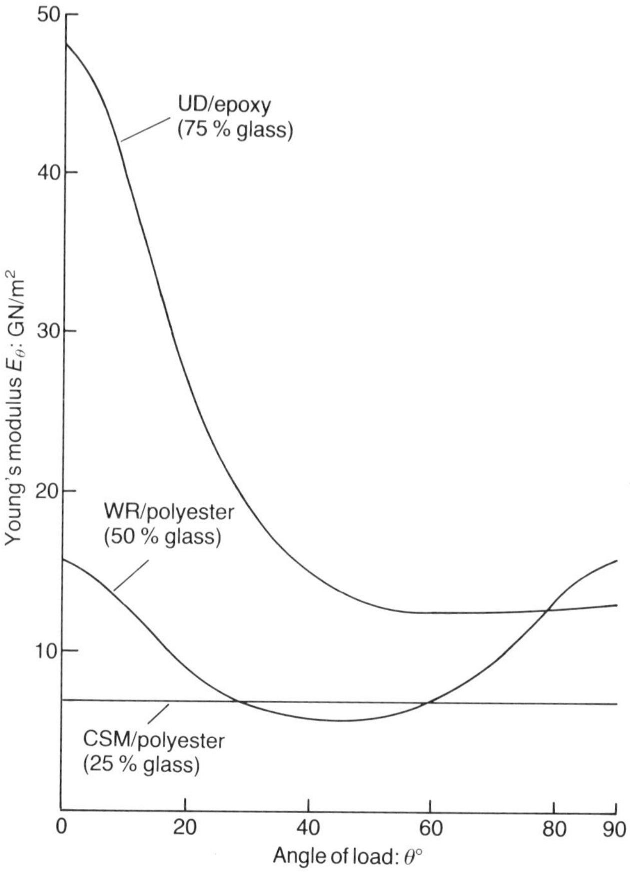

Fig. 3.5. Dependence of tensile modulus of elasticity E_θ on angle of load θ (relative to the principal fibre direction) for different types of reinforcement in GRP

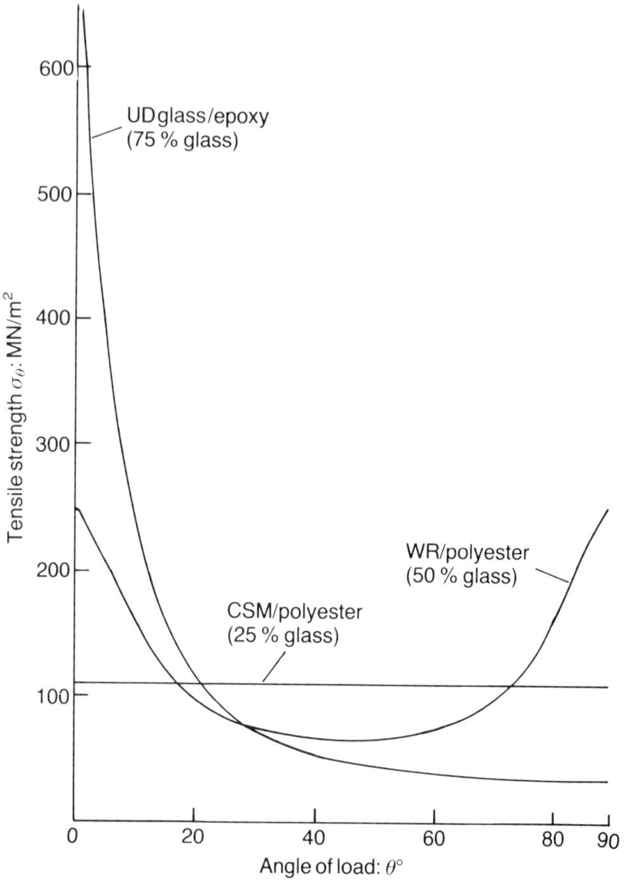

Fig. 3.6. Dependence of tensile strength σ_θ on angle of load θ (relative to the principal fibre direction) for different types of reinforcement in GRP

The significance of material anisotropy in polymer composites can be judged by graphs of E_θ and σ_θ as a function of orientation angle θ. Figs 3.5 and 3.6 show such functions for typical GRP composites of UD glass/epoxy, WR/polyester and CSM/polyester taken from reference 3.4 and which were computed from equations (3.8) and (3.11) using typical data for these materials. The marked orientation-dependence of properties in the UD composite, particularly the strength values, is clear from the figures, in contrast with the isotropic behaviour of CSM/polyester laminates. There is also seen to be a significant reduction in the properties of the WR composite when loaded at 45° to the weave direction, compared with the values at $\theta = 0°, 90°$ which are along the fibre directions in the fabric.

3.3.2. General laminates

Most polymer composites consist of several plies of resin-impregnated fibres laminated together (Fig. 3.2). In general the composite laminate will consist of more than one type of fibre reinforcement; for example a GRP material may contain CSM mixed with WR or UD fibres. Here the main loads are carried by the continuous UD fibres or by the WR plies, and the CSM/polyester, with its high resin content, is used as a corrosion barrier to protect the main load-bearing plies. CSM is also used as an interlayer with WR material to improve interlaminar adhesion and strength. Alternatively, an all-UD composite will have plies oriented at different angles, in order to alleviate the low transverse modulus and strength properties shown in Figs 3.5 and 3.6. In both types of laminate the designer is required to determine the laminate stiffness properties [**A**], [**D**] and failure loads in terms of laminate structure, ply orientations and ply material properties. The calculations may be performed using the analysis methods of laminated plate theory.[3.2,3.3]

In order to simplify laminate analysis, several microcomputer programs are now commercially available for composites materials design.[3.8 to 3.10] These have similar objectives and methods of operation, and are all based on laminated plate theory. They differ in details of operation, with MIC-MAC using a spreadsheet compared with the conventional input-output format in CoALA and the ESDU programs. There are also differences in the materials modelling used for predicting laminate failure loads. The programs are available in several floppy disc versions for running on microcomputers such as the Macintosh, IBM PC and compatibles.

The main input requirements for using the programs are ply thicknesses, ply orientations relative to the plate element axes, ply elastic constants and failure stresses. With this information, the programs calculate the element in-plane and bending stiffnesses, from which engineering constants such as effective elastic moduli and flexural rigidities may be calculated. If the external loads of interest are specified, the programs will then compute ply stresses and strains under these loads, and hence determine whether ply failure occurs and what are the reserve strength factors.

Both multiaxial failure criteria described above, and variants thereof, are used in the software for the determination of laminate failure loads. In some cases it is possible to calculate both first ply failure, when the laminate is damaged, and last ply failure which is the ultimate load for the laminate. The programs permit analysis of the influence on laminate failure of temperature, e.g. during cooling of a laminate after hot-press moulding, and of moisture uptake which causes ply swelling. Certain microstructural analyses to determine ply properties from fibre and matrix properties may also be undertaken, as discussed in section 3.3.1.

Because of the complexity of the mechanical behaviour of a general composite laminate, the designer is recommended to make use of these micro-

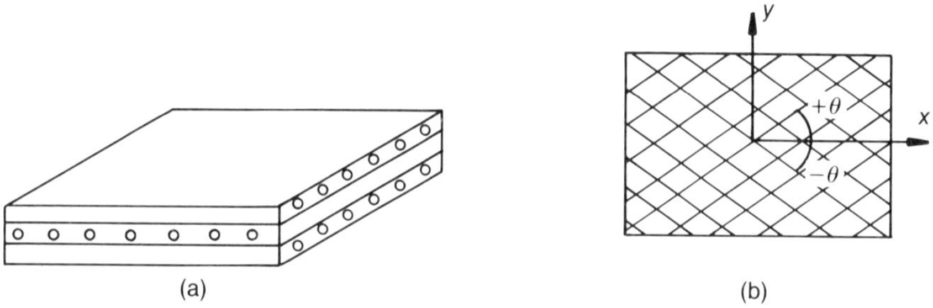

Fig. 3.7. Schematic diagram of laminates: (a) cross-ply; (b) angle-ply

computer programs for the analysis of all but the simplest laminates. However, for certain laminate constructions in wide use, it is possible to devise relatively simple laminate design formulae which are convenient for materials design. These are for UD fibre reinforced composites with cross-ply and angle-ply construction.

3.3.2.1. Cross-ply laminates

In a cross-ply composite alternate UD plies are oriented at angles of 0° and 90° to the plate element axis as indicated in Fig. 3.7(a). If the cross-ply laminate is symmetric about the midplane, then it has orthotropic properties. Application of laminated plate theory to calculate the in-plane stiffness properties then gives[3.1]

$$\begin{aligned} A_{11} &= (h/\mu)\,[VE_{11} + (1 - V)E_{22}] \\ A_{22} &= (h/\mu)\,[(1 - V)E_{11} + VE_{22}] \\ A_{12} &= V_{12}hE_{22}/\mu,\ A_{66} = hG_{12} \end{aligned} \tag{3.12}$$

where h is the laminate thickness and V the volume fraction of 0° plies.

In order to determine the flexural rigidities more information is needed on the ply lay-up, since the contribution of a ply to the flexural rigidity depends on its distance from the laminate neutral axis. In the commonly used case of a symmetric balanced laminate composed of n plies of equal thickness h/n, with $V = 0{\cdot}5$ so that there are an equal number of plies in the 0° and 90° directions, the flexural rigidities can be calculated as described in reference 3.2 to give

$$\begin{aligned} D_{11} &= (h^3/12\mu)\,[\tfrac{1}{2}(E_{11} + E_{22}) + (3/2n)(E_{11} - E_{22})] \\ D_{22} &= (h^3/12\mu)\,[\tfrac{1}{2}(E_{11} + E_{22}) - (3/2n)(E_{11} - E_{22})] \\ D_{12} &= \nu_{12}E_{22}h^3/12\mu \\ D_{66} &= h^3G_{12}/12 \end{aligned} \tag{3.13}$$

The flexural properties depend on the number of plies n, and for cross-ply composites with only a few plies there will be a significant difference in the principal flexural rigidities D_{11} and D_{22}. However, for a large number of plies, as n increases the terms of $0(1/n)$ become negligible and $D_{11} = D_{22} = (h^3/24\mu)(E_{11} + E_{22})$. It is then apparent from equations (3.12) and (3.13) that the cross-ply material behaves as a homogeneous material with modulus $\frac{1}{2}(E_{11} + E_{22})$ in both flexure and tension (*cf* (3.12) with $V = 0{\cdot}5$).

In a general orthotropic laminate there are no simple design formulae for accurately predicting laminate strengths. Even when a uniaxial stress is applied to a laminate, the individual plies will in general be subjected to a multiaxial stress field. The failure analysis then requires a check on the stresses in each ply and comparison with an appropriate multiaxial failure criterion such as those discussed in section 3.3.1, in order to assess ply integrity. Such calculations are most easily carried out using the available laminate analysis software,[3.8–3.10] which may be used for failure analysis under any combination of in-plane loads and bending moments applied to the laminate element. For preliminary analysis of laminates, simpler, approximate strength analysis formulae are useful. These may be derived on making certain simplifying assumptions about the laminate stress fields and modes of failure.[3.1,3.4]

For a cross-ply laminate under axial tensile load in the x direction, it is found that the initial damage usually consists of cracking in the transversely oriented plies. This occurs when the x component of stress in the transverse plies equals the transverse ply strength σ_{22}^*. Considering the influence of the uniaxial stresses only, and considering only the case of a balanced, symmetric laminate with a large number of plies, for the loads at transverse ply failure are found to be[3.4]

$$N_{xx}^* = N_{yy}^* = h\sigma_{22}^* \bar{E}/E_{22} \tag{3.14}$$

where $\bar{E} = \frac{1}{2}(E_{11} + E_{22})$ is the effective modulus for the cross-ply laminate, as given in equation (3.12) with $V = 0{\cdot}5$. When the load is increased beyond this transverse ply failure stress, ultimate failure will occur when the longitudinal plies fracture. If it is assumed that the cracked transverse plies take no further load and that the load is carried entirely by the longitudinal fibres, it follows[3.1] that the ultimate tensile failure load of a cross-ply laminate is

$$N_{xx}^* = N_{yy}^* = \tfrac{1}{2}h\sigma_{11}^* \tag{3.15}$$

which is a simple rule of mixtures result.

An analogous result to equation (3.14) for the laminate damage load under flexural loading may be obtained based on failure of a surface transverse ply, or the transverse ply adjacent to the surface. For a balanced symmetric cross-ply laminate with a large number of plies the bending moments at first ply failure are given by

$$M_{xx}^* = M_{yy}^* = h^2\sigma_{22}^* \bar{E}/6E_{22} \tag{3.16}$$

where $\bar{E} = \frac{1}{2}(E_{11} + E_{22})$ is now the effective flexural modulus. It is not possible to derive a simple result for the ultimate flexural load analogous to equation (3.15), since in flexure many of the transverse plies will still be intact and carrying load and thus contributing to the laminate ultimate strength. The flexural strength is thus expected to be higher than the ultimate tensile strength of $\frac{1}{2}\sigma_{11}^*$ implied by equation (3.15). Since it depends on details of the ply lay-up, it will generally be necessary to calculate it using laminate software.

3.3.2.2. Angle-ply laminates

In an angle-ply laminate (Fig. 3.7(b)) alternate plies are oriented at angles $+\theta$ and $-\theta$ to a plate element axis, the x axis here. Angle-ply laminates which are symmetric about their midplane and balanced, that is they have an equal number of $+\theta$ and $-\theta$ plies, are orthotropic. In the case when they consist of a large number of thin plies they are specially orthotropic with homogeneous stiffness properties. In this case the in-plane bending stiffneses are given[3.1,3.2] by

$$(A_{11}, A_{12}, A_{22}, A_{66}) = h(Q_{11}, Q_{12}, Q_{22}, Q_{66})$$

$$(D_{11}, D_{12}, D_{22}, D_{66}) = \frac{h^3}{12}(Q_{11}, Q_{12}, Q_{22}, Q_{66}) \tag{3.17}$$

where the functions Q_{11} depend on ply angle θ and ply properties, and are defined by the following equations.

$$\left.\begin{aligned} Q_{11} &= (1/\mu)[E_{11}c^4 + (2\nu_{12} + 4\mu G_{12})s^2c^2 + E_{22}s^4], \\ Q_{22} &= (1/\mu)[E_{22}c^4 + (2\nu_{12} + 4\mu G_{12})s^2c^2 + E_{11}s^4], \\ Q_{12} &= (1/\mu)[\nu_{12}E_{22}(s^4 + c^4) + (E_{11} + E_{22} - 4\mu G_{12})s^2c^2], \\ Q_{66} &= (1/\mu)[\mu G_{12}(s^4 + c^4) + (E_{11} + E_{22} - 2\mu G_{12})s^2c^2] \end{aligned}\right\} \quad (3.18)$$

in which $c = \cos\theta$, $s = \sin\theta$. Note that angle-ply laminates with a small number n of plies are not specially orthotropic since there are additional stiffness coefficients such as D_{16} which are $0(1/n)$ and cause bending/twisting coupling. Similar bending-extension coupling terms of $0(1/n)$ may also be present in antisymmetric angle-ply laminates. Both these aspects are discussed further in reference 3.2.

The failure of angle-ply laminates is complex and failure modes depend significantly on the ply angle.[3.1] For a laminate loaded uniaxially in the x direction, the initial damage is frequently observed to be shear cracking in the plies, especially when $\theta < 45°$. The ply shear stress σ_{12} in the 1–2 axes corresponding to a uniaxial stress σ_{xx} along the x axis is obtained from the stress transformation relations as

$$\sigma_{12} = \sigma_{xx}/\sin\theta\cos\theta$$

It follows that at first ply failure due to shear cracking, an estimate of the in-plane load N^*_{xx} and bending moment M^*_{xx} is

$$N^*_{xx} = h\sigma^*_{12}\sin\theta\cos\theta,\ M^*_{xx} = \tfrac{1}{6}h^2\sigma^*_{12}\sin\theta\cos\theta \quad (3.19)$$

where σ^*_{12} is the UD ply shear strength. At ultimate failure, fibre fracture occurs and the ultimate loads can be estimated by netting analysis,[3.1] in which it is assumed that the load in the composite is carried only by the fibres. This model then leads to the following estimates of the ultimate loads in an angle-ply laminate

$$N^*_{xx} = h\sigma^*_{11}\cos^2\theta,\ M^*_{xx} = \tfrac{1}{6}h^2\sigma^*_{11}\cos^2\theta \quad (3.20)$$

which is again expected to be valid for smaller ply angles where the netting analysis assumption is valid.

3.3.3. Sandwich materials

Sandwich materials consisting of polymer composite skins with a low density core are widely used for low weight panels, where the main loads are flexural. Typical sandwich materials used in civil engineering applications have GRP skins of CSM/polyester or WR/polyester, with cores of polyurethane (PU) foam, PVC foam, end grain balsa or plywood. Examples of sandwich materials used in other industries are carbon fibre/epoxy laminated skins, with aluminium honeycomb or resin-impregnated paper honeycomb cores, for aerospace components, and integral skinned PU/PU foam for automotive panel applications. These materials allow expensive polymer composites to be used efficiently as thin load-bearing skins, with other properties such as thermal insulation and good energy absorption characteristics being provided by the cheaper, lighter core materials.

Sandwich beams and panels consisting of thin stiff skins with low density cores are particularly efficient structural materials for flexural loading. The stiff skins are at the plate surface where stresses in flexure are high whilst the core, which is a low stress region, may consist of a lower modulus, low density material to reduce panel weight. For design purposes, sandwich panels may be divided into two main types.[3.12] In one type, core materials have low shear modulus so that the sandwich is subjected to significant shear deformations under transverse loads. Relevant design methods for such materials are discussed in reference 3.13. In this chapter attention is restricted to the second main type in which the core is stiff enough for shear effects to have only a small influence on panel deflections. This includes materials such as GRP/balsa and GRP/PU foam which are widely used in civil engineering applications. For such materials conventional elastic design methods for beams and plates may often be used for the analysis of panel deflections and failure loads, provided that the appropriate sandwich panel stiffness and strength properties are known. The materials design analysis described in this section determines these sandwich material properties in terms of skin and core properties and thicknesses. The methods discussed above for general laminate analysis and particularly some of the software[3.10] may also be used to analyse general sandwich laminates. However, some of the simpler design formulae given here for symmetric sandwich materials are useful for a preliminary design study of many polymer composite sandwich components.

The materials design analysis formulae given here assume a symmetric sandwich material with isotropic elastic skin and core materials. They are also valid in the more general case of sandwich materials with orthotropic skins or cores, in which case the modulus and strength values used would take the principal value in the 1 or 2 direction as appropriate. In practice, many sandwich materials are unsymmetric with one skin thicker than the other. More detailed design formulae for unsymmetric sandwich materials are given in reference 3.13. Alternatively, the present design procedure may be applied approximately to a symmetric sandwich with skin thickness equal to the thinnest skin.

Consider the symmetric three-ply sandwich beam shown schematically in Fig. 3.8. The beam has unit width thickness h and consists of a central core with thickness h_c and skins of thickness h_s, so that $h = h_c + 2h_s$. The core and skin moduli of elasticity along the beam axis are E_c and E_s respectively and it is assumed that transverse strains arising from differences in Poisson's ratio are negligible. The tensile modulus of the sandwich material E_T is given by the rule of mixtures formula

$$E_T = E_c(h_c/h) + E_s(1 - h_c/h) \qquad (3.21)$$

which shows that the skins and core contribute to the tensile stiffness in proportion to their thicknesses. When the beam is subjected to bending moments its behaviour can be determined in terms of an effective flexural modulus E_F given by

$$E_f = E_c(h_c/h)^3 + E_s[1 - (h_c/h)^3] \qquad (3.22)$$

Fig. 3.8. Schematic diagram of sandwich beam

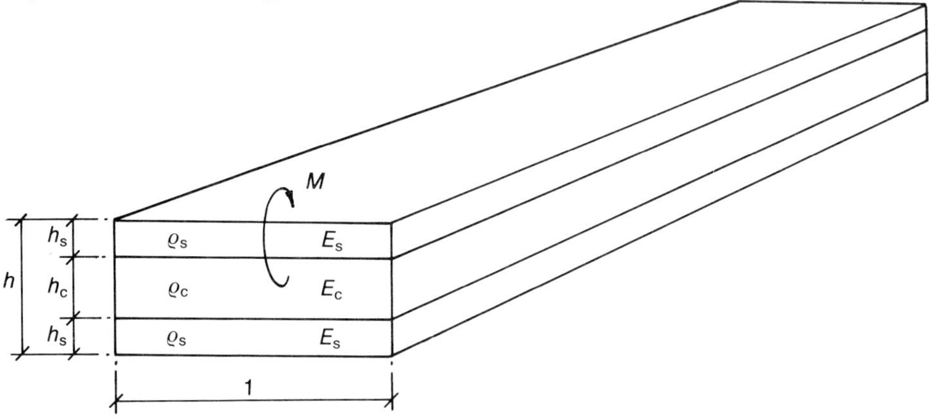

see reference 3.12 for details of the derivation. The main geometrical parameter is the core/sandwich thickness ratio h_c/h. For most sandwich materials $E_c \ll E_s$ so that the skin modulus term in equations (3.21) and (3.22) gives the main contribution to the sandwich stiffness properties and the core contribution is often neglected. Sandwich tensile and flexural moduli are different, with the flexural modulus being higher provided that $E_s > E_c$, which will usually be the case. Having determined the sandwich flexural modulus from equation (3.22), the flexural rigidity per unit width D of a sandwich panel may be defined in the usual way, as

$$D = h^3 E_F / 12(1 - \nu^2) \qquad (3.23)$$

where ν is taken as the Poisson's ratio of the skin material.

Next consider the strength properties of a sandwich material. If it is assumed that under tensile loads ultimate failure is the fracture of the skins, after the core is cracked and no longer carrying load, then the tensile failure load N^* is given by

$$N^* = 2h_s \sigma_s^* \qquad (3.24)$$

where σ_s^* is the tensile failure stress of the skin material in the load direction. If it is assumed that, under an applied bending moment, the sandwich material fails by a tensile or compressive failure of the skin material, the ultimate bending moment M^* is given by equation (3.11)

$$M^* = h^2 E_F \sigma_s^* / 6E_s \qquad (3.25)$$

Other modes of failure may be important in sandwich materials and these are discussed further in reference 3.12. For example, under bending moment the sandwich may fail by skin wrinkling on the compression face. This will occur when the skin-wrinkling stress, given approximately from a local buckling analysis as

$$\sigma_s^* = \tfrac{1}{2} E_s^{1/3} E_c^{2/3} \qquad (3.26)$$

is lower than the tensile or compressive strengths of the skin material. The lowest of the wrinkling stress and tensile or compressive strength is then used in equation (3.25) to determine the ultimate bending moment. Under transverse shear load S applied to the sandwich, it is usually assumed that all the shear stress is transmitted through the core, when core shear failure may become the critical failure mode. If the through-thickness shear strength of the core material is τ^*, the transverse load at failure S^* per unit width is then given approximately by

$$S^* = h_c \tau^* \qquad (3.27)$$

This equation may be used to determine the minimum core thickness required. Other important aspects of sandwich materials analysis such as minimum weight design are discussed in references 3.6 and 3.12, and in greater detail in reference 3.13.

3.4. Analysis procedures for structural elements

The preceding sections discussed how to determine, by calculation or materials tests, the materials design properties of a composite material element. This section considers how this materials design information is used in component structural design analysis. Where finite element analysis (FEA) is used for component stress analysis, the materials design analysis is used to determine the required materials input data for the FEA software. In the case of programs such as NASTRAN[3.14] based on an orthotropic plate element, this will consist of effective orthotropic elastic material properties for the composite laminate or sandwich material. For composites with a single type of reinforcement, as discussed in section 3.3.1, the materials property data required will be the modulus and strength properties shown in Table 3.1. For the more general laminates described in section 3.3.2, the effective average composite properties are required for FEA based on an assumed homogeneous orthotropic behaviour. In this more general case the materials design formulae of section 3.3.2, or the laminate software,[3.8–3.11] are used as a preprocessor to provide the materials input data for the FEA program.

There are many situations where a detailed FEA is not required and design analysis is based on a combination of design formulae, design charts, codes of practice and previous experience. This is the case with many civil engineering structures, such as cladding panels, water tanks, pipework and lighting columns, where polymer composites are in current use. For more critical structures such as bridge decks, walkways, ladders, frame structures, these simpler design methods are valuable for a preliminary design analysis. This is required in order to determine the suitability and cost effectiveness of new structural materials and to select optimum material thickness and fibre-orientation directions.

Existing information for design analysis based on formulae, tables, etc. has been mainly developed for metals below yield (see for example reference 3.15) and is only suitable for isotropic elastic materials. These design methods may thus be used for the isotropic polymer composites, such as CSM/polyester and SMC materials. For the anisotropic polymer composites, these design procedures require modification to include material anisotropy. This section summarizes selected design formulae for orthotropic elastic materials which may be used, in conjunction with the materials design analysis, for the design analysis of polymer composite components. It is beyond the scope of this chapter to cover in detail design analysis with orthotropic materials: attention is restricted to design information on panels in section 3.4.1 and hollow section beams in section 3.4.2, since these are elements of many engineering structures. Section 3.4.3 concludes with a summary of current design codes of practice for polymer composite components, along with a brief note on available design software for composite component analysis.

3.4.1. Design analysis of panels

A polymer composite panel is modelled for design analysis as a flat rectangular plate with sides of length a, b and uniform thickness h, as shown schematically in Fig. 3.9. It is assumed that the panel material is specially orthotropic with material symmetry axes parallel to the x–y co-ordinate axes of the plate. Under in-plane loads panel properties are defined in terms of the in-plane stiffnesses A_{11}, A_{12}, A_{22}, A_{66} as indicated by the plate load–strain relations given in equation (3.2). The subscripts 1-2 for the material symmetry axes now coincide with the x–y plate axes, since the plate is specially orthotropic. Under bending loads, the panel properties are determined by the

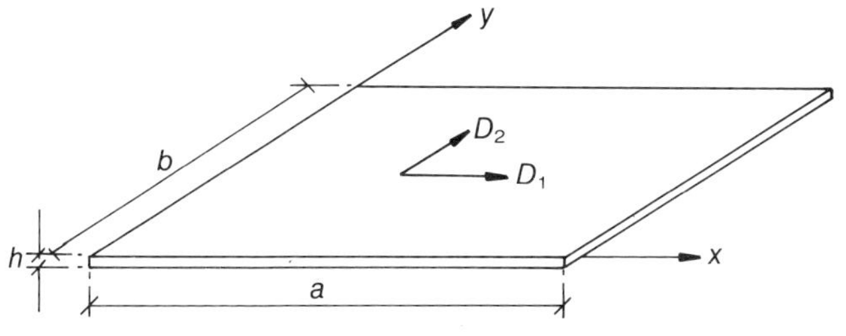

Fig. 3.9. Orthotropic plate geometry

flexural rigidities D_{11}, D_{12}, D_{22}, D_{66} given in equation (3.3). These stiffness properties will in general be dependent on laminate structure and it is assumed that they have been determined in terms of the structure by a materials design analysis, or alternatively by mechanical testing of composite plate specimens with a similar structure. For convenience an alternative notation for the flexural rigidities is introduced

$$D_1 = D_{11}, \quad D_2 = D_{22}, \quad D_3 = D_{12} + 2D_{66} \tag{3.28}$$

3.4.1.1. In-plane loads

The behaviour of the panel under in-plane loads is obtained directly from the load–strain relations (equation (3.2)). For example, under a uniaxial tensile load N_{xx} in the x direction, for a panel with unconstrained edges $N_{yy} = N_{xy} = 0$ and (3.2) gives for the plate extension u

$$u = aN_{xx}/(A_{11} - A_{12}^2/A_{22}) \tag{3.29}$$

since $e_{xx} = u/a$ is the x component of strain. The tensile failure load N_{xx} will depend on the laminate construction and the failure condition adopted and could be given, for example, by equations (3.7), (3.14), (3.15), (3.19) or (3.20).

Under in-plane compression loads, failure in thin panels will usually occur by panel buckling before the material compressive strength is reached. For uniaxial compressive loads applied in the x direction, the buckling load N_{xx}^1 will depend on panel support conditions and aspect ratio a/b. For a rectangular orthotropic plate with simply supported edges, the following formula for the critical buckling load at the onset of buckling may be obtained (see for example reference 3.16)

$$N_{xx}^1 = \frac{\pi^2}{b^2}\left[\frac{D_1}{(a/b)^2} + 2D_3 + D_2(a/b)^2\right] \tag{3.30}$$

which is valid for plates with small or moderate aspect ratios, i.e. $a/b < 2$. For isotropic materials where $D_1 = D_2 = D_3 = D$, the plate flexural rigidity, equation (3.30) reduces to the classical plate buckling formula.[3.15]

For long plates in which $a \gg b$ under compressive end load, local buckling may take place which is independent of the ratio a/b and is given by the design formula

$$N_{xx}^1 = 4\pi^2 D_1 K/b^2 \tag{3.31}$$

where

$$K = \tfrac{1}{2}[(D_2/D_1)^{\frac{1}{2}} + D_3/D_1] \tag{3.32}$$

The failure load in a composite panel under compression is now given by the lowest of the buckling load N_{xx}^1 as given by equations (3.30) or (3.31) and the uniaxial panel strength N_{xx}^* which depends on panel material structure as discussed above.

Design formulae such as these are useful for making a preliminary assessment of the influence of anisotropy in composite panels. For isotropic plates the factor $K = 1$ and equation (3.31) then give the local buckling load for an isotropic plate with flexural rigidity D_1. In a composite plate with longitudinal fibre reinforcement in the loaded x direction, the plate anisotropy ratio $D_2/D_1 < 1$, and it is usually found that $D_3/D_1 < 1$. It follows from equation (3.32) that $K < 1$, and hence the critical buckling load N_{xx} in equation (3.31) is lower than that for an isotropic plate with flexural rigidity D_1. It follows that the longitudinal fibre reinforcement has a destabilizing effect and a long UD composite plate is more likely to fail by local buckling than a corresponding isotropic plate. It is also apparent that use of the classical isotropic design formula, equation (3.31) with $K = 1$, could lead to errors in the design of a composite plate.

3.4.1.2. Flexural loads

Attention is given next to design procedures for the rectangular orthotropic panel shown in Fig. 3.9 under transverse flexural loads. The panel is assumed to have supported edges and to be loaded by a transverse load P, which could be for example a uniform pressure load or a concentrated point or line load.

The main requirement for the design analysis of plates is to determine the maximum plate deflection and the maximum bending and twisting moments. The maximum deflections characterize the plate stiffness and the maximum moments may be compared with plate-bending moment at failure M^*_{xx}, M^*_{yy} defined in section 3.3, or used to determine maximum stresses for comparison with material strength properties. They may then be used in conjunction with suitable failure criteria to specify plate thickness or type of reinforcement. Consider an orthotropic rectangular plate under flexural loading, as depicted in Fig. 3.9. The detailed analysis of reference 3.17 shows that the maximum deflection w may be expressed in the general form

$$w = \alpha P a^2 / D_2 \tag{3.33}$$

where α is referred to as the plate stiffness parameter. For orthotropic plates α is a dimensionless parameter which depends on plate geometry ratio a/b, the flexural rigidity ratios D_1/D_2, D_3/D_2, the plate edge conditions and the loading type and position. If the plate contains a free edge, α depends on Poisson's ratio v_{12}, but it is independent of v_{12} for combinations of clamped and simply supported edges. It may also be shown that the maximum bending and twisting moments in the plate are expressible in the form

$$M_{xx} = \beta_1 P, \quad M_{yy} = \beta_2 P, \quad M_{xy} = \beta_{12} P \tag{3.34}$$

where β_1, β_2, β_{12} are stress concentration parameters which again depend on a/b, D_1/D_2, D_3/D_2, v_{12} and the plate edge and loading conditions. Computation of the full solution to orthotropic plates shows that the twisting moment M_{xy} is usually significantly lower than the bending moments M_{xx}, M_{yy}. Thus β_1, β_2 are expected to be the critical stress parameters for orthotropic plate design and so design information for α, β_1, β_2 is concentrated on.

For isotropic plates $D_1 = D_2 = D_3 = D$ and it is commonly assumed that $v = 0{\cdot}3$, so that the design parameters depend only on plate geometry a/b and loading conditions. Roark and Young (reference 3.15, Table 26) give an extensive table of values for α and β (the maximum of β_1, β_2) for isotropic elastic plates, showing the effect of varying a/b, plate edge and loading conditions. This table is widely used for design calculations on isotropic plates. The main objective here is to supplement these standard results with design information on the influence of material anisotropy on the plate stiffness and stress parameters. In order to provide information valid for a wide range of orthotropic materials, we must select values of the material parameters D_1/D_2, D_3/D_2 and v_{12} which are representative of the main reinforced plastics materials. Note that the α and β parameters are independent of the absolute values of flexural rigidities D_1 or D_2, which are only required when calculating plate deflections from equation (3.33). This simplifies considerably the presentation of design data for orthotropic plates since a wide range of materials can be covered by a few values of the ratios D_1/D_2, D_3/D_2.

For many composite materials, for example, with UD fibres in the 1 direction, it is found that the ratio $D_3/D_2 \simeq 1$ (see for example reference 3.17). It is also found that when $D_3/D_2 \neq 1$, as in woven fibre reinforced composites where typically D_3/D_2 is in the range 0·5–1·0, that the stiffness and stress parameters are relatively insensitive to D_3/D_2. It follows that the main materials parameter which characterizes anisotropy for plate design is the ratio D_1/D_2.

Design data for the stiffness and stress paramters α, β_1, β_2 are given here as functions of the plate aspect ratio a/b and anisotropy ratio D_1/D_2 and for $\nu_{12} = 0{\cdot}3$. Simply supported rectangular plates are considered under uniform pressure and centre point loads. Further design data for clamped edged plates are given in reference 3.10. Reinforced plastics plates with woven fibre reinforcement will have anisotropy ratios $D_1/D_2 = 1$, but will not be isotropic. Their behaviour is determined by the second material parameter D_3/D_2 related to the plate torsional properties. Design data for α, β_1, β_2 for such plates are also presented in reference 3.18. It is found that they are reasonably well approximated by the isotropic values and this type of plate symmetry is not considered specifically here. Further design information in the form of approximate formulae for α, β_1, β_2 as functions of D_1/D_2, a/b are given in reference 3.17 and found to be in good agreement with the tabulated values given here.

Uniform pressure loading p. The applied load is defined by $P = abp$. Maximum deflections and stresses are at the plate centre. Fig. 3.10 shows the stiffness parameter α as a function of anisotropy ratio D_1/D_2 for $1 < D_1/D_2 < 15$, for a range of plate geometries $0{\cdot}25 < a/b < 4$. These α values together with corresponding design data for the stress parameters β_1, β_2 are listed in Table 3.3.

Centre point load P. The maximum deflection and stresses are at the central loading point. Table 3.4 gives the stiffness parameter α as a function of anisotropy ratio D_1/D_2 for $1 < D_1/D_2 < 15$ with plate geometries in the range $0{\cdot}25 < a/b < 4$ and for $\nu_{12} = 0{\cdot}3$. Corresponding design data for β_1, β_2 are also listed in Table 3.4. For point loads the stress concentrations β_1, β_2 depend on the loading diameter d. Table 3.4 is based on computed solutions to the plate equations in which it is assumed that $d = b/32$. Values of β_1, β_2 for other loading diameters d may be estimated on multiplying the value from the table by the factor $0{\cdot}19\,(0{\cdot}435 + 1{\cdot}3 \log 4b/\pi d)$ which has been obtained from reference 3.14.

Use of the basic design formulae equations (3.33) and (3.34), together with

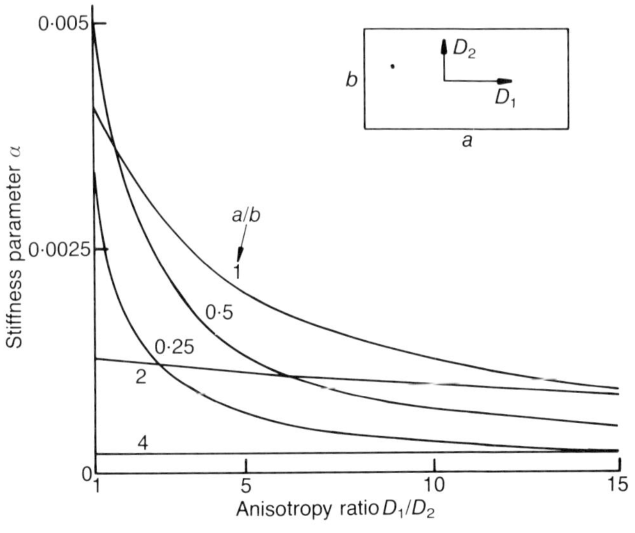

Fig. 3.10. Orthotropic plate stiffness parameter α as function of D_1/D_2 for simply supported plates under pressure loading

Table 3.3. Stiffness and stress parameters for rectangular orthotropic plates under uniform pressure loading, with simply supported edges

D_1/D_2		a/b				
		0·25	0·5	1	2	4
1	α	0·00321	0·00504	0·00406	0·00127	0·0002
	β_1	0·031	0·051	0·048	0·023	0·011
	β_2	0·011	0·023	0·048	0·051	0·031
2	α	0·00163	0·00303	0·0032	0·00122	0·0002
	β_1	0·031	0·059	0·069	0·033	0·014
	β_2	0·0073	0·015	0·038	0·05	0·031
4	α	0·00083	0·00163	0·00232	0·00117	0·0002
	β_1	0·032	0·063	0·094	0·052	0·02
	β_2	0·005	0·0098	0·026	0·047	0·032
7	α	0·00046	0·00093	0·00158	0·00103	0·0002
	β_1	0·032	0·064	0·112	0·078	0·026
	β_2	0·0035	0·0073	0·017	0·042	0·032
10	α	0·00033	0·00067	0·00123	0·00097	0·0002
	β_1	0·032	0·064	0·012	0·098	0·032
	β_2	0·0032	0·0062	0·014	0·039	0·032
15	α	0·00022	0·00045	0·00088	0·00083	0·0002
	β_1	0·032	0·064	0·128	0·125	0·039
	β_2	0·0023	0·0048	0·011	0·034	0·032

Table 3.4. Stiffness and stress parameters for rectangular orthotropic plates under centre point loading, with simply supported edges

D_1/D_2		a/b				
		0·25	0·5	1	2	4
1	α	0·0172	0·0166	0·0116	0·00417	0·00108
	β_1	0·397	0·445	0·435	0·365	0·297
	β_2	0·295	0·365	0·435	0·445	0·397
2	α	0·0111	0·011	0·00933	0·00378	0·00098
	β_1	0·467	0·535	0·553	0·458	0·368
	β_2	0·23	0·29	0·363	0·395	0·357
4	α	0·00707	0·00701	0·00674	0·00337	0·00088
	β_1	0·548	0·635	0·692	0·587	0·463
	β_2	0·178	0·232	0·293	0·345	0·318
7	α	0·00486	0·0048	0·00485	0·00298	0·0008
	β_1	0·62	0·725	0·813	0·725	0·558
	β_2	0·147	0·192	0·243	0·307	0·288
10	α	0·00383	0·00377	0·00384	0·0027	0·00075
	β_1	0·67	0·787	0·892	0·892	0·63
	β_2	0·128	0·17	0·217	0·282	0·272
15	α	0·0029	0·00285	0·00292	0·00234	0·0007
	β_1	0·73	0·862	0·983	0·817	0·575
	β_2	0·11	0·148	0·188	0·213	0·213

design data for α, β_1, β_2 in Tables 3.3 and 3.4, permits a rapid assessment to be made of different types of fibre reinforcement.

For example, Fig. 3.10 may be used to assess the importance of anisotropy ratio D_1/D_2 on plate stiffness. For square plate when $a/b = 1$, α (and hence maximum deflection w) falls most rapidly for low values of D_1/D_2. It follows that use of a high performance UD composite with high values of D_1/D_2 will not lead to a corresponding increase in plate stiffness. This is more apparent when plate aspect ratio a/b is also varied. Fig. 3.10 shows that there is negligible change in stiffness for high aspect ratio plates ($a/b = 4$) with UD fibre reinforcement along the main plate axis. However, for long plates with transverse fibre reinforcement ($a/b = 0{\cdot}25$, $D_1/D_2 > 1$), there is a rapid increase in plate stiffness (i.e. low α and hence low w) as higher performance UD fibres, leading to a high D_1/D_2 value, are used. This demonstrates the advantage of selectively reinforcing long supported plates by fibres transverse to the main plate axis direction.

3.4.2. Design analysis of hollow beams and columns

Thin-walled structures provide an efficient method of using polymer composites, which are expensive materials compared with other structural materials. Fabrication methods such as filament winding, pultrusion and rotational moulding (section 2.8) are suitable for the fabrication of hollow thin-walled beam structures such as cylinders, box beams, U-channels etc. Typical applications of these hollow cylindrical components in civil engineering are filament wound GRP pipe, storage tanks, chimneys and lighting columns; GRP box beams are being used as elements of frame structures, walkways and ducting. The design of pipes and storage tanks is covered by codes of practice and is briefly discussed in section 4.3.

In this section the design analysis of hollow cylinders and box beams is briefly described. Compared with structural metallic materials, polymer composites generally have a low stiffness/strength ratio. It follows that for structural applications design is often stiffness-limited by excessive deflections or buckling failures, rather than stress-limited which is frequently the case with metals. Attention is thus given here to buckling failures in thin-walled sections.

3.4.2.1. Hollow circular cylinders

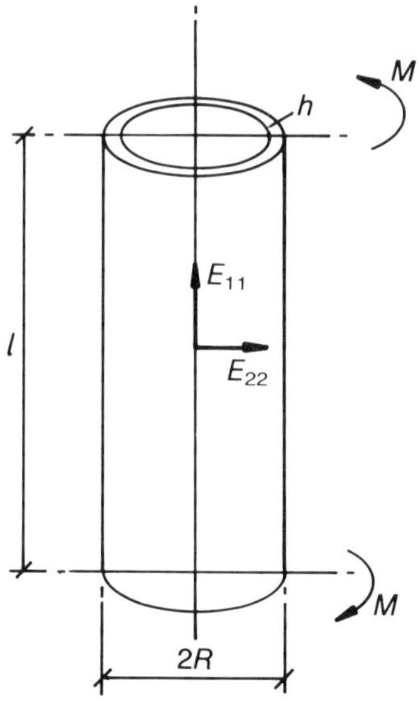

Fig. 3.11. Orthotropic circular cylinder geometry

Consider the thin hollow circular cylinder shown in Fig. 3.11, with radius R, wall thickness h and length l. The cylinder is composed of orthotropic elastic material with the 1 and 2 material symmetry axes in the axial and circumferential directions, respectively. The cylinder is thin-walled so that the material stiffness properties are characterized by the longitudinal elastic modulus E_{11}, the circumferential modulus E_{22}, the wall in-plane shear modulus G_{12} and the longitudinal Poisson's ratio ν_{12}.

Under axial compression load, failure of a long cylindrical column for which $l \gg R$ may take place through compression strength failure of the wall material, through Euler buckling of the column, or through local buckling in the cylindrical wall. Compression strength failure and Euler buckling may be analysed using conventional isotropic design formulae[3.15] provided that the appropriate longitudinal failure stress σ^*_{11} and modulus E_{11} are used in the analyses, which are essentially one dimensional. However, for local buckling some modification to the isotropic design formulae is required.

A detailed theoretical analysis[3.19] gives the following design formula for the critical buckling stress σ'_{11} at the onset of ring and "chessboard" buckling in a long orthotropic cylinder under axial compression

$$\sigma'_{11} = (h/R)(E_{11}/(3\mu)^{\frac{1}{2}})K_1 \qquad (3.35)$$

Here K_1 is a dimensionless parameter dependent on the orthotropic elastic constants of the wall material, and is given by

$$K_1 = [2(1 + \nu_{12}(E_{22}/E_{11})^{\frac{1}{2}})(E_{22}/E_{11})^{\frac{1}{2}}(G_{12}/E_{11})]^{\frac{1}{2}} \qquad (3.36)$$

The factor K_1, the anisotropy factor, takes the value $K_1 = 1$ for isotropic materials, in which case equation (3.35) reduces to the standard design formula for the local buckling stress in an isotropic cylinder.[3.15] It follows that when $K_1 < 1$ buckling is more likely, and when $K_1 > 1$, buckling is less likely, than in a column of isotropic material with modulus $E = E_{11}$. Equation (3.35) can thus be used to assess the influence of different fibre reinforcement geometries on the compressive failure load of a long composite column, such as a chimney.

The design of orthotropic circular cylinders under an applied bending moment M, as shown schematically in Fig. 3.11, has been considered.[3.20] Failure is assumed to take place by material fracture in the cylinder wall, or by local buckling in the column wall. A unified design formula for M^*, the bending moment at failure, is derived in reference 3.20 which takes account of these two failure modes. It is found that for long thin-walled circular cylinders

$$M^* = \pi R^2 h \phi_1 \sigma_{11}^* \qquad (3.37)$$

where σ_{11}^* is the longitudinal strength (tensile or compressive) of the wall material, and ϕ_1 is a buckling co-efficient. This is defined by the formula

$$\phi_1 = (\eta K_1/(3\mu)^{\frac{1}{2}})(h/R)(E_{11}/\sigma_{11}^*) \qquad (3.38)$$

or $\phi_1 = 1$, if this is smaller than the value calculated from equation (3.38). In equation (3.38) K_1 is defined in equation (3.36), and a conservative estimate for design purposes of the parameter η is $\eta = 0{\cdot}65$. When $\phi_1 < 1$, column failure is controlled by buckling, and the factor ϕ_1 in equation (3.37) shows the reduction in column bending strength due to buckling. If $\phi_1 > 1$ as calculated from equation (3.38), then failure is controlled by the longitudinal material failure stress and $\phi_1 = 1$ is set in equation (3.37), which is then the standard design formula for column bending strength.

Equations (3.37) and (3.38) may be used for the design analysis of hollow GRP lighting columns and telegraph poles under bending moments arising from wind loading,[3.18] along with a similar design procedure for octagonal section columns. The influence of material anisotropy on column bending strength is dependent on the factor K_1 defined in equation (3.36) which controls buckling failures. Typical data for GRP materials[3.20] show that: $K_1 = 0{\cdot}32$ for a UD glass/polyester column with axial fibre reinforcement;

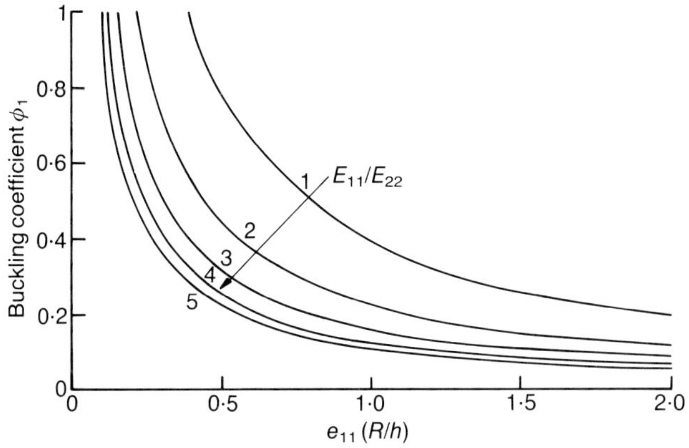

Fig. 3.12. Buckling coefficient for orthotropic circular cylinders is bending ($\nu_{12} = 0{\cdot}3$, $G_{12}/E_{22} = 0{\cdot}38$)

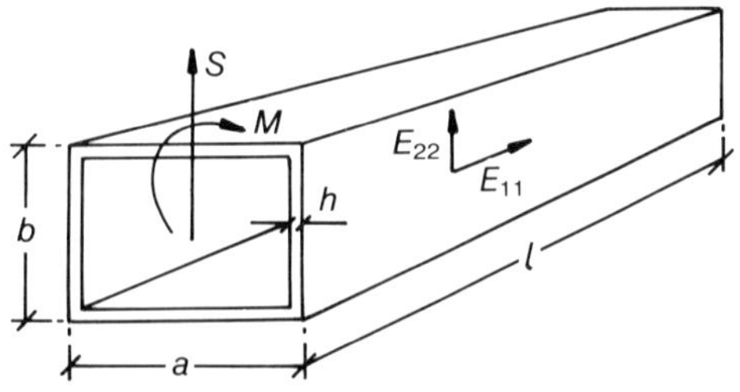

Fig. 3.13. Orthotropic box beam geometry

$K_l = 1{\cdot}62$ with hoop reinforcement; $K_l = 1{\cdot}32$ for $\pm 45°$ angle-ply reinforcement; and $K_l = 0{\cdot}79$ for a balanced woven glass reinforcement, with warp direction along the column axis. It is apparent that the isotropic value $K_l = 1$ would be a poor approximation for design with these GRP materials. It is also apparent that axial reinforcement, which is preferred for higher column flexural rigidity, has a destabilizing influence on local buckling, whilst hoop reinforcement is stabilizing.

As a further aid to the designer, it is convenient to show the buckling factor ϕ_l on a design chart. Fig. 3.12 shows such a chart for ϕ_l at different values of the material anisotropy ratio E_{11}/E_{22} with $\nu_{12} = 0{\cdot}3$ and a typical value $G_{12} = 0{\cdot}38E_{22}$ assumed. Replacing σ_{11}^*/E_{11} by the failure strain e_{11}^*, the parameter ϕ_l is plotted against $e_{11}^*(R/h)$. The figure shows how rapidly ϕ_l decreases with increasing E_{11}/E_{22} and increasing R/h. For a given choice of composite material, the design chart enables the critical value of cylinder radius/thickness R/h to be determined at which buckling will just take place, corresponding to $\phi_l = 1$ in Fig. 3.12. Hence the designer can specify column wall thickness.

3.4.2.2. Box beams

A hollow thin-walled box beam is shown in Fig. 3.13, with length l, sides a, b and uniform wall thickness h. The beam is composed of orthotropic elastic materials with the 1 and 2 material symmetry axes in the beam axial direction, and in the transverse direction of each sidewall respectively. It is assumed here that the same orthotropic material with elastic moduli E_{11}, E_{22}, G_{12}, ν_{12} is used for each sidewall, and for convenience its flexural stiffness is characterized by the flexural rigidities D_1, D_2, D_3 defined in equation (3.28).

The box beam is subjected to transverse loads, which give rise to a bending moment M and transverse shear force V on a beam cross-section, as indicated in Fig. 3.13. For thin-walled box beams the axial flexural rigidity D is given by

$$D = \tfrac{1}{6}b^3h(1 + 3a/b)E_{11}/\mu \qquad (3.39)$$

For beams long enough for transverse shear deflection effects to be neglected, as discussed in reference 3.21, classical beam theory[3.15] may be used for determining beam deflections under transverse load, with flexural rigidity given in equation (3.39). When considering beam failure under transverse loads, account must be taken of possible local buckling failures in the thin walls, and classical beam theory based on material fracture requires modification.

A unified failure analysis of composite box beams is described in reference 3.22 and compared with measured properties of pultruded GRP box beams in reference 3.21. Under the applied bending moment M and transverse shear force S shown in Fig. 3.13 it is assumed that possible failure modes are: tensile or compressive material fracture on the beam lower and upper faces; local buckling failure at the upper (compression) face; shear fracture in the sidewalls which carry the main transverse shear stress; and shear buckling failure in the sidewalls. The ultimate bending moment M^* for the box beam section is given as

$$M^* = \tfrac{1}{3}b^2h(1 + 3a/b)\phi\sigma_{11}^* \qquad (3.40)$$

where σ_{11}^* is the longitudinal material failure stress and ϕ is the compressive buckling coefficient defined by

$$\phi = \tfrac{1}{3}\pi^2K(h/a)^2(E_{11}/\mu\sigma_{11}^*) \qquad (3.41)$$

or $\phi = 1$, if this is lower than the value defined in equation (3.41). In the derivation of equations (3.40) and (3.41), equation (3.31) for the local buckling of a long orthotropic plate is used so that the parameter K is the anisotropy factor defined in equation (3.32).

The ultimate transverse shear load S^* in the box section is given by the design formula

$$S^* = \tfrac{4}{3}bh\psi\sigma_{12}^* \qquad (3.42)$$

where σ_{12}^* is the shear strength of the wall material and ψ the shear buckling coefficient. This is defined as

$$\psi = 6{\cdot}4K_2(h/b)^2(E_{11}/\mu\sigma_{12}^*) \qquad (3.43)$$

or $\psi = 1$ if this is lower. The anisotropy factor in shear K_2 is given by

$$K_2 = (D_2/D_1)^{\frac{1}{2}}(D_3/D_1)^{\frac{1}{2}}f(\delta) \qquad (3.44)$$

where

$$\delta = (D_1D_2)^{\frac{1}{2}}/D_3 \qquad (3.45)$$

and the function $f(\delta)$ is defined as

$$f(\delta) = \begin{cases} \delta^{\frac{1}{2}}(0{\cdot}62 + 0{\cdot}38/\delta), & \delta \geqslant 1 \\ 0{\cdot}89 + 0{\cdot}04\delta + 0{\cdot}07\delta^2, & \delta < 1 \end{cases} \qquad (3.46)$$

Again the anisotropy factor K_2 is defined so that $K_2 = 1$ for an isotropic box beam and hence the influence of anisotropy on shear buckling behaviour is readily assessed by calculating K_2.

Equations (3.40)–(3.43) are the basic design formulae for the failure analysis of box beams under transverse load. When $\phi < 1$, failure in bending is controlled by compression-face buckling and the ultimate bending moment M^* is reduced by a factor ϕ. If $\phi > 1$ when calculated from equation (3.41) then failure is controlled by the lowest of the material longitudinal tensile and compressive strengths. Then $\phi = 1$ in equations (3.40) which becomes the box beam bending-strength design formula. Similarly, when $\psi < 1$ shear failure is controlled by sidewall buckling, with the beam ultimate transverse force S^* reduced by the factor ψ. These design formulae thus show clearly the influence of material strength and stiffness properties, anisotropy and beam-section geometry on the failure loads of a fibre reinforced box beam. Computation of ϕ and ψ shows immediately whether local buckling will reduce the beam bending and shear strength. Section 3.5.1 considers as a design example the analysis of a GRP box beam with different types of fibre reinforcement using these design procedures.

3.4.3. Design codes

This section indicates briefly other sources of design information on polymer composites. Because polymer composites are still relatively new structural materials, only a few codes of practice have been established. Those that exist apply mainly to glass fibre reinforced polyester resins (GRP), which are the most widely used polymer composite materials, for applications such as GRP pipes and storage tanks. In order to use these materials in more general civil engineering applications, the designer should work from first principles using methods such as those described in this chapter. A number of design datasheets on specific analyses such as composite plate buckling, calculation of vibration modes, etc. have been produced to help in these calculations. The composite materials design software[3.8–3.11] is also a valuable tool for the designer working from first principles, and this is now being augmented by selected composite component analysis software.

Current BS design codes for GRP materials are BS 4994[3.7] on GRP vessels and tanks, BS 5480[3.23] on GRP pipes and fittings for water supply and Bs 6464[3.24] on GRP pipes and fittings for process plants. These standards are very detailed and include: specifications for GRP material properties; design failure criteria; safety factors; effect of the environment (temperature, water and chemical immersion); design formulae and tables for calculating the influence of design loads on pipe and tank components; and information relevant to the industry concerned, such as details of construction, erection and inspection. The main materials covered by these standards are CSM/polyester, WR/polyester, combined CSM/WR reinforced polyester, and filament-wound glass/polyester and glass/epoxy, which are widely used materials in pipework and process plant components. The design procedures recommended in these codes take account of the laminated structure of GRP materials, but differ in detail from the materials design procedure described in section 3.3.

The design procedure was first proposed in reference 3.7 for the design of GRP vessels and tanks, and has since been adopted in the pipes standards. The calculation of an appropriate laminate construction is based on the allowable unit loading and unit modulus for the polymer composite material. The unit load is the laminate load expressed per unit width of laminate. GRP material properties such as the ultimate unit strength are then defined as the laminate strength per unit width, per unit weight of fibre reinforcement. The unit modulus is defined similarly, as the unit load per unit strain, per unit weight of reinforcement. This approach to materials design recognizes the fact that the loads in a GRP laminate are carried by the glass fibre reinforcement and, for the same weight of reinforcement, the laminate unit properties are independent of thickness, which is determined mainly by the resin/glass ratio. Since most vessels and pipes are subjected to in-plane stresses, this approach enables the allowable laminate unit load to be easily calculated by summing the unit loads in each ply. Materials data on GRP materials are also simplified by presenting unit strength and unit modulus per unit weight of reinforcement.

In the design procedure the allowable strain in the laminate is limited to 0·2%, or to one tenth of the resin-failure strain if this is lower. On multiplying this strain by the unit modulus an allowable strain limited unit load N_S is determined. A load-limited unit load N_L is obtained on dividing the laminate ultimate unit strength by a design factor k. The allowable design unit load is then the lowest of N_S and N_L. The design factor k is included to cover such variables as degradation in GRP strength properties over long times and with cyclic fatigue loads, the reduction in properties with temperature and environment, and variations in properties due to methods of manufacture. The standard requires a minimum design factor for k of 8. As a result of the chosen strain limit and the design factor which allow for environment, GRP laminates are designed in BS 4994 so that the fibre/resin bond should not break down in the structure and failures such as environmental stress cracking should not occur in aqueous environments.

A valuable source of design information on polymer composites has been built up over many years by ESDU International.[3.25–3.36] This information was derived initially for aerospace composites, but it is also relevant to composite materials in many other branches of engineering. The design datasheets listed in references 3.25–3.28 are all concerned with materials design analysis, i.e. the prediction of laminated plate properties, and were used as the basis for the ESDU microcomputer software.[3.8] A selection of datasheets on composite component analysis are listed in references 3.29–3.32. These consider: the calculation of buckling loads in orthotropic plates (ESDU 80023); the natural frequencies of orthotropic and sandwich plates (ESDU 83036 and 85037); and the design of bonded joints (ESDU 81022), which is a topic very relevant for composite structures.

Many of these ESDU datasheets have been used as the basis of mainframe and microcomputer programs. Examples of microcomputer software are listed in references 3.33–3.36. These include: the calculation of stress concentrations in laminated plates containing a circular hole under in-plane loads (ESDU 2034); compressive buckling loads in composite plate (ESDU 2021); and the natural frequencies and damping factors of composite plates (ESDU 3007 and ESDU 3013).

In using this composite design software it is first necessary to determine composite material properties, using for example the laminate analysis software.[3.8–3.11] The use of microcomputers for composite design is facilitated further if the materials design and component design software are integrated into suites of programs. An example of such integrated design software is given in MIC–MAC[3.10] which includes, in addition to composites laminate analysis, design procedures for idealized structures such as tubes, beams and pressure vessels. After first carrying out the materials design analysis, it is possible with MIC-MAC to use a different spreadsheet to input component design data such as length and diameter of a cylindrical vessel, applied pressure or twisting moment, and then to carry out the composite component design analysis. Thus the basic software tools for polymer composite material and component design are being established. With the increasing use of these materials in structural applications, the designer can expect further design software to become available.

3.5. Design examples

3.5.1. Box beam analysis

This section gives a design analysis example for a polymer composite box beam, which shows how to combine the materials design of section 3.3 with the structural analysis of section 3.4.2. The example considers a GRP box beam in three-point bending fabricated from UD glass fibres with four types of fibre reinforcement: UD longitudinal fibres (fibres at 0° to beam axis); UD transverse fibres (90° fibres); a cross-ply laminate (0°/90° fibres) and an angle-ply laminate (±45° fibres). To show the influence of these different fibre reinforcement geometries on the box beam properties, the flexural rigidity, the failure load and the failure mode are calculated in each case.

The box beam geometry is shown in Fig. 3.13 and in this example a square cross-section thin-walled beam is considered with

$$l = 600\,\text{mm}, \quad a = b = 50\,\text{mm}, \quad h = 3\,\text{mm}$$

The beam is assumed to be simply supported at its ends and loaded transversely by a transverse centre load P. It follows that the maximum bending moment M in the beam, which is at the beam centre is $M = Pl/4$, and the transverse shear force $S = P/2$ is constant along the beam.[3.15] Let P_F^* and P_S^* be the ultimate loads in the beam corresponding to failure in bending and shear, respec-

Table 3.5. Box beam material properties for different fibre reinforcement geometries

Glass fibre reinforcement	Material properties			Rigidity ratios		Anisotropy factors	
	E_{11}/μ: GN/m^2	σ^*_{11}: MN/m^2	σ^*_{12}: MN/m^2	D_2/D_1	D_3/D_1	K	K_2
UD longitudinal	43·3	700	72	0·29	0·33	0·43	0·34
UD Transverse	12·4	30	72	3·5	1·12	1·5	2·17
0/90 cross-ply	27·8	350	72	1	0·5	0·75	0·81
±45 angle-ply	24·0	350	350	1	1·56	1·28	1·18

tively. On combining these beam relations with equations (3.40) and (3.47) for the box beam bending moment and shear force at failure, it follows that

$$P^*_F = (4b^2h/3l)(1 + 3a/b)\phi\sigma^*_{11}$$

$$P^*_S = 8bh\psi\sigma^*_{12}/3 \qquad (3.47)$$

On assuming there are no interaction effects between bending and shear failures, the ultimate load in the beam will be the lowest of P^*_F and P^*_S.

In order to calculate these beam failure loads from equation (3.47), materials data on the wall material longitudinal and shear strengths σ^*_{11}, σ^*_{12} and the buckling coefficients ϕ, ψ are required. The buckling coefficients depend additionally on longitudinal modulus E_{11}/μ in the beam wall and on the wall flexural rigidity ratios D_2/D_1, D_3/D_1. The required wall properties were first calculated from a materials design analysis and the results are given in Table 3.5. It is assumed that the beam is fabricated from UD glass fibre reinforced epoxy resin with fibre-volume fraction $V_f = 0{\cdot}6$ and the basic materials data given in Table 3.1 are therefore used. For the UD longitudinal fibre box beam the 1 axis of the beam coincides with the 1 axis in the material, so that the required wall properties, E_{11}, σ^*_{11}, coincide with the materials properties in Table 3.1, and the required values are listed again in Table 3.5. For the UD transverse fibre reinforced beam, the beam 1 axis now corresponds to the transverse 2 axis in the material, so that for example the longitudinal wall modulus $E_{11}/\mu = 12{\cdot}4\,\text{GN/m}^2$ which is the transverse modulus of the UD glass/epoxy material. Equations (3.6) and (3.28) are then used to calculate the flexural rigidity ratios from the relevant materials properties.

In the cross-ply laminate it is assumed that there are equal numbers of fibres in the 0° and 90° directions and there are a large number of individual plies so that the terms of $0(1/n)$ in equation (3.13) are negligible. The cross-ply longitudinal modulus A_{11}/h is then calculated from equation (3.12) with $V = 0{\cdot}5$, and the flexural rigidities from equation (3.13). The cross-ply material longitudinal strength N^*_{xx}/h is assumed to be the ultimate strength at fibre fracture and is thus obtained from equation (3.15). The laminate shear strength is equal to the shear strength σ^*_{12} of the UD ply material. In a similar manner the angle-ply material properties in Table 3.5 are calculated from equations (3.17) and (3.18) with $\theta = 45°$. The failure properties of the angle-ply laminate are assumed to be at ultimate fibre fracture and are estimated from the netting analysis method. The required longitudinal failure stress is $N^*_{xx}/h = \sigma^*_{11}\cos^2\theta$ as given in equation (3.29), and the corresponding wall material shear strength as given by netting analysis in reference 3.1 is $N^*_{xy}/h = \sigma^*_{11}\sin\theta\cos\theta$.

Having determined the effect of different reinforcement geometries from the materials analysis, the box beam properties are now calculated. First the anisotropy factors K, K_2 are obtained from equation (3.32) and (3.34) using the appropriate flexural rigidity ratios D_2/D_1, D_3/D_1, and the results are shown in Table 3.5. It is seen that K and K_2 have similar values for each reinforcement type. It follows from their definition that wall buckling is more likely in UD longitudinal and cross-ply laminates where $K, K_2 < 1$. The UD transverse and angle-ply laminates for which $K, K_2 > 1$ are stabilized against buckling, compared with isotropic wall materials. This is confirmed by Table 3.6 which gives the buckling coefficients ϕ, ψ for the box beams, calculated from equations (3.41) and (3.43) with the appropriate materials data from Table 3.5. The table shows that the flexural buckling coefficient $\phi = 0{\cdot}32$ and $0{\cdot}73$ for UD longitudinal and cross-ply beams, whereas $\phi = 1$ for UD transverse and angle-ply beams. The shear buckling coefficient $\psi = 1$ for all reinforcement types, indicating that shear buckling failures will not occur.

The beam failure loads P_F^*, P_S^* are now calculated from equation (3.47) and the results are given in Table 3.6. In each case it is seen that $P_F^* < P_S^*$, showing that box beam failure occurs through the applied bending moment, rather than the transverse shear load. It follows that beam failure is predicted to be on the upper or lower faces as a material tensile or compression strength failure in the UD transverse and angle-ply beams (where $\phi = 1$), or as a compressive buckling failure in the UD longitudinal and cross-ply beams (where $\phi < 1$). Table 3.6 also shows that the strongest box beam in three-point bending is the $\pm 45°$ angle-ply with a predicted failure load of 23·3 kN, followed by the cross-ply beam with a failure load of 17 kN. The UD longitudinal beam, whose strength is reduced considerably by buckling, has $P_F^* = 14{\cdot}9$ kN. The lowest strength beam is the UD transverse beam, which is the most stable against buckling, but has the lowest material strength in the beam axial direction.

Table 3.6 also shows the beam flexural rigidity calculated from equation (3.39). It is seen that the UD longitudinal beam has the highest stiffness with flexural rigidity $D = 216{\cdot}5$ Nm^2, which is nearly double the angle-ply beam value of 120 N mm^2. The cross-ply beam is about 15% stiffer than the angle-ply beam, with the UD transverse beam having a significantly lower stiffness than the others. In selecting the best type of reinforcement for an application, the designer has thus to balance the requirements for box beam stiffness and strength. None of these four types of reinforcement combines the highest stiffness with the highest strength, although the UD longitudinal beam would

Table 3.6. Calculated flexural rigidities and box beam failure loads

Glass fibre reinforcement	Flexural rigidity D : Nm^2	Buckling factors		Calculated failure loads		Predicted failure mode
		ϕ	ψ	P_F : kN	P_S : kN	
UD longitudinal	216·5	0·32	1	14·9	28·8	Compression buckling
UD transverse	62	1	1	2·0	28·8	Tensile/ compression material fracture
0/90 cross-ply	139	0·73	1	17·0	28·8	Compression buckling
±45 angle-ply	120	1	1	23·3	140	Tensile/ compression material fracture

achieve this if it were stabilized against buckling. An optimum design may well consist of an angle-ply beam for its strength and stability, reinforced with UD longitudinal fibres to increase its flexural rigidity. Such hybrid constructions could also be rapidly assessed using the materials and structural design methods given here. This design example shows the complexity of the interaction between fibre reinforcement geometry and structural behaviour, and demonstrate the value of simplified design methods for assessing these factors in the preliminary design of composite structures.

3.5.2. Design case study—building panel

3.5.2.1. Specification of building panel

It has been decided by the building architect to have a modular building structure employing a steel frame as the primary structure and secondary structural wall panels. The terms primary and secondary are used to catagorize the structure into that which, if it fails will cause the building to collapse (primary) and that which, if it fails, will cause only local damage (secondary). The modular construction will enable the building user to replace the wall panels with glass and vice versa should he wish to alter the internal layout during the life of the building.

As the wall panels need to be aesthetically pleasing, low in maintenance costs and light to handle, the designer has recommended the use of a sandwich construction incorporating fibre reinforced polymer skins and a synthetic foam core. This choice of material exploits the previously mentioned advantage of composites—good environmental behaviour, low thermal conductivity and light weight. The designer would also be conscious of the disadvantages when carrying out his design, particularly with respect to cost—the higher cost of materials, the need for tooling for manufacture and the reduced robustness.

The architect has set the parameters that the panels must give a minimum U value of 0·60 W/m^2 °C, that they should be self-coloured to reduce maintenance costs, and that the finish to both the outside and the inside surface be smooth and of a high quality. The panels must also provide Class 0 fire propagation rating when tested to BS 476.[3.28] The panel size is 1·80 m wide × 3·0 m high.

The designer's first consideration must be that of the manufacturing process as this will affect the required surface finish and the panel cost, which in turn

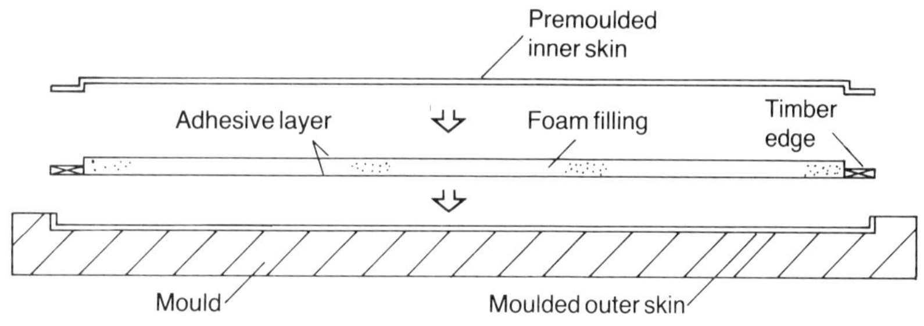

Fig. 3.14. Fabrication method 1—separate mouldings bonded together

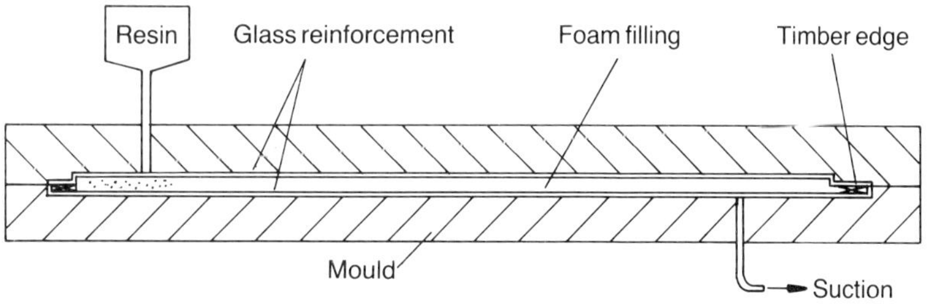

Fig. 3.15. Fabrication method 2—resin-injected panel

is affected by the required quantity of 100 panels. There are two manufacturing methods that could be considered—contact moulding in a female mould and resin injection in a closed mould. The former method would necessitate the bonding of the inner skin to the foam (Fig. 3.14) while the latter method enables the complete panel to be made in one operation (Fig. 3.15). Injection is best undertaken by vacuum assistance as this reduces the pressure on the moulds. The advantages of this method are that the panel is made to a known geometry and in one operation. The disadvantage is the difficulty of quality control of the bond between the resin-impregnated skin and the foam. However, as the vacuum-assisted injection method provides a superior product, and with only marginal increase in tooling cost, the designer, in conjunction with the manufacturer, makes the decision to use this method.

The load on the panel is primarily pressure induced by wind, although there will be handling considerations. It has been assumed that the panels do not contribute to the building's structural integrity. The designer, in conjunction with the architect, agrees the method of fixing the panel to the main frame, utilizing a plastic extruded mullion secured to the steel frame of the building (Fig. 3.16). This method of fixing determines the boundary conditions of the panel, i.e. simply supported with freedom both to rotate and to translate, enabling the structural engineer to undertake the design analysis calculations. These determine the required skin and core thicknesses and hence the detailed specifications, which enable the manufacturer's drawing to be produced.

The vacuum-injection manufacturing process necessitates the use of random glass fibre mat as the reinforcement in a continuous, as opposed to chopped filament form, as the latter tends to wash out when the resin is brought into the mould. With advice from the manufacturer the resin/fibre ratio is set at 2·5 : 1 which enables preliminary levels of ultimate stress to be selected for the surface skins by using either manufacturer's data or previously gained test data. Polyurethane foam is selected for the core material. The resin specification is a function of the manufacturing methods, the need for Class 0 fire resistance and the required colour pigmentation. If necessary, advice is sought from the resin supplier to ensure all the parameters can be satis-

Fig. 3.16. Panel to frame joint

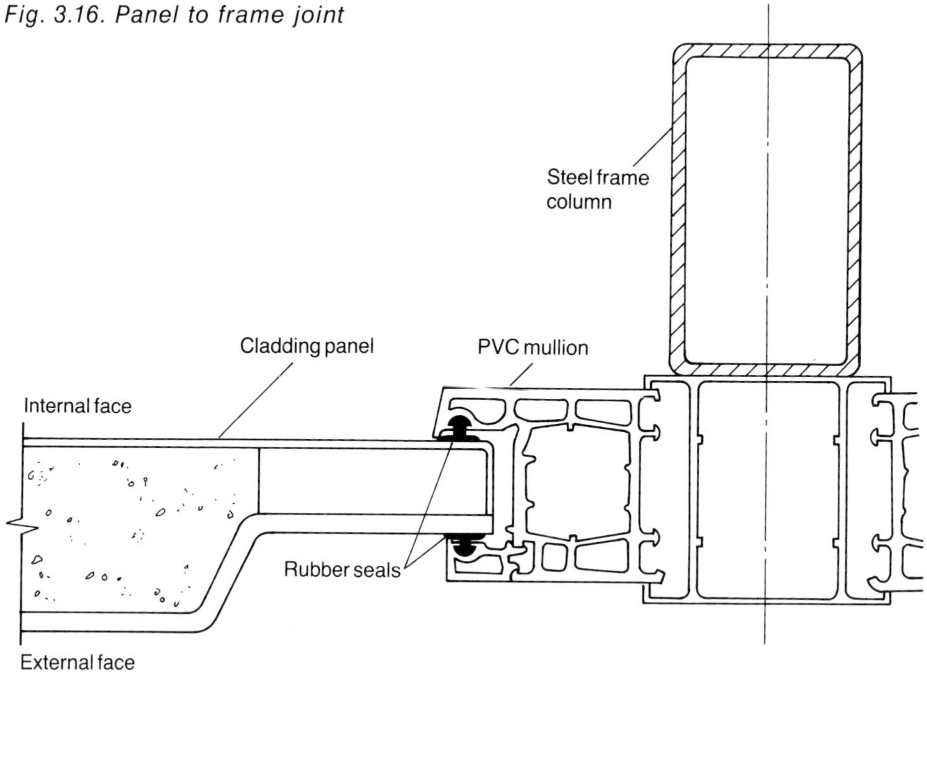

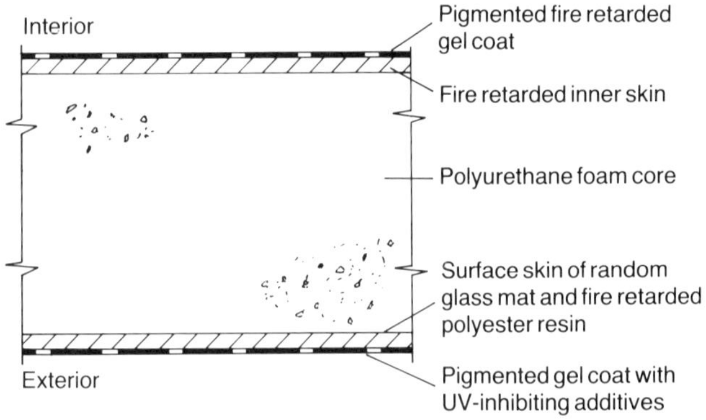

Fig. 3.17. Final panel specification

factorily met. Regarding colour it is important to select a resin that has additives to resist UV attack. If possible, dark colours should be avoided as these tend to fade in time. It is normal practice not to use fire-retardant additives to the gel coat exposed to the atmosphere as they will decrease the surface resistance to the environment and aggravate colour retention. The backing resin reinforcing the surface skin and that of the inner skin are specified to give the necessary fire resistance. The finalized panel specification is as shown in Fig. 3.17.

After manufacturing the first panel it is put aside for inspection and samples are cut out to determine whether the bond between the skin and the foam core is adequate. The sandwich is mechanically tested which confirms the stiffness of the sandwich, the integrity of the bond and the mechanical strength.

The completed building is shown in Fig. 3.18. Note the use of glass reinforced polyester fascia panels at the top of the building.

3.5.2.2. Design analysis of panel

Summary of panel design requirements.

Fig. 3.18. The completed building—the Training Centre for Exxon Chemicals (courtesy Trafalgar House Building and Construction Engineering Holdings Ltd)

(*a*) Flat panel to be designed, 1·80 × 3·0 m, simply supported on all edges.
(*b*) Panel is loaded by wind pressure assumed uniformly distributed over the outer surface.
(*c*) Basic wind speed assumed at 45 m/s. BS CP3[3.37] gives a surface pressure ranging from +440 Pa to −628 Pa depending upon the location of panel on the walls of building.
(*d*) Panel to be designed as a lightweight composite sandwich with GRP facings on a foam core.
(*e*) Facings to be continuous filament mat CFM/polyester with a resin/fibre weight ratio about 2·50 : 1.
(*f*) The foam core is required to achieve adequate shear strength and also to provide a panel with U at least equal to that of a brick cavity wall, say $U \leqslant 0{\cdot}60\ \mathrm{W/m^2\,°C}$.
(*g*) Mechanical properties of CFM facings are to be typical mean values appropriate to the resin/fibre ratio and good standards of workmanship.
(*h*) A design factor of safety of 4·0 to be applied to the properties to give permissible working values.
(*i*) Permissible mid-panel limit deflection to be not more than span/400.

3.5.2.3. *Design calculations*

A suitable sandwich construction is shown schematically in Fig. 3.19. The core is a rigid PU foam with a density of 32 kg/m^3 and thickness h_c = 30 mm. The facings are manufactured from layers of 300 g/m^2 CFM with R = 2·5 : 1. The thickness of individual layers will be approximately 0·75 mm and, with 4 layers on each face, will produce a symmetric construction with h_s = 3 mm.

Material mechanical properties. From Table 3.1 for CSM (equivalent to CFM)

Tensile modulus, $E_{11} = E_{22} = E_S = 8\ \mathrm{GN/m^2}$

Shear modulus, $G_{12} = 3\ \mathrm{GN/m^2}$

Poisson's ratio, $\nu_{12} = 0{\cdot}30$

Ultimate tensile strength, $\sigma^*_{11} = \sigma^*_{22} = 95\ \mathrm{MN/m^2}$

Ultimate shear strength, $\sigma^*_{12} = 80\ \mathrm{MN/m^2}$

For the PU foam

Ultimate shear strength, $\tau^* = 138\ \mathrm{kN/m^2}$

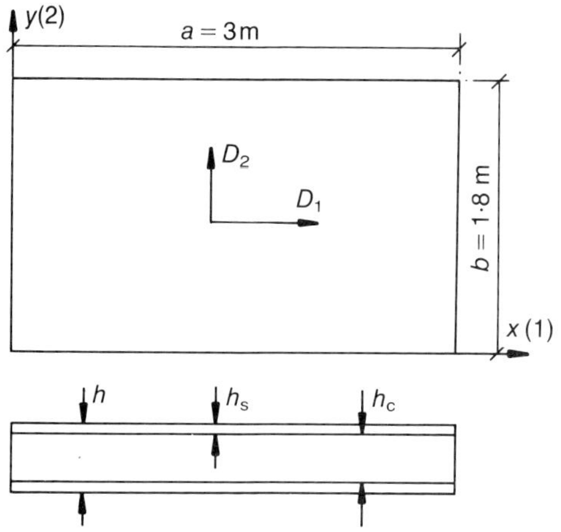

Fig. 3.19. Schematic diagram of sandwich panel

Compression modulus, $E_c \simeq 10\,\mathrm{MN/m^2}$ (anticipated)

The effective flexural modulus for the sandwich panel is obtained from equation (3.22)

$$E_F = E_C\left(\frac{h_c}{h}\right)^3 + E_S\left[1 - \left(\frac{h_c}{h}\right)^3\right]$$

The contribution from the low density, low stiffness, core will be insignificant consequently the first term can be neglected. Substitution gives

$$E_F = 8 \times 10^9\left[1 - \left(\frac{30}{36}\right)^3\right] = 3{\cdot}37\,\mathrm{GN/m^2}$$

The flexural rigidity per unit width of sandwich from equation (3.23) is given as

$$D = \frac{h^3 E_F}{12(1 - \nu^2)}$$

Substitution gives

$$D_1 = D_2 = \frac{(36 \times 10^{-3})^3 \times 3{\cdot}37 \times 10^9}{12(1 - 0{\cdot}30^2)} = 14398\ \mathrm{Nm^2/m}$$

Core shear. For a simply supported rectangular panel, with uniform pressure loading, Table 26 of reference 3.15 can be used to determine the shear load on the core. The maximum value occurs at the mid-point of the long edge and is given by

$$R = \frac{\gamma P}{a}$$

with a panel aspect ratio $a/b = 3{\cdot}0/1{\cdot}8 = 1{\cdot}667$. The interpolation of the data gives $\gamma = 0{\cdot}494$. P is the total load on the panel

$$P = 628 \times 1{\cdot}8 \times 3{\cdot}0 = 3391\,\mathrm{N}$$

thus giving a maximum value of R of

$$R = \frac{0{\cdot}494 \times 3391}{3{\cdot}0} = 558\,\mathrm{N/m}$$

With a core thickness $h_c = 30\,\mathrm{mm}$ and a core shear strength $\tau^* = 138\,\mathrm{KN/m^2}$, the allowable load at failure from equation (3.27) is

$$S^* = h_c\tau^*$$

$$= 30 \times 10^{-3} \times 138 \times 10^3 = 4140\,\mathrm{N/m}$$

and an applied design factor of safety of 4·0 gives

$$S^* = \tfrac{1}{4} \times 4140 = 1035\,\mathrm{N/m}$$

thus providing an adequate margin against the applied load of 558 N/m. Core thickness reduction is possible subject to adequate flexural strength and stiffness from the following calculations.

Equations (3.33) and (3.34) give the maximum panel deflection and the maximum bending moment respectively.

$$w = \frac{\alpha P a^2}{D_2}$$

$$M_{xx} = \beta_1 P, \quad M_{yy} = \beta_2 P$$

Note that for this particular loading condition the twisting moment M_{xy} is zero.

For CFM the anisotropy ratio $D_1/D_2 = 1$, and with a panel aspect ratio $a/b = 1{\cdot}667$ the data in Table 3.3 can be interpolated for the stiffness and strength parameters

$$\alpha = 0{\cdot}0022, \ \beta_1 = 0{\cdot}033, \ \beta_2 = 0{\cdot}051$$

With total panel load $P = 3391\,\text{N}$

$$\text{moment } M_{xx} = 0{\cdot}033 \times 3391 = 112\,\text{Nm/m}$$

$$\text{moment } M_{yy} = 0{\cdot}051 \times 3391 = 173\,\text{Nm/n}$$

The permissible bending moment is obtained from equation (3.25)

$$M^* = \frac{h^2 E_F \sigma_s^*}{6E_s}$$

where σ_s^* is the strength of the sandwich skin (either ultimate tension or compression, or local skin wrinkling depending upon the proportions and properties of panel components). Local skin-wrinkling stress can be approximated from equation (3.26) thus

$$\sigma_s^* = \tfrac{1}{2} E_s^{1/3} E_c^{2/3}$$

Substitution gives

$$\sigma_s^* = \tfrac{1}{2}(8 \times 10^9)^{1/3} \times (10 \times 10^6)^{2/3}$$

$$= 46{\cdot}38\,\text{MN/m}^2$$

This value is less than half the ultimate tensile strength, $\sigma_{11}^*, \sigma_{22}^*$ ($= 95\,\text{MN/m}^2$) and will therefore be used in equation (3.25) for calculation of permissible moment. Substitution gives

$$M^* = \frac{(36 \times 10^{-3})^2 \times 3{\cdot}37 \times 10^9 \times 46{\cdot}38 \times 10^6}{6 \times 8 \times 10^9}$$

$$= 4220\,\text{Nm/m}$$

and an applied design factor of safety of 4·0 gives

$$M^* = \tfrac{1}{4} \times 4220 = 1055\,\text{Nm/m}$$

providing a more than adequate margin against the maximum applied moment

$$M_{yy} = 173\,\text{Nmm}$$

The maximum deflection, at the panel centre, (by substitution into equation (3.33)) gives

$$w = \frac{0{\cdot}0022 \times 3391 \times 3^2}{14398}$$

$$= 4{\cdot}66 \times 10^{-3}\,\text{m} \quad \text{or } 4{\cdot}66\,\text{mm}$$

From this the deflection/span ratio w/b can be calculated

$$w/b = \frac{4{\cdot}66}{1800} = 1/386$$

This value is marginally higher than the specified requirement of 1/400 and the flexural stiffness is seen to dictate design.

It is necessary to check U value for the proposed sandwich panel. For the PU foam core with a density of $32\,kg/m^3$, the value of K_c is

$$K_c = 0{\cdot}0231\,W/m\,°C$$

For CFM/polyester facing laminates

$$K_s = 0{\cdot}20\,W/m\,°C$$

The U value for sandwich $= 1/\Sigma R$ where $\Sigma R = h_c/K_c + h_s/K_s$. With $h_c = 30\,mm$ and $h_s = 3\,mm$

$$\Sigma R = \frac{30 \times 10^{-3}}{0{\cdot}0231} + \frac{2 \times 3 \times 10^{-3}}{0{\cdot}20}$$

$$= 1{\cdot}299 + 0{\cdot}030 = 1{\cdot}329$$

Hence

$$U = \frac{1}{1{\cdot}329} = 0{\cdot}752\,W/m^2\,°C$$

This is greater than the specified value of 0·60 and will require an increase in core thickness (hence try $h_c = 40\,mm$). The revised calculation for the U value is

$$\Sigma R = \frac{40 \times 10^{-3}}{0{\cdot}0231} + \frac{2 \times 3 \times 10^{-3}}{0{\cdot}20}$$

$$= 1{\cdot}732 + 0{\cdot}030 = 1{\cdot}762$$

Hence

$$U = \frac{1}{1{\cdot}762} = 0{\cdot}568\,W/m^2\,°C$$

This is satisfactory, hence the 40 mm core is acceptable.

The revised core thickness will modify the earlier calculations for panel strength and stiffness, particularly the stiffness calculation which found the 30 mm core to be marginal. Revise stiffness calculations

$$E_F = 8 \times 10^{-3}[1 - (40/46)^3] = 2{\cdot}74\,GN/m^2$$

$$D_1, D_2 = \frac{(46 \times 10^{-3})^3 \times 2{\cdot}74 \times 10^9}{12(1 - 0{\cdot}30^2)} = 24423\,Nm^2/m$$

$$w = \frac{0{\cdot}0022 \times 3391 \times 3^2}{24423} = 2{\cdot}75\,mm$$

$$\text{Ratio } w/b = 2{\cdot}75/1800 = 1/646$$

This is now more than adequate. The facing thickness can be reduced without significant effect on U value. It will be practical to reduce the thickness of the inner face only leaving the outer face at 3 mm to provide protection against local damage.

If 3 layers are tried on inside face

$$h_s = 3 \times 0{\cdot}75 = 2{\cdot}25\,mm$$

4 layers on outside face give

$$h_s = 4 \times 0{\cdot}75 = 3{\cdot}0\,mm$$

Further revise stiffness calculation

$$E_F = 8 \times 10^{-3}\,[1 - (40/45{\cdot}25)^3] = 2{\cdot}47\,GN/m^2$$

This is not strictly correct as the equation for E_F is for a symmetric sandwich construction. However, for the degree of asymmetry present the error is negligible and acceptable for this calculation

$$D_1, D_2 = \frac{(45{\cdot}25 \times 10^{-3})^3 \times 2{\cdot}47 \times 10^9}{12(1 - 0{\cdot}30^2)} = 20957\,\mathrm{Nm^2/m}$$

$$w = \frac{0{\cdot}0022 \times 3391 \times 3^2}{20957} = 3.20\,\mathrm{mm}$$

$$\text{Ratio } w/b = 3{\cdot}20/1800 = 1/562$$

The panel remains more than adequate and it can therefore be accepted.

References

3.1. Hull D. *An introduction to composite materials*. Cambridge University Press, Cambridge, 1981.
3.2. Tsai S.W. and Hahn H.T. *Introduction to composite materials*. Technonic, Westport, 1980.
3.3. Tsai S.W. *Composites design*. Think Composites Software, Dayton, 1987.
3.4. Johnson A.F. *Engineering design properties of GRP*. British Plastics Federation, London, 1984.
3.5. Christensen R.M. *Theory of viscoelasticity*. Academic, New York, 1971.
3.6. Johnson A.F. and Sims G.D. *Design procedures for plastics panels*. National Physical Laboratory, Teddington, 1987.
3.7. British Standards Institution. *Design and construction of vessels and tanks in reinforced plastics*, BS 4994. BSI, London, 1981.
3.8. Engineering Sciences Data Unit. *Stiffnesses and properties of laminated plates*. ESDU 2022. ESDU International, London, 1987.
3.9. Engineering Sciences Data Unit. *Failure of composite laminates*, ESDU 2033. ESDU International, London, 1987.
3.10. Think Composites Software. *MIC-MAC*. Think Composites Software, Dayton, 1987.
3.11. Cranfield Institute of Technology. *CoALA*. Cranfield Institute of Technology, 1987.
3.12. Johnson A.F. and Sims G.D. Mechanical properties and design of sandwich materials. *Composites* **17**, No. 4, 1986.
3.13. Allen H.G. *Analysis and design of structural sandwich panels*. Pergamon, Oxford, 1969.
3.14. McNeal-Schwendler. *NASTRAN*. McNeal-Schwendler Co. Ltd, Ewell Surbiton, Surrey.
3.15. Roark R.J. and Young W.C. *Formulae for stress and strain*. McGraw-Hill, New York, 1975.
3.16. Allen H.G. and Bulson P.S. *Background to buckling*. McGraw-Hill, London, 1980.
3.17. Johnson A.F. and Sims, G.D. Simplified design procedures for composite plates under flexural loading. In: *Proc. 2nd Int. Conf. Composite Structures*, I. Marshall (ed), Applied Science, London, 1983.
3.18. Johnson A.F. Design analysis for GRP plates in flexure. In: *Proc 13th Reinforced Plastics Congress*. British Plastics Federation, London, 1982.
3.19. Dow N.F. and Rosen B.W. Structural efficiency of orthotropic cylindrical shells subjected to axial compression, *AIAA Journal* **4** 1966, No. 3, 481–485.
3.20. Johnson A.F. Design of RP cylinders under buckling loads, In: *Proc. 40th Reinforced Plastics/Composites Institute Annual Conference*, Paper 10-E. SPI, New York, 1985.
3.21. Sims, G.D. *et al.* Mechanical and structural properties of a GRP pultruded section, *Composites structures*, **8**, 1987, 173–187.
3.22. Johnson, A.F. *Simplified buckling analysis for RP beams and columns*. ECCM-1, Bordeaux, 1985.

3.23. British Standards Institution. *Specifications for GRP pipes and fittings for water supply*. BS 5480. BSI, London, 1985.
3.24. British Standards Institution. *Specification for reinforced plastic pipes, fittings and joints for process plants*, BS 6464. BSI, London, 1984.
3.25. Engineering Sciences Data Unit. *Stress analysis of laminated flat plates*. ESDU 74039. ESDU International, London.
3.26. Engineering Sciences Data Unit. *Stiffness of laminated flat plates*. ESDU 75002. EDSU International, London.
3.27. Engineering Sciences Data Unit. *Laminate stacking sequence for special orthotropy*. ESDU 82013. EDSU Internation, London.
3.28. Engineering Sciences Data Unit. *Failure modes of fibre reinforced laminates*. ESDU 82025. ESDU International, London.
3.29. Engineering Sciences Data Unit. *Buckling of rectangular specially orthotropic plates*. ESDU 80023. ESDU International, London.
3.30. Engineering Sciences Data Unit. *Natural frequencies of rectangular, specially orthotropic laminated plates*. ESDU 83036. ESDU International, London.
3.31. Engineering Sciences Data Unit. *Natural frequencies of simply-supported sandwich papers with laminated face plates*. ESDU 85037. ESDU International, London.
3.32. Engineering Sciences Data Unit. *Guide to the use of data items in the design of bonded joints*. ESDU 81022. ESDU International, London.
3.33. Engineering Sciences Data Unit. *Buckling of rectangular orthotropic plates*. ESDU 2021, ESDU International, London.
3.34. Engineering Sciences Data Unit. *Stress concentrations in composite plates*. ESDU 2034. ESDU International, London.
3.35. Engineering Sciences Data Unit. *Natural frequencies of laminated flat plates*. ESDU 3007, ESDU International, London.
3.36. Engineering Sciences Data Unit. *Damping in composite plates*, ESDU 3013. ESDU International, London.
3.37. British Standards Institution. *Wind loads CP3:* chapter V, part 2. BSI, London, 1972.
3.38 British Standards Institution. *Method of test for fire propagation for products*. BS 476, part 6. BSI, London, 1981.

4 Application of limit state design principles to composite structural systems

P. R. Head and R. B. Templeman

4.1. Introduction

An understanding of composite material properties and manufacturing methods and knowledge of design analysis methods are not sufficient on their own to enable reliable cost-effective structures to be designed and built. They enable strengths and stiffnesses to be evaluated for assumed composite configurations so that strengths and deflections of structures can be predicted. However, they do not address the questions of reliability and performance, which are fundamental to engineering design. For example they do not provide answers to the following questions.

(*a*) Is a thin-walled composite section as reliable as a thicker one and which manufacturing processes produce the most reliable performance?
(*b*) Is fire resistance more important than environmental resistance in a particular application? Improving fire resistance often leads to a reduction in durability.
(*c*) Can the use of a more expensive material with better quality control be more cost-effective than the use of a cheaper, more variable material?
(*d*) What stress and strain levels are acceptable under normal working load conditions?
(*e*) Can the extra tooling and production costs associated with fine dimensional tolerances be justified because of the advantages in structural reliability which would result from their adoption?

These aspects are usually covered in a specification for manufacture and construction and a code of practice for design, but standard specifications and codes of practice do not yet exist for composites for use in civil and structural engineering applications. This chapter describes limit state design principles which can provide a basis for these which is able to embrace materials, materials testing, manufacturing processes, quality control, design procedures and reliability.

Limit state design principles have been developed in Europe on both a national and international basis by the civil and structural engineering industries and are widely recognized as providing a sound basis for the design of steel and concrete structures. However there has been some opposition to the introduction of limit state codes of practice from engineers familiar with the previous allowable stress methods. This opposition has focused particularly on the complexity of the new codes and the difficulty of using them for design. There has also been criticism of the empirical adjustments that have been made to ensure that structures designed to the new codes are not much different from those designed to the old ones. These adjustments have meant that consistency in reliability between different loading scenarios has not always been maintained.

The approach to defining the materials criteria for design of composites is complicated when compared with steel, concrete and aluminium. The designer needs to become intimately involved with the details of stress–strain and strength behaviour and changes in this behaviour with time, temperature, stress environment and process. Limit state design principles are ideally suited to dealing with a large number of interacting design aspects and it should be possible to develop design codes of practice for composites which simplify rather than increase the complexity of the designer's task. Furthermore, there are no existing allowable stress codes for composites, and therefore there is nothing to stop a completely rational limit state design philosophy being adopted throughout.

Limit state design provides a logical design procedure which identifies the limit state at which a structure ceases to fulfil its design functions. The aim of limit state design is to achieve acceptable probabilities that the relevant limit states will not be reached during the intended life of the structure. The assessment of these probabilities sets up a framework within which the uncertainties in design data, loading, stress analysis, processing, geometry etc, can be quantified and understood. Competitiveness and reliability should flow from careful application of the procedures. Indeed, limit state design may be the key to unleashing the full potential of the structural use of composites in the construction industry.

The chapter begins with an introduction to limit state design principles and then goes on to describe those characteristics of composites which make the application of the principles different from those relating to other structural materials.

A design procedure is then discussed under the same headings that would appear in a code of practice, i.e. design life, methods of analysis, specification for loads, limit state requirements, properties of materials, and analysis of the strengths of laminates. No design is complete without a specification, fully integrated with the assumption and philosophy of the design method, and therefore this is then discussed, paying particular attention to quality assurance and control.

Finally some simple design examples are given to illustrate the application of limit state design methods to composite materials.

4.2. General principles of limit state design

The development of limit state principles began in the 1960s and in 1974 it was agreed by the Joint Committee on Structural Safety (JCSS—an inter-association committee sponsored by six major international associations concerned with structures and materials) that the principles form part of an international system of unified standard codes of practice for structures. The first volume of this work, entitled *Common unified rules for different types of construction and materials*, was published in 1978 and deals with a common basis for setting up codes to ensure adequate and appropriate treatment of safety and reliability whatever the material or type of construction. It is these principles which can be adopted for design of fibre reinforced polymer (FRP) structures. Further details can be found in references 4.1, 4.2 and 4.3. FRP is used in this chapter to embrace all fibre/matrix composites.

4.2.1. Aims of limit state design

In the JCSS unified rules the aim of design is defined as

> "the achievement of acceptable probabilities that the structure being designed will not become unfit for the use for which it is required during some reference period and having regard to its intended life. Thus, all structures or structural elements should be designed to sustain, with an appropriate degree of safety, all loads and deformations liable to occur during construction and proper use, to perform adequately in normal use and to have adequate durability during the life of the structure."

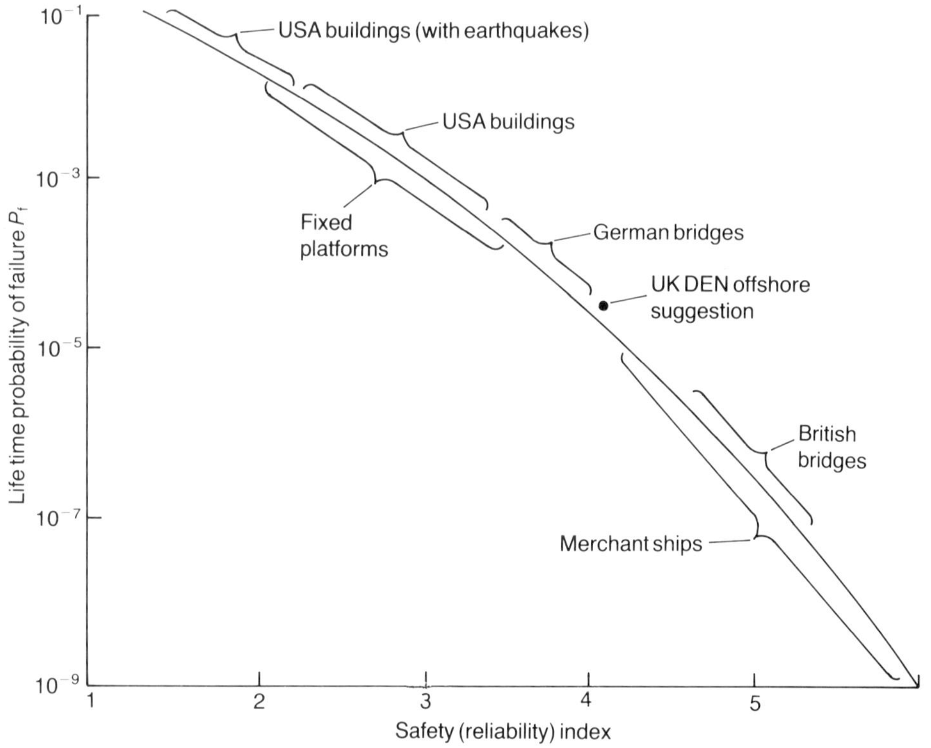

Fig. 4.1. Safety index plotted against probability of failure and values for certain codes and structures

The phrases "acceptable probabilities" and "appropriate degree of safety" beg the question of what is acceptable or appropriate. The document does not attempt a definition, this being a task for the drafters of regulations, taking due account for all conditions, whether they be technical or economic. The qualitative area of "safety" is replaced by the quantitative term "reliability", which is defined as the complement to the probability of an adverse event (e.g. failure). Each industry that has adopted limit state design has had to decide on acceptable probabilities of failure, or reliability criteria, and much research has been done into actual reliability levels in different industries. Fig. 4.1 shows a range of reliability levels which are accepted for structures in different industries in different countries. FRP structures could be expected to require similar reliability levels.

4.2.2. Levels of sophistication

Three levels of sophistication are recognized in checking the degree of safety in structures at a given limit state.

Level 3: An exact, probabilistic design method involving integration of the representation of the various uncertainties to derive optimum failure probabilities.

Level 2: A probabilistic design method in which the basic variables are represented by their known or postulated distributions and a defined reliability level is accepted. The distributions are assumed to be uncorrelated.

Level 1: A semi-probabilistic design method in which appropriate levels of reliability are provided by the specification of a number of partial safety factors related to nominal values of the basic variables.

Level 3 concepts are used primarily in research. Design codes generally adopt a Level 1 approach, although Level 2 concepts will normally have been used in the design of the code itself. Level 2 methods may be used directly in design if sufficient statistical data for the basic variables are available (or can be obtained) and the required level of reliability justifies the cost. However it is envisaged that for the majority of FRP applications Level 1 design methods will be the most appropriate, once appropriate codes of practice have been developed.

4.2.3. Limit states

The performance of all or part of an FRP structure or structural component can be described with reference to a limited set of limit states beyond which the design requirements are no longer satisfied. Limit states can be placed into three categories

(*a*) the ultimate limit states (ULS), corresponding to the maximum load-carrying capacity
(*b*) the serviceability limit states (SLS), relating to the criteria governing normal use
(*c*) the conditional limit states, corresponding to an infrequent major random event, e.g. fire; conditional limit states are frequently placed in one of the other categories (ULS or SLS).

Examples of ultimate limit states include loss of stability (buckling) and rupture of critical sections due to the material strength being exceeded (strength may be reduced by environmental effects or by repeated loading). Serviceability limit states may include excessive deformation, excessive vibration and local damage which reduces durability or affects efficiency or appearance.

4.2.4. Design

All the relevant limit states would need to be considered in the design of an FRP structure, although it would generally be acceptable to design on the basis of the limit state judged to be critical and then to check that the remaining limit states would not be reached.

The consideration of each limit state would require the setting up of a calculation model incorporating the appropriate basic variables allowing for the influence of direct and indirect loads, the response of the structure and the behaviour of the material. The uncertainties associated with the variables and with the calculation model itself would be allowed for either by a method of partial coefficients (Level 1 approach) or by a probabilistic method (Level 2). The basic variables would be considered to be independent, random variables.

4.2.5. The method of partial coefficients (Level 1)

The relationship between test loads and in-service loads and also the relevance of test data on materials to the performance of structural elements is dealt with in limit state design by partial coefficients.

In the method of partial coefficients the variability of these basic parameters would be taken into account by the selection of characteristic values of each of them. The uncertainties associated with the characteristic values and the calculation models would be taken into account by the application of partial coefficients to the variables or their effects. Where the necessary data were available the characteristic value for a variable would normally be based on a statistical interpretation of those data. Thus for a varying load the characteristic value would be defined as that value which had a prescribed probability of not being exceeded within the reference period. When characteristic values for loads cannot be established from statistical data they can be estimated on the basis of available information and, possibly, forecasts of future developments.

The variation of strength and other properties of the materials would be treated by defining characteristic properties which would be related to standard test specimens and procedures. Standardized test procedures would be vital for the successful application of the method. The characteristic value of a material property would normally be defined as that value which had a prescribed probability of not being reached in a hypothetical unlimited test series.

Two types of partial coefficient would be used, one type for the loads or load effects and the other for the strength of materials or elements. The partial coefficients would vary depending on the load type, the material, the type of FRP structure or component, the importance of the application, and the safety and economic consequences of the limit state under consideration being exceeded. When partial coefficients are applied to a characteristic value, the resulting value is termed a design value. The design value for a particular variable might vary for different limit states (whereas the characteristic value would normally be constant). Thus, for a particular limit state, the design value of the different types of variable would be given by

$$F_{\mathrm{d}} = \gamma_{\mathrm{f}} F_{\mathrm{k}}$$

$$f_{\mathrm{d}} = f_{\mathrm{k}} / \gamma_{\mathrm{m}}$$

where F_{d} is the design value of a load, F_{k} is the characteristic value of the load, γ_{f} is the partial coefficient for the load, f_{d} is the design material property (e.g. strength), f_{k} is the characteristic property, and γ_{m} is the partial material coefficient for the property.

The partial coefficients could be further subdivided to enable rational and consistent values for the individual coefficients to be defined. Thus the partial coefficient γ_{f} could be considered as a function of three coefficients γ_{f1}, γ_{f2} and γ_{f3}, where γ_{f1} takes account of the possible unfavourable deviation of the loads from the characteristic values; γ_{f2} takes account of the reduced probability that combinations of loads will simultaneously reach their characteristic value; and γ_{f3} takes account of possible inaccurate assessment of the load effects.

The partial material coefficient γ_{m} could also be broken down, and could be considered as a function of two coefficients γ_{m1} and γ_{m2}, where γ_{m1} takes account of unfavourable deviations of the material properties from the specified characteristic values, and the possible differences between the material property in the structure or element and that derived from test specimens; and γ_{m2} takes account of possible local weaknesses arising from the manufacturing process, and unfavourable geometric deviations resulting form manufacturing tolerances. A greater number of subdivisions could be chosen if necessary.

A further coefficient, γ_{n}, could be introduced to adjust either γ_{f} or γ_{m}. The coefficient γ_{n} could be considered as a function of two coefficients, γ_{n1} and γ_{n2}, where γ_{n1} takes account of the nature of the failure associated with the limit state e.g. whether ductile or brittle and whether collapse can occur without warning; and γ_{n2} takes account of the consequences of failure, from economic considerations to those of human safety. However, γ_{n} would normally be used simply as a modifying factor to the other factors rather than explicitly as a separate factor. Fabrication tolerances are a problem to both moulders and designers. In certain cases where the variability of a geometric parameter could have a significant effect on the strength of an element, for example the thickness of a thin composite skin, it would be appropriate to apply a coefficient explictly rather than by inclusion in γ_{m}. The design values a_{d} of geometrical parameters would be obtained from the characteristic values a_{k} (normally the nominal specified values) and an additive element Δ_{a}, since additive elements would generally be more suitable than factors for geometric parameters

$$a_d = a_k + \Delta_a$$

where Δ_a takes account of the importance of variations in a, and the given tolerance limits for a.

4.2.6. Choice of values for the partial coefficients

The values of the partial coefficients would depend on the limit state under consideration and should be based on statistical data when available. Where such data are not available the coefficients would be derived by calibration with pre-existing practice or by experience and judgement. Some further guidance is given in section 4.3.

4.2.7. Summary of the partial coefficient method

The object of the partial coefficient method outlined above for FRP structures would be to achieve, by means of suitably derived partial coefficients applied to the characteristic values of the relevant basic variables, acceptable probabilities that for each limit state the resistance of the structure or element would exceed the local effects. This is illustrated in simple form in Fig. 4.2 where a typical resistance function f and a typical load function F are plotted.

4.3. Characteristics of FRP materials

4.3.1. General

The great variety of plastics, additives and modifiers means that there are no universal methods describing the behaviour of all plastic materials. Nevertheless plastics do exhibit many similar characteristics, with differences only in magnitude, and for the most part it is these characteristics which distinguish the structural behaviour of plastics from that of conventional materials. The more pronounced viscoelastic behaviour and influence of temperature and environmental factors are examples of this. The many types of fibre reinforcement would appear to introduce further complexity. However in some respects the behaviour of FRP is less complex than unreinforced and short-fibre reinforced plastics, because typically both the fibre reinforcement and the plastic matrix are more elastic (i.e. less viscous) than thermoplastics, resulting in less dependence of properties on temperature and duration of loading. In other respects, of course, the behaviour is more complex, arising from the wide disparity in properties of the constituents and the possible variations in the agreement of the fibres. All of these factors would need to be embraced by the limit state design method.

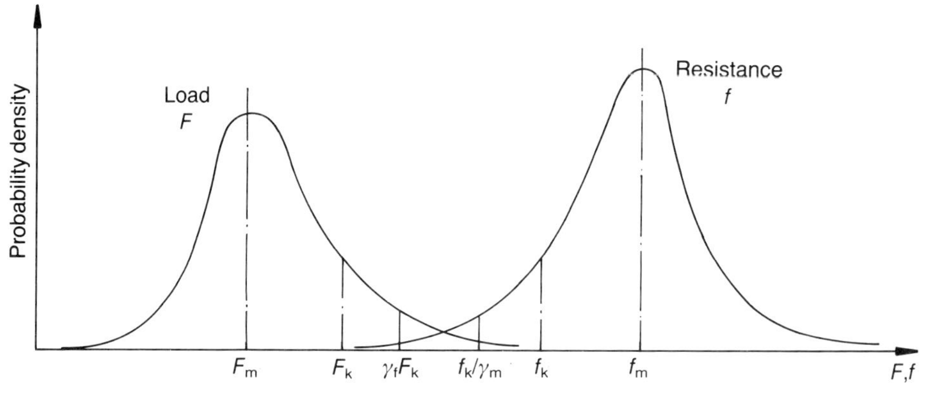

Fig. 4.2. Load and resistance distributions

In the following sections the main characteristics of FRP materials are described, drawing particular attention to where these differ from conventional structural materials (e.g. concrete, metals, masonry, wood) so that those aspects that need to be covered by the design method are summarized.

4.3.2. Structural characteristics of FRP materials

4.3.2.1. Fibres

Fibre reinforcement not only comes in a variety of materials with different strengths and stiffnesses, but also in a variety of forms e.g. mats, straight rovings and woven fabrics (see sections 2.3.1, 2.3.2, 2.3.3). In some forms the fibres are grossly kinked to conform to a weave pattern and this can reduce the strength of the composite material. Unlike most conventional materials, the strength and stiffness of the material can be varied by adjusting the fibre content.

4.3.2.2. Matrices

The wide variety of plastics with different characteristics is further complicated by the addition of fillers and plasticizers which can significantly alter the composite properties (see section 7.4.2). This can result in published properties being of little value because the exact composition is not stated or cannot be reproduced. Some matrices may exhibit poor bonding with the fibre reinforcement and are thus unable to develop the full strength capacity of the fibres.

4.3.2.3. Processes

The characteristics of materials with identical constituents can be influenced by the process used in their production. Factors such as unplanned fibre orientation, insufficient compaction, inadequate curing, inaccurate fibre placement and lap joints in preformed reinforcement can all adversely affect the material properties.

4.3.2.4. Short-term stress–strain relationships

Provided fibres are oriented in the stress direction, most FRP materials exhibit mainly linear elastic behaviour to failure. Characteristic stress–strain curves are shown in Fig. 4.3. Type I is the most typical for materials with fibres aligned in more than one direction, with the "knee" indicating the onset of first damage, i.e. crazing or microcracking at the transverse fibres. Type II is typical for unidirectional laminates stressed parallel to the fibres and Type III is typical for cases where the influence of the matrix is significant, i.e. random in-plane reinforced laminates or aligned fibre laminates stressed off the principal fibre axes.

Unlike most structural materials, FRP materials do not exhibit yielding (except for aramid fibre composites in compression) and therefore are not ductile. However they do not generally display the other characteristics of classic brittle materials, i.e. low tensile strength and sensitivity to impact and stress concentrations. This is because the fibres act as crack inhibitors, although where there are no fibres to limit the spread of cracks, e.g. transverse to a unidirectionally reinforced laminate or in interlaminar layers, the material behaves in a brittle manner.

4.3.2.5. Directional properties

FRP materials are neither homogeneous nor, generally, isotropic. Depending on the fibre alignment, an FRP laminate can exhibit in-plane elastic properties which are planar-isotropic, orthotropic, or anisotropic. In aligned fibre laminates, properties can alter significantly when the direction of applied

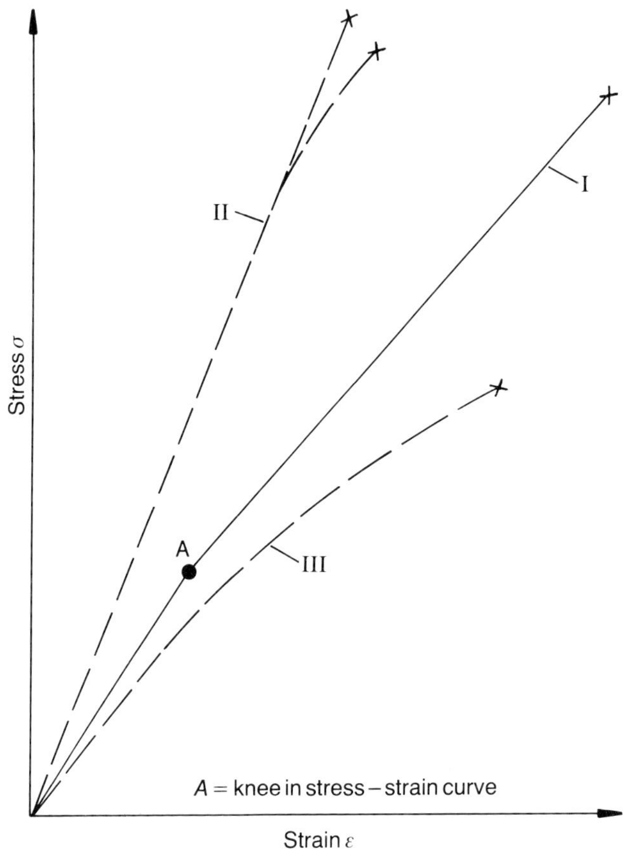

Fig. 4.3. Characteristic stress–strain curves for reinforced plastics

stress departs from a principal fibre direction. Unbalanced and unsymmetric multi-ply laminates show even more complex behaviour.

4.3.2.6. Long-term loading

Under sustained stress, the strain in FRP continues to increase while the magnitude of stress needed to produce rupture diminishes with time (see chapter 7, part B). These phenomena are referred to as creep and creep rupture respectively. The behaviour is complex and is influenced by temperature and environment. Although the effect in FRP is less pronounced than with unreinforced plastic, it is generally more significant than with other structural materials.

4.3.2.7. Fatigue

As with static loading, fatigue behaviour under repeated or cyclic loading is highly directional. Unlike most conventional structural materials, some FRP materials, notably glass-reinforced materials, do not have a fatigue limit. High frequency stress variations can generate internal heat in the material.

4.3.2.8. Environmental factors

The interaction of time, temperature, stress and environment can have a profound effect on the structural performance of FRP materials, to a far greater extent than with any other major structural material except possibly wood (see chapter 7, part A). Even normally occurring agents such as water (or water vapour) and ultraviolet rays from the sun can seriously affect FRP materials.

4.3.2.9. Characterization of FRP materials

Other structural materials such as wood and concrete display widely varying properties and characteristics. However, design in these materials is facilitated by classification of the material, related to established specifications in the case of concrete, and by generally accepted simplifications of the overall behaviour. No such characterization exists in FRP materials.

4.4. Design codes and criteria for FRP materials

4.4.1. General

The broad range of applications of FRP materials makes it inadvisable to attempt to develop a single code and set of design criteria to treat all possible applications. A similar approach is adopted for other materials: for instance there are different design criteria for steel buildings and steel offshore structures. Design methods and national codes will tend to be application-oriented and the following paragraphs endeavour to set out the general principles on which limit state design criteria for FRP structures could be based. The importance of this is that a consistent and rational approach to design could be adopted despite the diversity of applications, giving confidence to designers and end users.

4.4.2. Design life

The design life of composite structures needs to be specified not only to evaluate fatigue performance but also to evaluate the cumulative effects of creep under load and environmental effects. The assumption of a design life does not necessarily mean that the structure will no longer be fit for its purpose at the end of that period. However some repair and refurbishment may be expected at that time and also inspection and maintenance will still be required regularly. One of the most important reasons for using composites is to reduce maintenance costs but it is still likely that the frequency of inspection will need to be similar to that for other materials in strategic structures. If the decision is taken to recoat the structure, say every 40 years in order to renew the eroded resin surface, then the environmental impact on strength and stiffness may be able to be reduced in the design. Recoating composite structures is much cheaper than coating steel because there are no rust deposits to remove and coating thicknesses are generally thinner.

4.4.3. Limit states

For the design of any structure or component it would be necessary to select all the relevant limit states for consideration. The actual limit states selected would obviously depend on the nature of the FRP structure and its intended use, and the following is intended only as an indication of those limit states which would have to be considered in view of the particular characteristics of FRP described previously.

4.4.3.1. Serviceability limit states (SLS)

(*a*) Deflection/deformation. This is likely to be a critical limit state in many applications because of the generally lower magnitude and directional nature of FRP stiffness compared to conventional structural materials such as steel and concrete.

(*b*) Residual deformation. The viscoelastic behaviour of some FRP materials may result in excessive permanent deformation under sustained loading.

(*c*) First damage. Crazing or microcracking of the material may constitute a limit state for various reasons, e.g. it accelerates environmental degradation or causes leakage from a fluid container.

(*d*) Buckling or wrinkling (local). This may be a serviceability or ultimate limit state depending on the nature of the application and the post-buckling behaviour of the material.
(*e*) Vibration.
(*f*) Environmental damage. Effect on performance (reduced properties) or appearance without causing failure.
(*g*) Impact damage—as (*f*).

4.4.3.2. Ultimate limit states (ULS)

(*a*) Collapse. Loss of equilibrium of whole or part of structure.
(*b*) Rupture of Section. This can cause failure of structure or element of structure. It may take several forms, e.g. tension, shear, flexure or interlaminar shear and may require definition of failure criteria for the material. Rupture may be influenced by the interaction of time, temperature, stress and environment.
(*c*) Buckling. Loss of stability: may be critical on account of lower stiffness and directional nature of properties.
(*d*) Fatigue.
(*e*) Resonance.
(*f*) Fire resistance.

4.4.4. Methods of analysis

4.4.4.1. Analysis at ultimate load

Methods of analysis at ultimate loads would have to take account of the lack of ductility in most FRP materials, in contrast to most conventional structural materials. This would have two main effects.

(*a*) Non-linear structural analysis involving redistribution of stresses in redundant structures would not be valid.
(*b*) Internal redistribution of the stresses in a cross-section cannot take place. This means that load effects such as temperature gradients and shear lag would have to be considered at the ultimate limit state (unlike ductile materials where these effects need only be considered at the serviceability limit state).

However, the almost linear stress–strain curves to failure of most FRP materials means that elastic methods of analysis would be adequate for most applications.

4.4.4.2. Global analysis

Global analysis would generally be undertaken using linear elastic computer models of the structure with appropriate allowances for loss of stiffness due to creep. Allowance for in-plane shear stiffness might be found to be important in some structures and complex three-dimensional finite element analyses might be necessary in order to evaluate stress concentration and shear lag effects.

4.4.4.3. Analysis for deflections

Live load deflections would generally dominate designs in composites and it is therefore important to calculate them accurately using appropriate short- or long-term stiffnesses depending on the loading regime. The analysis would generally be the same global analysis that is used to evaluate stresses.

4.4.4.4. Analysis of stresses

The calculation of stresses from the global analysis for serviceability or ultimate limit state conditions would need to include the following.

(*a*) Stresses due to axial forces and global bending moments, both longitudinal and transverse including the influence of shear lag.
(*b*) Shear stresses including those due to torsion.
(*c*) Warping stresses due to torsion and distortion of box members.
(*d*) Transverse stresses due to distortion of the members.
(*e*) Stresses due to creep.
(*f*) Stresses occurring in the vicinity of major stress concentrations around local loads or near supports.

4.4.5. Specification for loads

It would generally be found that the loads specified in limit state codes of practice for other materials would be able to be used for composite structures. Dead loads would generally be much lower and therefore care should be taken to recognize that the live load would represent a larger proportion of the total load effect than is normal with other materials.

Loading specifications would additionally need to define the duration of loading as composite behaviour is time-dependent. This would increase the complexity of load combination effects, since both long- and short-term loads would need to be considered, and load combinations should therefore be re-evaluated. Temperature effects might be more significant in some respects because of the larger coefficients of thermal expansion with composites but temperature restraint might be less significant because of the lower modulus of elasticity.

4.4.6. Material parameters

4.4.6.1. Characteristic material properties

The required characteristic material properties for FRP materials differ in many respects from those required for most conventional structural materials. Firstly, standard grades of material are not available. Steel, for example, is produced under stringent quality control to a nominal guaranteed yield stress, which is then used as the characteristic strength in limit state design, and the elastic modulus and Poisson's ratio vary little, whatever the composition. The material is homogeneous and isotropic, so no other elastic constants are required and the properties vary little under normal temperature ranges and environmental conditions.

By contrast, standard data on FRP composites are generally either not available or cannot be relied upon except for preliminary design. Up to 21 elastic constants may be required for a rigorous analysis of a complex laminate. Stiffness as well as strength varies with the composition, and the properties vary with temperature and production process and may degrade with time. Material design data are therefore normally derived from the testing of specimens. For maximum reliability the specimens and test conditions should represent as closely as possible the materials and conditions of use of the final product.

Derivation of characteristic design properties from test data is a specialist subject and is dealt with elsewhere.[4.4,4.5] In the construction industry characteristic material properties are commonly taken as the lower 5% fractile, but other values may be appropriate for different applications. The quoted characteristic values, which for FRP materials will include stiffnesses as well as strengths and possibly first damage as well as ultimate strengths, should be associated with the governing limitations, e.g. constituent materials, production process, limits of temperature or environmental conditions. This implies, of course, that separate characteristic values should be used when any of the conditions is changed e.g. a different temperature range. An alternative method would be to use a single characteristic value for each property, for a given set of conditions, and use factors to derive representative values for other conditions. These would not be true partial coefficients, however, since they do not represent uncertainties.

4.4.6.2. Partial material coefficients

The partial material coefficient, γ_m, would allow for uncertainties in the assumed properties of the material in the final structure. Ideally it should be derived from statistical data but this requires specialist techniques[4.5] and there is rarely sufficient data available (since specimen testing does not deal with all the uncertainties). The coefficient is normally derived by a combination of statistical methods and calibration with pre-existing practice, or simply by judgement where there is no relevant pre-existing practice.

Most existing limit state design codes do not subdivide the γ_m factor, although different values are frequently given for different limit states, indicating that the derivation involved subdivision. FRP material properties used in design tend to have uncertainties not associated with more conventional materials however.[4.6] One such uncertainty is in the method of deriving properties. For the properties of a multi-ply laminate, for example, there are three alternative methods of deriving the laminate characteristics, with reducing level of uncertainty

- (*a*) properties of constituent materials (fibre, matrix) derived from test specimen data; properties of individual plies derived from theory; properties of laminate derived from theory
- (*b*) properties of individual plies derived from test specimen data; properties of laminate derived from theory
- (*c*) properties of laminate derived from test specimen data (not always practicable).

Clearly different partial coefficient values should be applied to the characteristic values depending on which derivation method is used, and this may justify a further subdivision of the partial material coefficient. Different production processes may also justify different partial coefficient values (in addition to different characteristic values) in view of the different level of uncertainty associated with each process. Materials produced by automated processes generally demonstrate a lower spread of test results and are produced to smaller dimensional tolerances than manual processes, and this would be reflected in lower partial coefficient values and a resulting economy in design.

A simple illustration is the flexural stiffness of a 3 mm thick laminate. The mechanical properties are provided by a reinforcing mat close to each laminate surface. Any variation in wall thickness changes the lever arm between the two mats and hence the mechanical properties. Automated processes such as pultrusion can produce laminate thicknesses within a tolerance of $\pm 0{\cdot}2$ mm. This thickness variation, though small, still gives a partial coefficient of 1·25 on stiffness. If the tolerance were doubled to $\pm 0{\cdot}4$ mm, because a less accurate production method was used, the partial coefficient would increase to 1·50.

Partial material coefficients for stiffness and strength would generally take account of variations in laminate thickness and reinforcement position, inaccuracies in the theory used, creep, weathering and variation in constituent glass weight. Also other variations in conditions may be taken account of where appropriate. A separate coefficient would be evaluated for each of the following for both stiffness and strength

- (*a*) membrane stresses in two directions
- (*b*) flexural stresses in two directions

(*c*) membrane shear stress
(*d*) out-of-plane shear stress.

The only British Standard Code of Practice for design of composites is the British Standard for the design of reinforced plastic vessels[4.7] which incorporates a design factor K composed of six separate strength reduction factors. No partial load coefficients are specified so this code is not in strict limit state format. Of the six strength reduction factors only two, both relating to the production process, incorporate partial material coefficients according to limit state philosophy (note that automated methods are given lower values). The other factors, apart from one described as a safety factor, are really condition factors applied to the characteristic strength as described in section 4.4.6.1.

4.4.6.3. Brittle failure modes

Higher values of the partial material factor are normally applied when the failure mode is brittle. Thus tensile failure of a unidirectional composite, failure by interlaminar shear of tension, and failure of certain bonded connections would be examples of limit states where the higher values would apply.

4.4.7. Buckling instability

Two distinct methods of dealing with buckling instability can be found in existing limit state design codes. In those dealing with concrete structures, slender compression members must be designed for an additional eccentricity of load, the magnitude depending on the slenderness ratio of the member. In those dealing with metal structures, local plate buckling and overall buckling instability are dealt with by strength reduction factors dependent on slenderness ratios, and this would appear to be the more appropriate method for FRP structures. Additionally, for FRP structures and plates in compression, uncertainties in the stiffness parameters would need to be allowed for by the partial material coefficient.

Buckling would usually be considered at both the serviceability and ultimate limit states. Composites, particularly cellular box sections, have good post-buckling reserves of strength because the large out-of-plane deformations can occur elastically, permitting redistribution of load to stiff corner sections of the members. Hence the onset of elastic buckling could be treated as a serviceability limit, and this could often be more critical than the ultimate post-buckling strength.

4.4.8. Fatigue

For fatigue assessment two distinct methods have been formulated in other limit state codes. In one draft code it is proposed that the calculated fatigue life should exceed the design life factored by a partial coefficient. However a more rational approach[4.8] is to use characteristic S–N curves (i.e. curves joining lower fractile points rather than the mean values) to calculate fatigue life, which is then required to exceed the design life.

For FRP materials, fatigue behaviour is influenced by temperature, environment and direction of stress (for non isotropic laminates), and therefore characteristic S–N curves should be associated with governing limitations as suggested for other material properties (see section 4.4.6.1).

4.5. Specification for FRP materials

The design criteria outlined earlier in the chapter cannot be defined without making assumptions about test methods, materials, processing requirements, material and fabrication tolerances and quality control. Any code of practice for design of composites must therefore have an associated specification which sets out the detailed requirements assumed in preparing the design criteria. The following sections briefly outline important specification headings.

4.5.1. Classification of laminates

The responsibility for selection of laminate materials and proportions of materials could depend on whether the specification is a "performance" type or not. In a typical performance specification the short- and long-term strength, stiffness, durability and fire resistance of laminates and components might be specified, but the choice of final materials would be left to the manufacturer to suit the processing aspects. This is similar to the approach taken with concrete structures, with respect to the concrete mix design, but on the other hand it puts some responsibility onto the manufacturer for the design of the reinforcement. Alternatively the full laminate make up and materials might be specified but this would rarely be appropriate since some variations would almost certainly be needed to suit process requirements.

4.5.2. Constituent materials

Reinforcement, resin, veil and gel coat materials would need to be fully specified in relation to appropriate standards which include quality control aspects. Particular specifications might have to be written where appropriate standards are not available. Resin additives would also need to be specified including pigments, ultraviolet absorbers, fillers, flame-retardant substances and low profile (low shrinkage) additives.

4.5.3. Type approval

With performance specifications it would generally be desirable to require type approval of the final laminate configuration to be obtained prior to full production. A specification would be needed setting out the test requirements, typically covering short- and long-term strength and stiffness, durability and fire resistance. This would involve a programme including accelerated weathering tests using established test procedures.

4.5.4. Process requirements

The manufacturing processes would usually need to be defined and specifications would typically need to cover handling and storage of materials, splicing of reinforcement and cutting of sections in continuous processes, quality control, packaging and labelling. Most British manufacturers will have a Certificate of Registration of Firms of Assessed Capability to BS 5750: Part 2. It would be realistic to expect a manufacturer to establish full low traceability of materials from raw materials through to the finished structure in the same way as steelwork records are made.

4.5.5. Physical properties

The specification of physical properties would cover the requirements for the finished product and would define both destructive and nondestructive testing. Visual inspection and various forms of ultrasonic examination would make up the nondestructive tests while destructive batch tests on components might include strength and stiffness tests on coupons, glass content tests, Barcol hardness tests and flammability tests.

Full tolerances on finished dimensions would also need to be specified including cross-sectional and straightness values.

4.5.6. Fabrication and erection

The specification would need to cover fabrication and erection aspects in similar detail to steelwork specifications.

4.6. Conclusions and recommendations

A framework has been set out for the logical and rational design of FRP structures which would embrace materials, materials testing, manufacturing processes, quality control, design procedures and structural reliability. This framework is the use of limit state design principles combined with concepts of structural reliability, which have received international recognition as a common basis for Codes of Practice for design in any material or type of construction. The potential advantages to the industry would be significant in education and for material suppliers, manufacturing suppliers, moulders, designers and end users. The limit state framework could provide a common language for communication between these different disciplines, between different industries and between different countries.

4.7. Examples

The partial coefficients used in the following examples have not been derived by statistical or any other methods, and should not therefore be regarded as proposed values. The purpose of the examples is simply to illustrate limit state design methods.

4.7.1. Example 1

A composite beam is to be designed for use as a sideguard device for commercial vehicles (Fig. 4.4(a)). The span between supports is 3000 mm and the projection beyond the rear support is 500 mm. The regulations require the beam to be capable of withstanding a force of 2 kN applied transversely to the vehicle at any point on the beam, with the permitted deflections of the beam limited to 150 mm generally, but only 30 mm within 250 mm of the rear of the beam.

The chosen form of the beam is a pultruded, hollow glass-reinforced polyester box beam with lugs at the rear for attachment to the supports, as shown in Fig. 4.4(b). The calculated transformed section properties of the beam are

Moment of inertia I_{yy} = 884500 mm^4

Section moduli z = 26364 mm^3 (centre of front wall)

19787 mm^3 (extreme fibre at rear)

Transverse shear area A_x = 300 mm^2

The laminate forming the front wall is three-ply. The predicted properties of this laminate (calculated entirely from laminate theory) are as follows.

Fig. 4.4. Vehicle sideguard device

Plate (flexural) rigidity matrix

$$D_{11} = 77876\,\text{N-mm} \quad \text{(longitudinal)}$$

$$D_{22} = 40612\,\text{N-mm} \quad \text{(transverse)}$$

$$D_{12} = 12161\,\text{N-mm}$$

$$D_{33} = 15967\,\text{N-mm} \quad \text{(torsional)}$$

Equivalent longitudinal in-plane modulus

$$E_{11} = 29539\,\text{MN/m}^2$$

For the laminate forming the shorter walls, the predicted equivalent in-plane shear modulus is

$$G = 4500\,\text{MN/m}^2$$

The above values are adopted as characteristic material properties for preliminary design.

4.7.1.1. Limit states

The critical limit states are judged to be

- deflection at tip of rear cantilever
- local buckling (wrinkling) of front wall at midspan
- first damage at extreme rear fibre at midspan.

Since the front wall is expected to exhibit substantial post-buckling strength, the above limit states are all serviceability criteria.

4.7.1.2. Partial coefficients

The design loading is a one-off test loading applied under controlled conditions, therefore the partial load coefficient, γ_f is taken as 1·0.

The partial material coefficient allows for uncertainties arising from the production process and in the method of assessing the characteristic values. Large factors are not required at serviceability limit states, and the pultrusion process provides a highly consistent and accurate product. However, the characteristic properties are derived from theory rather than test and there is therefore uncertainty associated with this, particularly for shear rigidity values where experience has shown the predicted values to be optimistic. For SLS, γ_m is therefore taken as:

$$\gamma_m = 1{\cdot}1 \text{ (all values except shear, torsion)}$$

$$= 2{\cdot}0 \text{ (shear, torsion values)}$$

4.7.1.3. Material property reduction factors

There is no sustained or repetitive loading and the environment is normal. Only age degradation is applicable, but the pultruded beam incorporates a surface veil for durability and experience has shown little degradation of stiffness with time. No property reduction factors are therefore applied.

Fig. 4.5.

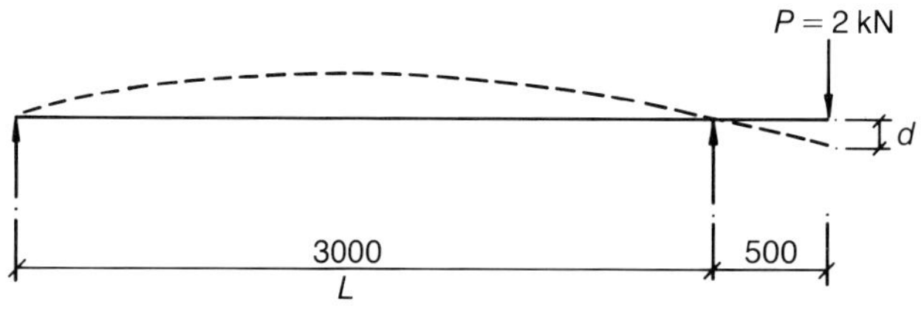

4.7.1.4. Design

(a) Deflection at tip of cantilever (see Fig. 4.5). Allowing for shear deformation, deflection

$$d = PC^2(C + L)/3EI + PC/GA$$

where

$$\begin{aligned} EI &= (E_{11}/\gamma_m)I_{yy} \\ &= (29539/1{\cdot}1)\ 884500 \\ &= 2{\cdot}375 \times 10^{10}\ \text{N-mm}^2 \end{aligned}$$

$$\begin{aligned} GA &= (G/\gamma_m)A_x \\ &= (4500/2{\cdot}0)\ 300 \\ &= 675000\ \text{N} \end{aligned}$$

therefore, deflection

$$\begin{aligned} d &= 2000 * 500^2(3000 + 500)/3 * 2{\cdot}375 * 10^{10} + 2000 * 500/675000 \\ &= 26{\cdot}0\ \text{mm} \end{aligned}$$

that is, less than the permitted 30 mm.

(b) Local buckling of front wall at midspan. In the view of the expected substantial post-buckling strength, and the limited application of the load, the stress will be permitted to exceed the onset of buckling stress by 25%. Within this range the buckling deformations are expected to be elastic and fully recoverable on removal of load. Assuming the plate is simply supported at the edges, the critical buckling stress f_{cr} is given by

$$f_{cr} = (2\pi^2/b^2 t)\,(\sqrt{D_{11}D_{22}} + D_{12} + 2D_{33})$$

where b is plate width (97·5 mm), t is plate thickness (3·5 mm), and D is plate flexure rigidities. Therefore, applying γ_m factors

$$\begin{aligned} f_{cr} &= (2\pi^2/97{\cdot}5^2 * 3{\cdot}5)\,(\sqrt{77876 * 40612/1{\cdot}1 * 1{\cdot}1} \\ &\quad + 12161/1{\cdot}1 + 2 * 15967/2{\cdot}0) \\ &= 46{\cdot}4\ \text{MN/m}^2 \end{aligned}$$

Bending moment on cross-section at midspan

$$\begin{aligned} M &= PL/4 \\ &= 1{\cdot}5 * 10^6\ \text{N-mm} \end{aligned}$$

therefore compressive stress at centre of plate

$$\begin{aligned} f &= M/z \\ &= 1{\cdot}5 * 10^6/26364 \\ &= 56{\cdot}9\ \text{MN/m}^2 \end{aligned}$$

which exceeds f_{cr} by 22·6%

(c) First damage at extreme rear fibre at midspan. The first damage criterion will be taken as a maximum strain limit of 0·3%. Bending moment at midspan is $1{\cdot}5 * 10^6$ N-mm, therefore tensile stress at extreme rear fibre

$$\begin{aligned} f &= M/z \\ &= 1{\cdot}5 * 106/19787 \\ &= 75{\cdot}8\ \text{MN/m}^2 \end{aligned}$$

Fig. 4.6. Testing of sideguard device

The strain is given by

$$e = f/(E_{11}/\gamma_m)$$
$$= 75{\cdot}8/(29539/1{\cdot}1)$$
$$= 0{\cdot}0028$$

which is less than permitted 0·003.

4.7.1.5. Testing

Subsequent full-scale testing (Fig. 4.6) verified the suitability of the design. A comparison of the measured and predicted properties (with $\gamma_m = 1{\cdot}0$) is given below.

Equivalent longitudinal in-plane modulus

$$E_{11} = 28010\,\mathrm{MN/m^2}\ \text{(predicted } 29539\,\mathrm{MN/m^2}\text{)}$$

In-plane shear modulus

$$G = 2259\,\mathrm{MN/m^2}\ \text{(predicted } 4500\,\mathrm{MN/m^2}\text{)}$$

Critical buckling stress

$$f_{cr} = 56\,\mathrm{MN/m^2}\ \text{(predicted } 59{\cdot}5\,\mathrm{MN/m^2}\text{)}$$

Failure occurred at approximately twice the buckling stress.

4.7.2. Example 2

An electronic scoreboard is to be suspended from the roof of a covered stadium by a number of hanger rods. The design criteria for each rod are listed below.

Length	8·0 m
Diameter	to be determined

Loading	110 kN dead load
	15 kN live load (occasional maintenance)
	33 kN (30% DL) accidental overload due to failure of one other hanger
Deflection	15 mm maximum during design life (due to dead load)
Design life	10 years

The selected material is graphite/epoxy composite with protection against environmental degradation. The characteristic short-term material properties (from tests) are

Longitudinal modulus $E = 180\,\mathrm{GN/m^2}$
Ultimate tensile strength $f_u = 1500\,\mathrm{MN/m^2}$

Sustained loading tests indicate that rupture occurs at 80% UTS after 10^5 hours (11·4 years).

4.7.2.1. Limit states

The following limit states will be considered

Ultimate (ULS)	1. Creep rupture
	2. Accidental collapse
Serviceability (SLS)	3. Deflection
	4. First damage

The rod will be designed for the creep rupture limit state and then checked against the remaining limit states.

4.7.2.2. Partial coefficients

		ULS	SLS
Partial load coefficient γ_f	dead load	1·5	1·2
	live load	2·0	1·5
	accidental	1·2	–
Partial material coefficient γ_m	strength	3·0	–
	stiffness	–	1·1

4.7.2.3. Material property reduction factors

(a) Creep. Graphite/epoxy composites exhibit little creep deformation, but a stiffness reduction factor of 1·1 will be used to allow for what may occur within the design life.

(b) Creep rupture. Strength reduction factor of 1·25 will be used to allow for the reduction in strength due to sustained loading over the reference period.

4.7.2.4. Design

(a) Creep rupture. Reduced characteristic material strength for sustained loading

$$f_k = 1500/1{\cdot}25$$
$$= 1200\,\mathrm{MN/m^2}$$

Design material strength (ULS)

$$f_d = f_k/\gamma_m$$
$$= 1200/3{\cdot}0$$
$$= 400\,\mathrm{MN/m^2}$$

Design load (ULS)

$$F_d = \gamma_f F_k$$
$$= 1{\cdot}5 * 110 + 2{\cdot}0 * 15$$
$$= 195\,\text{kN}$$

Required cross-section area of rod is given by

$$A \geqq 195000/400$$
$$\geqq 487{\cdot}5\,\text{mm}^2$$

therefore required diameter of rod

$$D = 25\,\text{mm}$$

(b) Accidental collapse. Design load for this limit state includes dead load only plus accidental load. Therefore design load (ULS)

$$F_d = \gamma_f F_k$$
$$= 1{\cdot}2\,(110 + 33)$$
$$= 171{\cdot}6\,\text{kN}$$

This is less critical than creep rupture.

(c) Deflection. For 10 year design life the material stiffness reduction factor is 1·1, therefore the characteristic viscoelastic modulus is given by

$$E_{vk} = E/1{\cdot}1$$
$$= 180/1{\cdot}1$$
$$= 163{\cdot}6\,\text{GN/m}^2$$

Design viscoelastic modulus (SLS)

$$E_{vd} = E_{vd}/\gamma_m$$
$$= 163{\cdot}6/1{\cdot}1$$
$$= 148{\cdot}8\,\text{GN/m}^2$$

Design load (SLS—dead load only)

$$F_d = \gamma_f F_k$$
$$= 1{\cdot}2 * 110$$
$$= 132\,\text{kN}$$

Therefore long-term deflection is given by

$$d = F_d L/(E_{vd} A)$$
$$= 132000 * 8000/(148800 * 490)$$
$$= 14{\cdot}5\,\text{mm}$$

This is less than the permitted value (15 mm).

(d) First damage. The criterion adopted here is a maximum elastic tensile strain of 0·2%. Design material modulus (SLS)

$$E_d = E/\gamma_m$$
$$= 180/1{\cdot}1$$
$$= 163{\cdot}6\,\text{GN/m}^2$$

Design loading (SLS)

$$F_d = \gamma_f F_k$$
$$= 1{\cdot}2 * 110 + 1{\cdot}5 * 15$$
$$= 154{\cdot}5\,\text{kN}$$

therefore the strain is given by

$$e = F_d/(E_d A)$$
$$= 154500/(16300 * 490)$$
$$= 0{\cdot}19\%$$

which is less than the permitted value (0·2%).

Conclusion. A 25 mm diameter rod with the specified properties satisfies all the critical limit states and is therefore acceptable.

References

4.1. CEB-FIP. Common unified rules for different types of construction and materials *International System of Unified Standard Codes of Practice for Structures* (3rd edn) **1**. CEB-FIP, Lucerne, 1978.

4.2. Joint Committee on Structural Safety. *General principles on reliability for structural design*. Report 35, part II. International Association for Bridge and Structural Engineering, Zurich, 1981.

4.3. Baker A.L.L. Criteria of structural design. *Proc. Inst. Civ. Engrs, Part 1*, November 1974.

4.4. Construction Industry Research and Information Association. *Rationalization of safety and serviceability factors in structural codes*. Report 63. CIRIA, London, 1977.

4.5. King R.L. Statistical methods for determining design allowable properties for advanced composite materials. In *Proc. Reinforced Plastics Congress*. British Plastics Federation, London, 1986.

4.6. American Society of Civil Engineers. *Structural plastics design manual*. ASCE Manual No. 63. American Society of Civil Engineers, New York, 1984.

4.7. British Standards Institution. *Specification for vessels and tanks in reinforced Plastics*. BS 4994. BSI, London, 1973.

4.8. British Standards Institution. Steel, concrete and composite bridges. In: *Code of practice for fatigue*. BS 5400, Part 10. BSI, London, 1980.

5 Architectural use and aesthetics of composite structures

A. J. Leggatt

5.1. Introduction

Polymer composites have attributes which cover an enormously wide field, as is evident from the coverage in this book. Such a wide spectrum of functions and applications must be an advantage when considered objectively, but the sheer wealth of features causes some difficulty in presenting the material to the uninitiated. The pure technologist will be easily fascinated by the stress analysis and load-time characteristics of the material as they open new fields of scientific and mathematical exploration. Compared with such heady stuff, the technologist might dismiss the "artistic" applications as trivial and consider that building cladding panels are not serious structures. The architect on the other hand might be deterred by the sophistication of composites technology and even frightened by the thought of "long-term creep under sustained load" and by the knowledge that many polymers can be made to burn. Both attitudes would be wrong and could lead to missed opportunities of enormous potential.

The architectural use of polymer composites requires the fusion of sophisticated technology with equally sophisticated creativity and original thinking. As the structural designer has to move away from the comparatively simple concepts of materials such as steel and concrete, so the architect or "visual" designer has to move away from the attributes of more familiar materials and ask some searching questions about the basic requirements of building elements. This chapter takes a rest from the technology and the more prosaic attributes of composites. It will concentrate instead on what would generally be called architectural matters, such as appearance, non-structural performance and durability. The polymer composite which has been used generally in architecture so far is glass-reinforced polyester (GRP). Other composites are used but they are generally non-polymer, for instance cement-based materials reinforced with glass or polymer fibres and gypsum plasters similarly reinforced. Such materials lie outside the strict definition of this book's coverage but some of the aesthetic considerations, especially those concerning shape, would have some application to other composites.

Polymer composites other than GRP exist—epoxy resins and polymer reinforcements are available—but have been rarely used in architecture so far. No doubt further research will evolve matrix and reinforcement materials which will enhance the properties already exhibited by GRP. For the purpose of this chapter only GRP will be discussed.

5.2. Surfaces

Although it is fairly easy to mould GRP into solid blocks and the like, the cost of the ingredients and the high strengths obtainable give more efficient use of the material in the form of thin sheets or laminae.

There is an obvious truth about surfaces which has been disregarded by a surprisingly high number of designers. In simple terms, it is that a flat sheet has virtually no rigidity. A sheet with single curvature, e.g. a cylindrical vault, has much increased rigidity but needs stabilizing at its ends, whereas a doubly curved sheet, e.g. a dome or a bulge, has immense rigidity. In a building, certain surfaces, those for travelling on, need to be flat. Nearly all the remaining surfaces (walls, roofs etc.) can be given curvatures to some degree or other. As the external surfaces of most buildings are either walls or roof elements, curvature can be used extensively, thus making thin sheet materials very appropriate and at the same time offering exciting scope to the imaginative designer. When combined with the excellent properties of GRP in regard to durability, colour retention, lightness and strength, the building designer is presented with a material which gives him outstanding opportunities in visual design to an extent unrivalled by any building material produced so far.

5.3. Thoughts for the designer

Unfortunately, GRP does impose some constraint on the designer. Care has to be taken in the structural design of GRP elements so that the effect of high continuous stress is allowed for and perhaps more significantly, proper attention must be paid to the fire behaviour of the material. The former of these problems is not normally a great constraint where GRP is used as a cladding element rather than part of the main structure. Fire behaviour is discussed in chapter 8 and although GRP can be made to burn, there are effective techniques and resin formulations which enable GRP to be used in compliance with existing building regulations. GRP has been used in buildings now for 20 years or more and there is no statistical evidence that GRP has presented a particular fire hazard. Where there have been instances of GRP performing unsatisfactorily in buildings, nearly always they have been due to poor performance in a mechanical sense, that is to say, the elements have distorted, joints have failed or, less commonly, there has been colour or surface degradation. In nearly all such cases of failure, the blame has to be put on inadequate or ill-informed design or specification.

Before deciding to use GRP, the designer should consider the "personality" of the material. By this is meant the architectural use, particularly with regard to shape, colour and texture, into which the material fits comfortably and economically.

The designer should consider whether the building he has in contemplation would allow for the visual personality of GRP to be expressed and whether he can honestly assess his own willingness to recognize and utilize in his design the very different shapes and colours which GRP can so easily provide. If the answer to this question is "no", then to proceed with GRP could well lead to difficult design and construction problems and possibly even to great disappointment in the end.

The point in the progress of the general design of a building at which the designer starts to consider GRP is all-important. In many ways, the earlier GRP is brought into the design process the better, although it should not be so early that the design of the GRP and of the building itself is distorted to fit a premature decision to use the material.

Perhaps the most successful GRP designs occur where the idea of using the material has fallen naturally into place in the early consideration of the primary components of the building design. It is as wrong to say "let's design something in GRP" as it is to translate into GRP a design concept which has been visualized previously in another material. Ideally, the architect or design leader should have a sufficiently wide knowledge of GRP (as he should of other materials—concrete, brickwork, timber, etc.) that the merits and demerits of the material are to hand in his visual vocabulary. The design leader need not burden himself with knowledge of detailed chemistry or stress analysis, provided he has a broad knowledge of what can be done economically.

For design in GRP, a strong sense of three-dimensional form is required, coupled with a knowledge of membrane structures, GRP technology and basic building construction. The theoretically ideal GRP designer would be architect, sculptor, structural engineer, building foreman, plumber and GRP technologist all rolled into one person. Such a multi-skilled genius would be hard to find, but normally successful GRP design can result from a small multi-disciplinary team led by a person with a firm sense of direction and purpose. The following plan is a suggested sequence for contemplating the numerous aspects of the design. It is not essential to follow this sequence although it has a certain logic about it, but it is important to check that all the listed elements are given full consideration at some stage, and it is even more important that there should be a free system of reaction of one element on the others.

(*a*) Size of unit
(*b*) Functions (i.e. those functions which the design must perform or which may conveniently be incorporated)
(*c*) Rigidity
(*d*) Visual design
(*e*) Stress flow
(*f*) Ease of manufacture
(*g*) Texture
(*h*) Colour
(*i*) Fixing details
(*j*) Thermal insulation
(*k*) Fire performance

5.3.1. Size of unit

Jointing between units forms a major part of the total cost of a GRP installation. Furthermore, if the GRP system is going to give trouble in the future, it will almost certainly be at the joints! For these reasons jointing should be kept to the minimum and this means the larger the unit the better. Consider the economic aspects of large and small units. There is no basic cost penalty in the material content and labour input per square metre between a large unit and a small unit. A single large mould is obviously more expensive than a single small mould but on a large project, several small moulds may be required in the factory in order to meet the production programme. In such a case the moulding cost for a larger unit may well work out cheaper.

GRP units are relatively light in weight so that handling within the factory and on the site itself does not normally present a problem. Limitations therefore are most commonly imposed by road transport constraints. Units up to 2·5 m wide and 12 m long should present no special problems especially if they can be nested within each other during transport. The Author's firm has used roofing units as long as 17 m by 2·5 m wide. In some cases a module of a certain size will be imposed on GRP design by visual or structural considerations. Even in these cases, the unit size need not be limited to such a module, as two, three or more modules may be incorporated in the same unit. The consequent saving in total length of jointing would be well worth achieving. Such a design decision would not prevent the manufacture of smaller units comprising a smaller number or even a single module. With a little ingenuity the mould designer can arrange for detailed adaptations to the mould so that it can be used for casting a smaller or larger number of modules in one unit.

All GRP units need to be carefully handled at all stages—in the factory, during transport, and particularly during site erection. Large units will need large cradles and these will add to the total cost. Nevertheless, within the scale of a particular project, the rule should be—the larger the unit the better.

5.4. Multiple uses

A material such as GRP gives the opportunity to combine a number of functions in the same material and in many cases within the same unit. Although it is not always practical and sensible to incorporate all possible functions into the design of a single GRP unit, the designer should run through the following checklist to see whether some advantage could be gained by incorporating some of these functions or at least the facilities for the functions.

Weatherproofing. The impermeable nature of GRP make this function self-explanatory.

Rainwater run-off. The ease with which free-formed shapes can be achieved with GRP makes it easy to incorporate channels and gutters in the geometry and thus avoid separate 'rainwater goods'.

Snow clearance. Low-powered electric heating elements embedded in the GRP could provide a useful facility in certain climates.

Solar heat rejection. The natural gloss finish of a polyester gel coat, combined with white or near-white pigmentation, will provide a high degree of solar reflection. Conversely a darkly pigmented textured surface produced by mineral inclusions in the gel coat, would provide a higher degree of solar heating where this is required.

Reduction of wind loading. The ease with which smooth curved shaping can be introduced into GRP moulding facilitates the reduction of localized wind forces on the building.

Thermal insulation. By insertion of insulating layers.

Sound insulation. Although normally associated with lightweight construction, GRP can, if necessary, incorporate dense infill to provide mass damping of sound. A more probable configuration would be the inclusion of a sound-absorbent material behind a perforated GRP skin facing into the building for the absorption of internal noise.

Natural lighting. Polyester resin is naturally translucent and it is easy to make panels, or parts of panels, with a high light transmissivity. The boundaries between translucent and opaque sections are easily made without disrupting the surfaces of the finished unit and without the need for frames or seals.

Artificial lighting. GRP units can be moulded to form the support for electric light sources as well as the support of, and conduits for, supply cables. The internal surfaces of the unit can be moulded to form reflecting and diffusing surfaces for the artificial light source.

Air conditioning. Diffusers and ducting for air supply and venting can be formed internally in GRP units.

Natural ventilation. Airways can be created through a GRP unit so as to provide natural ventilation in harmony with the visual design of the building. As separate ventilation units are rendered unnecessary, it is possible to disseminate the air inlet or outlet appertures throughout the total design. Suitable weather protection can be moulded into the unit for the externally exposed end of the air ways; movable louvres or register plates can be installed on the internal surfaces of the finished unit. The traditional combination of natural lighting and natural ventilation in the one device, i.e. the opening window, can be avoided together with the problems of draught dispersal and the architectural treatment of the heavy double frame required with an opening light.

Services support and ducting. Ducts and conduits for wiring and pipe services can be moulded into a GRP unit with great ease. It is preferable to have such moulded passages continuously open on the internal face to be sealed with a clip-on or screw-on cover strip. This arrangement facilitates installation and revision at a later date. The cover strip can be moulded in GRP so as to provide a visually continuous internal surface.

Access ways. Doorways, manholes, roof access traps, etc. can be moulded into GRP units without disrupting the primary architectural module and sometimes without even requiring a new mould, but merely an adaptation of the standard mould.

Smoke extraction. As with natural ventilation, airways can be formed in the GRP units and sealed with drop or hinged panels which open under gravity or spring action. Fusible links or other releasing devices can be incorporated.

Self-supporting structures. This is, of course, a fundamental function and the extent to which it is incorporated depends on the size of the unit envisaged and its position in the building.

The Author has not yet heard of a GRP unit which incorporates all the functions mentioned above. Nevertheless the incorporation of each one of these functions is entirely feasible and these ideas should be allowed to flow through the designer's mind during the initial stages of his work.

5.5. Rigidity and curvature

The general nature of the problem and the manner in which the designer's initial concept of shape can provide adequate rigidity without excessive use of material will now be considered. GRP is often looked upon as a fairly flexible material. Indeed it is. When handled in the form of a thin flat sheet, it may appear alarmingly so, although in fact its modulus of elasticity is somewhat higher than that of concrete and of most timbers. But concrete is never produced in such thin sections as GRP and the comparable thickness in timber would be the thinnest plywood grades. Because of its high strength and some of the other physical qualities, GRP can be used in very thin sections and so rigidity becomes a significant problem in the basic design. Rigidity is of more concern than is normally the case in the design of concrete, where sections have to be much more massive or in steel where the modulus of elasticity is 20 or 30 times greater.

For a structural engineer, GRP units are normally so thin in relation to their overall dimensions that they are best treated as shells or membrane structures. The most significant characteristic of a membrane structure is that it cannot carry stress otherwise than in the plane of the membrane itself. It follows that flat membranes are incapable of sustaining any load applied at right angles to the membrane. On the other hand, curved membranes, even though the curvature may be quite slight, can resist appreciable loads applied at right angles because the curvature of the membrane provides a certain component of its direction in the line of the applied force.

Curvature is the simplest and most economical means of providing rigidity to thin sheet materials (Figs 5.1, 5.2). In the case of GRP, it costs nothing, or very little, to provide, for in practical manufacture a mould with curved surfaces takes hardly any more effort to produce than one with plane surfaces.

For the purposes of design, two types of curvature can be considered— one-dimensional curvature and two-dimensional curvature. A sheet with one-dimensional curvature would produce a curved cross-section in one direction but a straight-line cross-section in the direction at right angles. The

Fig. 5.1. Three-dimensional curvature has been used within the framing of these panels to give considerable stiffness, yet a strict and formal geometry has been maintained: Police Sports Centre, Hendon, London

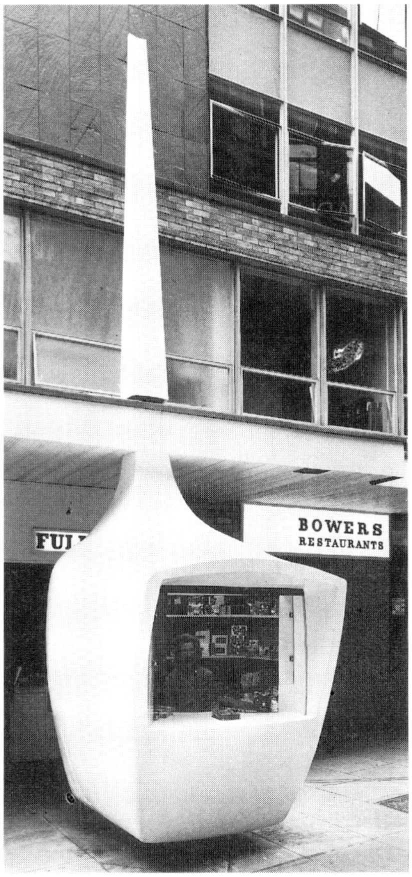

Fig. 5.2. A small building, but a good example of the use of free curvature to achieve a rigid and pleasing structure

surface of a cylinder or of a cone would be examples of one-dimensional curvature.

A surface with two-dimensional curvature would show curvature in one cross-section and also in a cross-section taken at right angles. The surface of a sphere would be an example of two-dimensional curvature where both curvatures are in the same sense (e.g. both convex). A saddle would be an example of two-dimensional curvature with the curvatures having opposite senses (i.e. one convex and other concave). A saddle shape is not so common as a spherical shape but both types of two-dimensional curvature are easily formed in GRP.

Despite the increased rigidity obtained with one-dimensional curvature, there are still problems under certain conditions with a tendency to buckle (localized instability) and sway (i.e. collapse in a direction at right angles to the applied load). Two-dimensional curvature can overcome these problems and brings a further leap upwards in rigidity. This form of surface is not easy for the desk-top experimenter to simulate by bending pieces of paper, but many forms of such surface are all around us. The egg-shell is probably the most dramatic illustration of the rigidity of two-dimensional curvature and there are commonplace derivatives such as the construction worker's hard hat and the pingpong ball. An interesting and valuable property of the two-dimensional curved surface is its ability to be rigid in itself, so fixing or constraint of the edges of the surface is not always necessary. This is a most useful attribute for the GRP designer to remember.

There are other devices available to the designer which will increase rigdity, such as troughing, ribbing and sandwich construction.

5.6. Troughing

The mechanism of troughing is self-evident. Corrugated iron sheets provide a common illustration of the rigidity that can be obtained. It should be realized that the corrugated sheet is an example of single-dimension curvature as a cross-section taken along the length of the corrugations would show a straight line. In GRP design, an advance on simple corrugations can easily be achieved by shaping the corrugations in the longitudinal sense so that an element of two-dimensional curvature is introduced. The troughs then become more like elongated dishes.

5.7. Ribbing

Ribbing is a form of troughing whereby troughed sections are added to one side of a flat sheet in order to give it rigidity. Corrugated cardboard is an illustration of the technique and stronger examples can be found in articles made from sheet metal. Although the technique is used in GRP design, there are a number of disadvantages to it. Firstly, it is not economical, as two components (the flat sheet and the trough) have to be separately manufactured and then fixed together either by adhesives or by further laying up of resin. Secondly, there is nearly always a tendency for the ribs to show through to the flat surface, which is normally the visible surface. Nevertheless it is a useful technique where it is absolutely essential to provide a flat finished surface to the GRP unit and where sandwich construction is inappropriate.

5.8. Sandwich construction

For a finished unit which has to have a flat surface on each side, increased rigidity can be achieved by making the panel thicker. Increasing the thickness simply by using more GRP would be an expensive method, so it is common practice to achieve the required thickness by sandwiching a low-density core between the two outer skins of GRP. Foamed plastic is often the core material, but timber (balsa wood) and void-formers fashioned from stiffened paper or cardboard, are also used.

5.9. Shape

Although troughing, ribbing and sandwich construction are widely used to achieve rigidity they are really only expedients, sometimes necessitated by

Fig. 5.3. Going too far with curvature? Perhaps, but it shows what can be done with GRP: house in Finland, 1968

functional constraints on the geometry of the finished unit, but perhaps more often they are an indication that the designer has not allowed his imagination to venture into the much more elegant realm of overall curvature or shape of the unit.

It is so easy to form GRP into almost any conceivable shape, that it is not surprising that sculptors are making increasing use of the material for the most imaginative pieces (Fig. 5.3). However, the inexperienced designer may be bewildered by such a wide freedom of choice of shape and thus welcome some constraints which will narrow down the choice. Although the constraints mentioned here can nearly always be loosened by some means or other, if the designer wants to give full rein to his imagination, they are useful starting points for evolving the shape of a unit.

Firstly, there is the constraint of the overall architectural design of the building. This may impose, or at least suggest, a design module in terms of size and perhaps also shape for the GRP unit. It will be remembered from section 5.3.1 that the unit size does not necessarily have to coincide with the design module size. A second constraint could be the achievement of rigidity as discussed in the preceding section. This is of course an optional constraint for even the designer who is enamoured with dead flat surfaces can obtain the desired rigidity of the finished unit. Thirdly, there are other functional constraints such as the need to shed rainwater, the avoidance of dirt traps and the need for it to be practical to make and use the mould which will form the shape.

The designer will surely obtain the maximum satisfaction from his work when he achieves a shape which is an elegant synthesis of visual pleasure, rigidity, functional performance and economy of material.

5.10. Texture

It has been seen that GRP can be formed into almost any shape and this characteristic applies right down the scale to the microscopic shape of the mould surface. Thus there is a very wide range of surface textures which can be obtained quite easily by simple contact moulding. As the mould itself can be taken direct from an original surface, exact facsimilies of material such as

Fig. 5.4. A bold GRP statement illustrating the crisp precision and uniform colour obtainable. The six nacelles on the terrace approach closer to the freedom of expression which polymer composites can give so easily and chcaply: Mondial Housc International Telephone Exchange on the banks of the River Thames

Fig. 5.5. Alderman Smith School, Nuneaton; architects Donald James and Partners (courtesy Brensal GRP Products)

timber, stone, brickwork and natural rock can be moulded with ease. Additional surface treatments to achieve a particular finish can be carried out before and after the moudling process. For example, the inclusion of mineral particles such as sand grains in the gel coat will produce a certain effect and post-moulding treatments such as light abrasion and, of course, painting, can all be used.

Such a wide variety of surface finishes is used to great advantage in reproduction GRP work. Some manufacturers have become particularly skilled at simulated finishes and can product GRP masonry, brickwork, timber, cast bronze and lead, which can deceive even the expert eye. Although there is a legitimate role to be played by such simulations, the majority of GRP designers will be using the material in its own right and will probably prefer a surface texture which befits the basic nature of the material.

The question of what is the natural textural appearance of GRP is a subject for debate. The basic liquid origin of the resin suggests the hard smooth shiny surface which is commonly employed. The hard gloss surface certainly has some advantages. It is likely to give the least possible problem with dirt retention and, in building design, it displays a unique property of the material. Although glass, glazed ceramics, and stove-enamelled or plastic-coated steel can give a hard gloss surface, none of those materials can combine that finish with a free-flowing large-scale sculptural shape so easily obtained with GRP. This combination allows the designer to introduce light and shade into a building facade to a degree perhaps even greater than that achieved with masonry in classical architecture. To the conventional repertoire of shadow and relief can be added the quality of reflection. The two-dimensional curvature, advocated in the section on rigidity, enables interesting transition zones to be achieved from light to shade. The precision of GRP moulding techniques enables highly disciplined repetition of shade patterns to be achieved over a building elevation (Fig. 5.4). Because of the high reflectivity of a hard gloss finish, such patterns are sometimes clearly defined even on an overcast day without direct sunlight.

Some designers prefer to depart from the hard gloss finish and use a slightly undulating or dimpled texture. The advantages claimed are that slight surface blemishes are not so evident and that the hard gloss surface will disappear over the years anyway so that a slightly textured surface makes the change less discernible. The loss of gloss from a polyester gel coat certainly does occur, depending on the exposure to sunlight, but in well-made GRP in normal conditions the loss is very slow.

Another approach to surface texture is to coat the external surface of the units with polyurethane after casting. This is more expensive than using a self-coloured gel coat but is likely to give a more durable finish. The polyurethane coating will intrinsically have a hard gloss finish but being a thin surface layer, will of course follow any moulded pattern in the base GRP.

5.11. Colour

Any colour desired can be used. But before indulging in this freedom, the designer ought to consider carefully a number of technical points.

Firstly, there is the question of colour matching if the designer wishes to have uniformity of colour throughout his building. If, as is usually the case, the colour is to be achieved by a pigmented gel coat, then it is best to ensure that sufficient pigmented resin is produced in one batch to provide the gel coat for all the units to be produced. If this is not practical, very careful colour matching of resin batches is needed and it is important to ensure that samples are viewed under scientifically controlled identical circumstances. The unskilled eye is easily deceived by colour and expert advice would be worth seeking.

Secondly, there is the matter of colour retention. The problem is not severe for internal GRP but, as most GRP is used on the outside of a building, the action of ultraviolet light is significant. Most pigments will change to some extent through the years but many colours can be produced nowadays by pigments whose rates of change are so slow as to be insignificant, although some blues and black remain a problem. It is essential to check that the pigment used has good light stability. The designer should ask to be shown actual samples of pigmented resin which have been exposed outside for a number of years. Most manufacturers now have such samples available. The Building Research Establishment has been conducting tests on colour retention. There are now sufficient GRP-clad buildings around which have been in existence for 10 to 15 years and a tour around to see some of those would be interesting for the designer. He should be encouraged by such a visit.

Colour change can be brought about by yellowing of the polyester resin itself. This was a well-known effect in earlier uses of the material but more light-stable resins are now available. Again, this point should be checked with the resin supplier and, if possible, samples with several years exposure inspected.

Most colour changes go unnoticed provided they are not too violent and more importantly, provided the change is reasonably uniform through the building. Differential colour change can be distressing, although it is a comparatively rare phenomenon in GRP buildings. It is best avoided by careful attention to quality control in the production of the GRP units. Aesthetically, unforeseen colour variation is considered much more objectionable in GRP than in other common building materials such as masonry, brickwork and (particularly) concrete. Perhaps this is due to the unique precision of form which is achieved in GRP which leads one to expect equal precision in colour behaviour; or perhaps the often-heard claim of permanent colour for GRP, leads one to expect just that. Certainly GRP has much stricter standards of appreciation imposed on it compared with, say, concrete where the most horrid patchiness is forgiven as a natural characteristic of the material; (in fact, with careful attention to specification and construction, very uniform concrete can be made).

It has been assumed so far, that the GRP units are intended to be the same colour. Indeed most architects, up to the time of writing, seem to have used GRP in monochrome, particularly in whitish colours and pale beige in the early days. More recently, adventurous use has been made of bolder colours such as orange, brown and the darker greens. Hardly any use has made so far of the ability to produce cheaply and easily the same basic unit in different colours. The same mould is used and the only change in production is to use a different pigment in the gel coat. By this means alternating colours, or a graded change of tint of the same colour can be used across a building facade.

5.12. Repetition of design

The GRP designer often will start his work under the threat of heavy costs if his design requires a number of different shapes and sizes of panel to be moulded. It is certainly true that each different mould required will add to the cost of the project. The designer therefore will probably resolve from the beginning to use only a standard panel, or at the most perhaps two or three standard designs. Even with these good intentions, he will eventually find that by the time he has catered for corners, internal and external, differing elevations or heights, various types of necessary openings in the facade, left-handed and right-handed units, he may find he has 50 or more different types on this final schedule. The problem can be reduced in two ways.

- By reducing the number of variations.
- By adapting the moulds to produce more than one shape.

Firstly, an excessive multiplicity of shapes usually results when the designer decides to wrap the whole of his building, or the whole of one elevation, in GRP. The aesthetic argument would be that GRP has such a strong individuality that it does not blend easily with other facing materials. Although a legitimate argument, it is perhaps based more on the comparative newness of GRP as an architectural element. Alternatively perhaps the designer is so fascinated with the artistic possibilities of the material that he is determined on a GRP building. Whatever the motives, it is certainly true that some of the most successful examples of GRP use have been based on the concept of a complete GRP wrap of the building. If the designer is prepared to face the extra moulding cost involved then so be it.

The alternative is to use GRP panels only in those parts of an elevation where the truly standard panels can be used. Commonly this would still result in 70 or 80% of the elevation being clad in GRP. Where odd corners and difficult geometric intersections occur, then more adaptable materials are substituted such as brickwork, blockwork or in situ concrete. There are strong aesthetic arguments for using different materials according to their suitability, in different parts of the building. One does not see facing brickwork used as a roof-covering material.

The second general approach to the problem of reducing mould cost is by designing the units so that variations can be cast from an extension or simple adaptation of one basic mould. Lengthening and shortening of a mould is an obvious example. Other adaptations can be made by the use of removable panels in the mould to form basic units or similar units with window openings, door openings and the like. The designer of the GRP unit should have these concepts firmly in mind and be prepared to adapt his design, using the advice of the mould-maker on what is practical to achieve.

5.13. The future

In building design, from both architectural and engineering aspects, design techniques and attitudes tend to follow the development of materials rather than the reverse. Although it is becoming increasingly possible to produce a material to fit a specified performance, particularly in the case of polymers, that procedure is followed in only a few special cases. But once a new or an improved material is made available, designers usually react promptly and imaginatively to exploit it. Therefore it is appropriate to look at possible developments in materials.

It is undeniable that many building designers consider GRP a difficult material in buildings from the point of view of fire behaviour. This is to a large extent unjustified. The intrinsic characteristics of the polyester resin contributes only a part of the total fire behaviour of a GRP panel in a particular building, but there would be undoubtedly a ready market for a polymer which would have even lower combustibility than those presently available. It is possible, therefore, that in the future a resin component of GRP will be available which, although perhaps not absolutely incombustible, would behave in a manner tantamount to an incombustible material. It seems likely that this would be achieved by admixtures to the polymer rather than by any particular arrangement of the carbon, hydrogen and oxygen structure within the molecule.

Greater resistance to ultraviolet light degradation might be achieved although present standards are good.

An existing, but relatively new development, is the use of catalysts which react only in the presence of ultraviolet light. This enables resin and catalyst to be mixed, stored and used in the manufacture of the laminate with no time restriction, provided that the mixture is not exposed to a light source containing the ultraviolet component. When the manufacture is complete and all is ready, a relatively short exposure time to ultraviolet light will set the resin. This is primarily an advantage in manufacturing but could lead to the use of more sophisticated and complicated construction techniques. It is already of considerable convenience in repair work.

There is plenty of scope for improvement in jointing techniques and materials. No doubt a great deal of research and development will go into this problem. Although glass fibre has proved to be a most successful reinforcement medium in polyester resin, it is not the only high-strength fibre available. Carbon fibre is already used where particularly high stresses occur and it is possible that composites will be developed using other polymers as the tensile reinforcement. Whether a completely homogenous polymer can be produced with physical properties of GRP and its relatively low cost, has yet to be established.

5.13.1. Design in the future

The prime requirement of a building is that it should provide an environment which is an improvement on the ambient external condition. In very many cases this primary object can be satisfied simply by the provision of a shell or membrane of GRP. Some buildings can produce a worsening of the ambient environment. An important example of this paradox relates to earthquakes. No one is likely to come to much harm during a major earthquake if he sits in the middle of an open field. Buildings of GRP, in so far as they are likely to be very much lighter than conventional construction, have an intrinsic advantage here.

The use of laminar materials is only beginning. The progress to be made depends largely on the inventiveness of the designer and his willingness to consider depatures from the conventional attitudes.

6 Adhesive and bolted joints

L. Hollaway

6.1. Introduction to adhesive joints

Joint design is dependent upon the nature of the materials to be joined as well as upon the method of jointing. Civil engineering composite structures are generally manufactured as units or members and these members then have to be joined to complete the whole structure. When jointing fibre/matrix composites to similar to dissimilar materials, there are three options open to the designer. These are to form

(*a*) an adhesive bonded joint (chemical jointing)
(*b*) a mechanical fastener (mechanical jointing)
(*c*) a combination of mechanical and chemical jointing.

Each of these options will be discussed separately in terms of lap joints. There are other types of joints, such as flange joints, which play a vital role in pressure pipes and pipework at chemical plants where pipes have to be joined to their like or to a structural component. The difficulty with these lies in the anisotropic nature of composite materials.[6.1]

The use of adhesively bonded joints for composites commenced in earnest with the start of aviation. Wooden structures (wood being a composite) which contained scarf joints were used as the primary members. Before that, the carpenter's joint had used glue but this was the minor component in forming a rigid mortice and tenon connection. Since then, the jointing of materials has been subjected to a long process of development with the fibre/matrix composite structural joint being a major area for investigation.

In load-bearing structural composite joints there exists an inter-relation between the materials to be joined, the nature of the stresses to be transmitted and the method of jointing adopted. It is therefore necessary for the designer to have a thorough understanding of the initiation and progress of the failure process if analytical models and methods of prediction are to be developed. There are pitfalls in attempting to correlate the results of bonded test coupons designed to force a failure in the adhesive with the performance of a structural bonded joint in which the objective is to ensure that the adhesive does not fail. There are a number of parameters upon which the exact nature of failure of the joint will depend; these are surface ply orientation, stacking sequence, joint geometry and loading. In addition, the matrix of the fibre/matrix composite may be much weaker than the structural adhesive and in this case failure will initiate in the material component of the composite system.

Adhesive bonding has other functions apart from transferring load from adherend to adherend. Bonding can be used to increase the structural efficiency of a laminate structure. Stiffeners, also, are generally bonded to thin plates to prevent local buckling, in which case very little load is transmitted through the glue lines, provided that the structure remains intact. However, the bonding here is important from the point of view of containing local damage.

The failure modes of fasteners for fibre-reinforced polymers are the same as in metals but the mechanism by which the damage commences and propagates is very different and extremely complex. Indeed, the mechanism of failure of the composite will depend upon the types of fibre and matrix being used. The failure mode is also highly dependent upon joint geometry; the width in single hole joints or pitch in multi-hole joints, edge distance, hole diameter and laminate thickness are all important parameters in understanding the failure modes of fastened joints.

A number of excellent reviews representing the state of the art for the analysis of adhesive bonded and mechanical fasteners have been given in references 6.2–6.14.

Throughout the discussions the following abbreviations will be made

GRP—glass fibre reinforced polyester resin.
GFRP—glass fibre reinforced epoxy resin.
KFRP—Kevlar fibre reinforced epoxy.
CFRP—carbon fibre reinforced epoxy resin.

6.2. Adhesively bonded joints

The variety of adhesives now available commercially is immense. These are usually supplied in liquid, paste or solid form, and are categorized according to their chemical type which may be further classified as inorganic or organic. Structural adhesives are organic and comprise thermosetting resins and elastomers or thermoplastics. These combinations of resins from different chemical groups are strong, tough infusable, insoluble and can be used over a wide range of temperatures. Currently, the two adhesives most commonly used for structural applications are the modified phenolic and epoxide resin systems.

The advantages and limitations for adhesive bonding which are fully documented in most texts on the subject indicate that engineering offers a potentially large market for manufacturers of adhesives. However, apart from the aerospace industry, their acceptance to date has been sparse and disappointing and this may be attributed to the following

(*a*) the potential user's lack of confidence in the material, and the shortage of information available
(*b*) adhesive-bonding technology being largely ignored in the training of engineers
(*c*) the low risk attitude of industry in times of recession
(*d*) the engineer's prejudice against replacing traditional jointing methods.

A strong and reliable joint is critically dependent not only on the correct adhesive being used but upon a number of other factors. These are

(*a*) the adherend surface pretreatment (cleaning and etching) must be undertaken conscientiously
(*b*) the adhesive preparation (proportioning, mixing and application) must follow the manufacturer's instructions precisely
(*c*) the component parts must be matched and a suitable pressure applied, with the curing schedule conforming to the manufacturer's requirements.

Having followed these procedures it is still difficult to estimate the degree of bond achieved and to detect a poorly constructed joint. Not surprisingly then, the on-site use of adhesives in the building sector, which traditionally uses unskilled labour, sometimes in adverse working and climatic conditions, has been restricted to essentially light loading applications. These include gap-filling sealants and the bonding of wall and flooring tiles; these items are discussed in chapter 11. Notable examples of load-bearing applications are in concrete bonding (by replacing mortars in segmental construction), in the restoration of cracked concrete and in the adhering of temporary or permanent steel plates to supplement the existing inadequate reinforcement.

Over the last 20 years, the use of composites for load-bearing structural applications in the construction industry has increased and attention has been directed to the plate continuum systems manufactured from chopped strand mat or bidirectional reinforcement and to polymer matrices and the skeletal systems fabricated from essentially unidirectionally aligned fibre sectional members.[6.15] The former system would generally be composed of a number of identical units joined by bonding and/or bolting along flanged edges. The members of the latter system are eminently suited to a space frame type system in which the loading of the members is mainly axial. Whilst these structural systems have great potential, it has been found that the fabrication and jointing of the members present difficulties. Currently, research effort is being directed towards this problem, using bonded metal or composite end caps which are received by the nodal joint.[6.16,6.17] Section 6.20 discusses this technique.

The structural components would normally be fabricated separately and then joined by the following jointing configurations, either separately or in combination

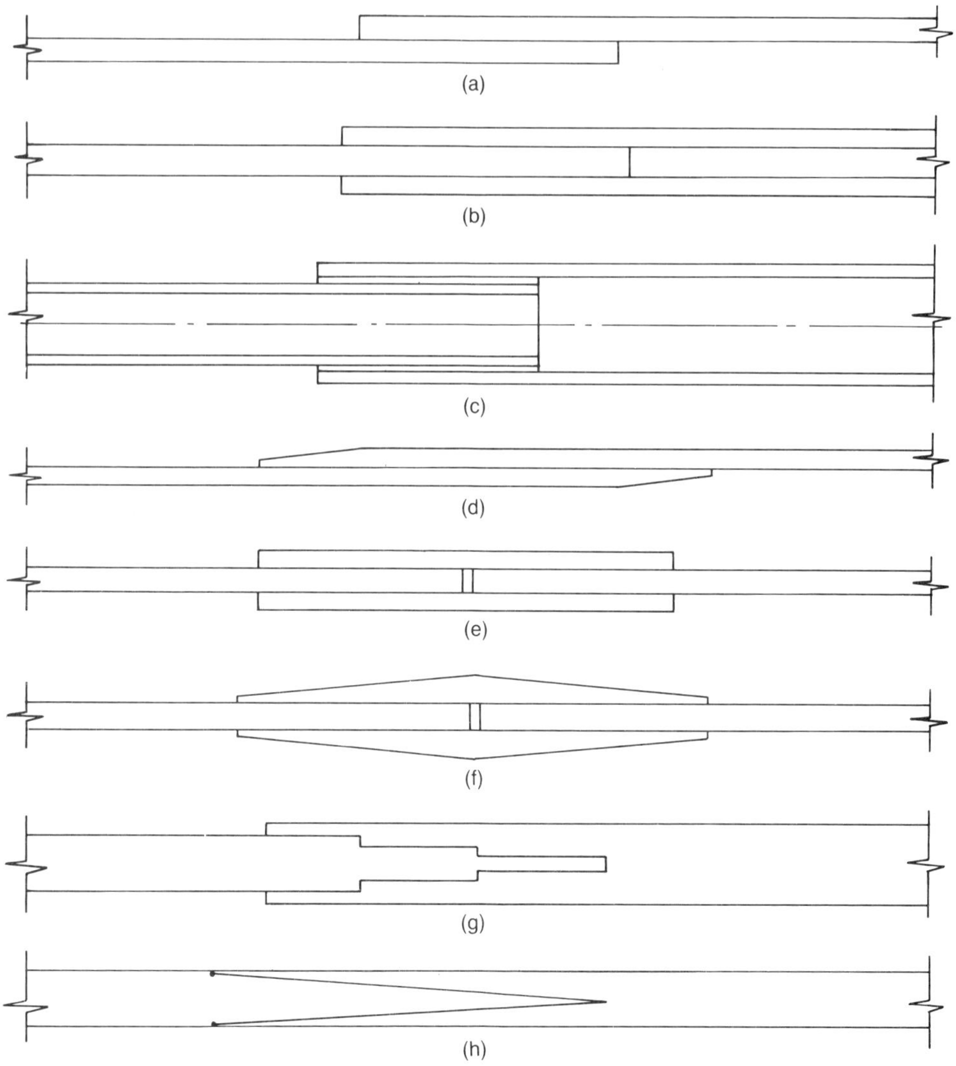

Fig. 6.1. Adhesively bonded joint types: (a) unsupported single lap joint; (b) double lap joint; (c) tubular lap joint axisymmetric loading; (d) tapered single lap joint; (e) double strap joint; (f) tapered strap joint; (g) stepped lap joint; (h) scarf joint

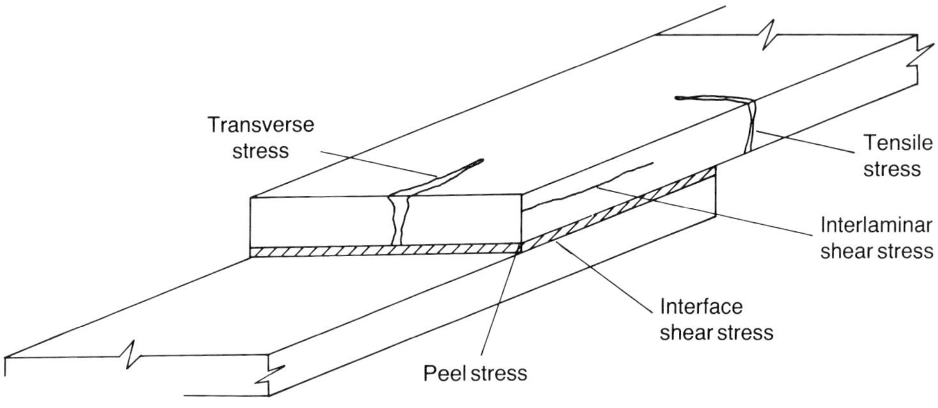

Fig. 6.2. Modes of failure of an adhesively bonded joint

(*a*) the single lap joint to enable overlapping plates to be bonded (Fig. 6.1(a))
(*b*) the double lap joint which uses both faces of the central adherend to transfer load (Fig. 6.1(b))
(*c*) the tubular lap joint which is capable of bonding overlapping co-axial tubes (Fig. 6.1(c)).

The double lap joint is symmetric about the centre line and is analogous to a plate bonded to a rigid support. The tubular lap joint would be employed in fabricating double layer skeletal structures using tubular fibre-reinforced polymer members.

The aerospace industry often replaces those configurations with more efficient joints which are usually produced by tapering the adherends to give the bevel, scarf and step joints. The various joint types are illustrated in Fig. 6.1.(d)–(h). The costly machining and performance of the adherends render these refinements less attractive to civil engineering structural systems unless a highly specialized composite jointing system is required. However, the simpler and cheaper configurations which would generally be used in the construction industry could not be expected to sustain high load levels, no matter how much quality workmanship was employed in their manufacture.

It is appropriate to introduce here the concept of the four distinct modes by which adhesively bonded joints can fail. These are

(*a*) a failure of the adherend
(*b*) a shear strength failure of the adhesive
(*c*) a mode associated with the failure of the adhesive under a peel load
(*d*) a failure by delamination of the fibrous composite adherends.

These modes are illustrated in Fig. 6.2. Item (*a*) above occurs in the strongest joint, in that the load level is 100% of the strength of the adherend in which failure takes place. In this case the strength of the joint is linearly proportioned to the thickness of the laminate.

Item (*b*) is an adhesive failure of the bond between the adherend and the adhesive. The shear strength of the bonded joint is proportional to the square root of the laminate thickness. This form of failure must not be confused with cohesive failure.*

The peel stress in item (*c*) is a tensile stress which develops its maximum value near the free end of a double or a single lap joint. The stress is developed in association with the shear stresses and can be particularly severe in thick

* Cohesive failure means a failure wholly within the adhesive material whereas adhesive failure implies the failure of the bond between the adhesive and adherend.

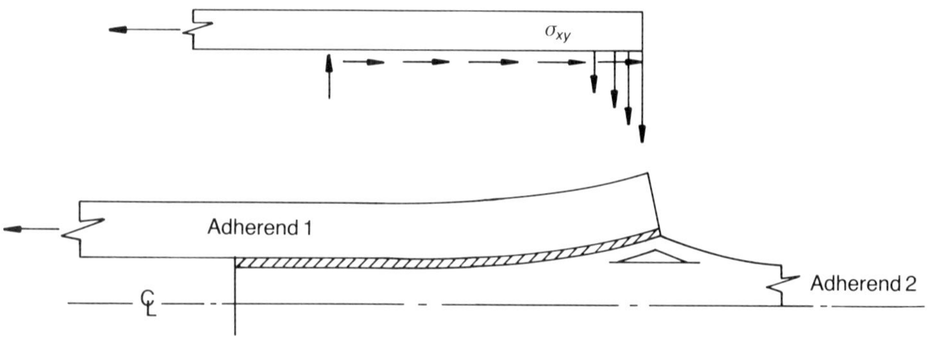

Fig. 6.3. Peel stress failure through thick joint

double lap joints (high peel stresses are developed in single lap joints). The low interlaminar tension strength of composite laminates limits the thickness of the adherends which can be bonded together efficiently by lap joints. The inner laminate of the adherend splits locally due to full stresses, thus destroying the shear-transfer capacity between the inner and outer plies. This overloads the outer filaments which break in tension and the failure progresses through the adherend thickness. The process is illustrated in Fig. 6.3. The peel stress induced failure occurs within the laminate as soon as the peel stress exceeds the interlaminar tension; this latter strength is far less than the peel strength of typical structural adhesives.

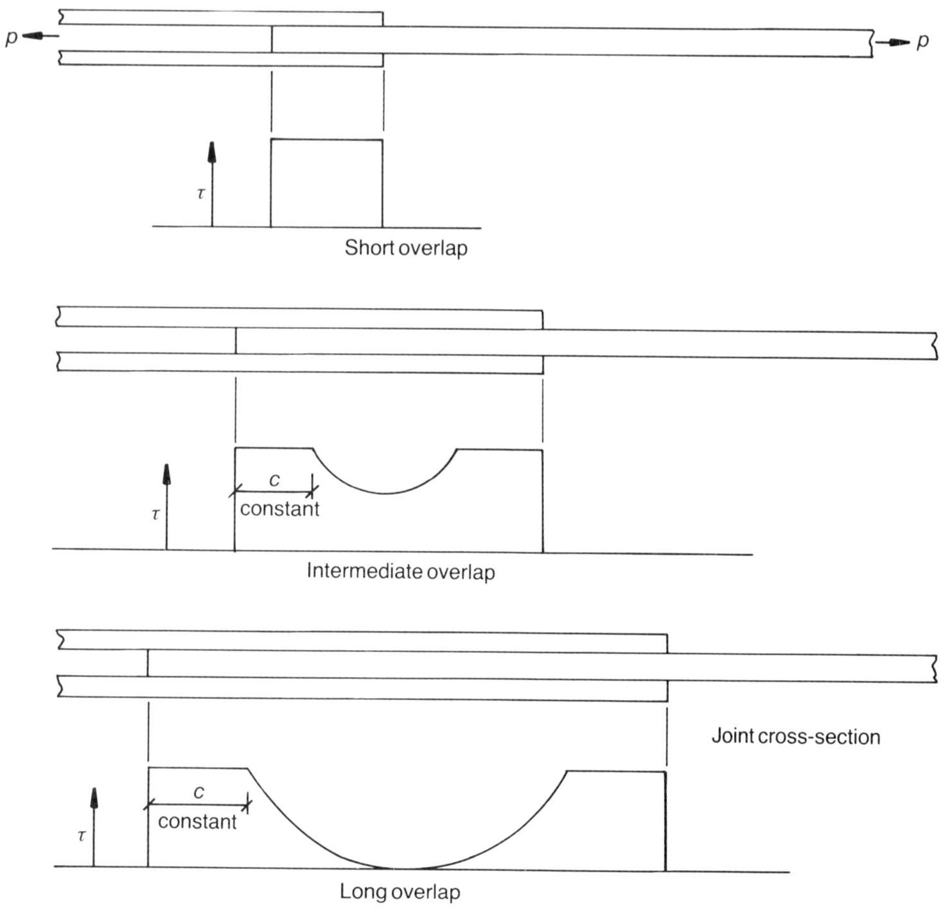

Fig. 6.4. Influence of lap length on bond stress distribution (after Hart-Smith)[6.10]

6.3. Single lap joints

The tensile single lap joint will not be as strong as the adherend being bonded. However, provided the overlap length/thickness of the adherend is of the order of 100 to 1, the structural efficiency of the joint should be high.

Single lap joints exhibit out-of-plane bending moments when under load. These moments are associated with the eccentricity in the load path and will produce high peel stresses, which are significant, particularly at high applied loads. Hart-Smith[6.10] has suggested that an effective way of preventing premature failure of adhesively bonded joints is to taper the outer adherends locally as shown in Fig. 6.1(d) (the tapered single lap joint), to a tip thickness of $0{\cdot}5 \pm 0{\cdot}25$ mm with a taper angle of 1 in 10. Such adhesive tapering prevents failures by induced peel stresses, but it will not increase the shear strength of the joint. If the latter strength needs to be increased, tapered slice plates could be used (Fig. 6.1(f)) and their centres could be thickened, thus simultaneously reducing the peel stresses and increasing the shear strength of the joint. Another possibility of preventing premature failure is to increase the overall length as shown in Fig. 6.4.

Thicker single lap joints require the ends of the adherends to be tapered, primarily to prevent a premature failure by induced peel stresses before the full potential shear strength of the adhesive can be developed.

6.4. Double lap joints

Unlike single lap joints, symmetric double lap joints have no bending moments. However, the joint does undergo internal bending as shown in Fig. 6.5. The outer adherends bend under load, causing tensile stresses within the adhesive at the end of the overlap and compressive stresses at the ends where the load is applied.

The magnitude of the tensile stresses (peel stresses) may be reduced considerably by the same techniques as for the single lap joints.

Whilst it is necessary to incorporate these sophisticated details in high technology composite design, they may not be relevant to some of the medium technology of the construction industry. However, it is vital for the designer to be aware of the high stresses associated with bonded joints and to be familiar with the possible methods of reducing these stresses.

6.5. Spew fillets

It is generally assumed that the adhesive layer within double and single lap joints ends in a square edge. If the adhesive layer has a square edge as shown in Fig. 6.6 it would be expected that high tensile and compressive stresses (of about 4 times the applied shear stress and in a direction at right angles to the sides on which the shear bond is applied[6.9]) would occur in the corners of this layer because of the free edges. However, real joints tend to form a fillet of

Fig. 6.5 (below). Internal bending moments within a double lap joint

Fig. 6.6 (right). Single lap joint: (a) without spew fillet; (b) with spew fillet

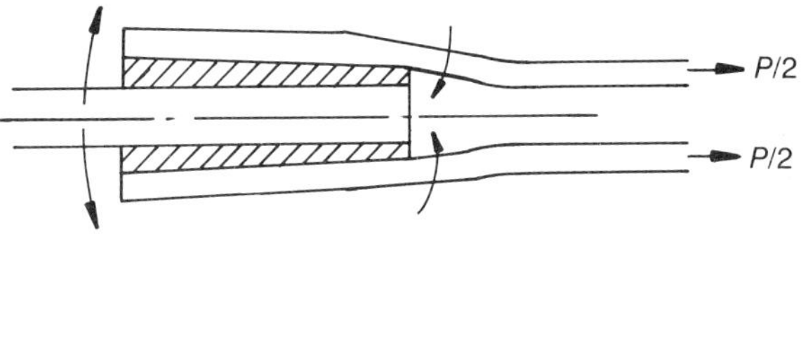

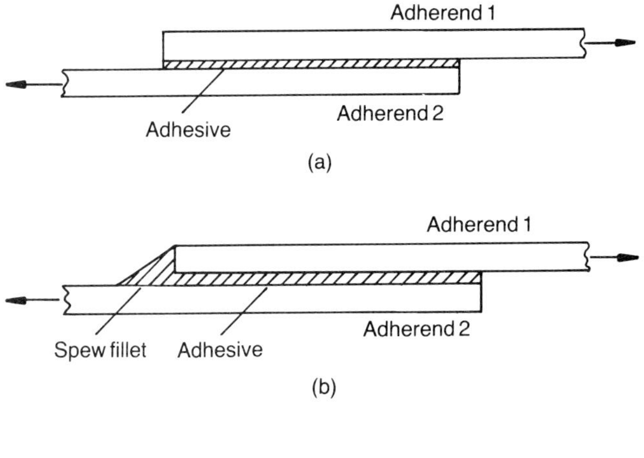

adhesion spew due to the squeezing of the adherends during bonding. Adams and Wake[6.9] have stated that, when a spew fillet does exist at the end of a joint, the maximum shear stress is reduced to 70–85% of the square-ended adhesive layer, depending upon the geometry of the fillet. Adams and Wake have also cited analytical analysis on various geometries of spew fillets. They have shown that the presence of fillets causes very large changes in the adhesive stress distribution in the highly stressed region of the joint end. These stress analyses were undertaken by the finite element methods which are currently the recognized techniques for stress analyses of fillets and bond lines.

It has been recognized that adhesive yielding is present in structural joints at appreciable levels of loading; yielding is regarded as a desirable feature to alleviate stress concentrations and to redistribute loads. Existing theoretical work has been undertaken using linear material assumptions and also non-linear material assumptions employing an elastic-plastic model for the adhesive (section 6.14).

6.6. Analysis of tubular joints

Figure 6.7 shows examples of tubular joints. The principles governing the analysis of lap joints may also be applied to these joints. The system can be conveniently analysed as an axisymmetric body.[6.18] The peel stresses are present in this type of joint but the load transfer is generally improved. It is difficult to know whether a good bond has been produced in many of these joints as no access is possible once the two adherends have been brought together.

The joint can be formed in several ways and assembled wet or dry. In the latter case, holes into which the adhesive resin is pumped are made in the adhesive region. In Figs 6.7(c) and 6.7(d) a hole at point x may be used either to allow air to escape in the wet joint or as an access for the resin to be pumped into the adhesive region of the dry joint.

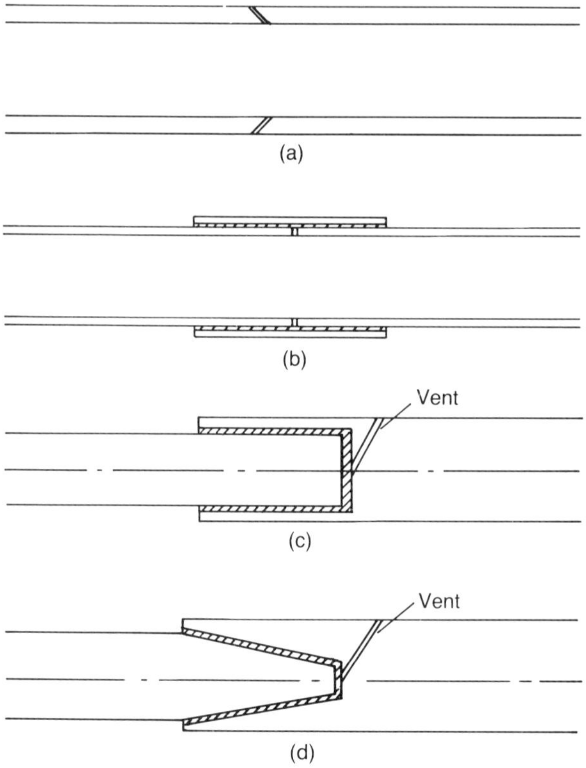

Fig. 6.7. Examples of tubular joints for axial or torsional loading: (a) bevelled tube joint; (b) sleeved tube joint; (c) parallel shaft joint; (d) tapered shaft joint; (e) crimped aluminum end cap on to tube (cross-section); (f) composite end cap on to tube (cross-section)

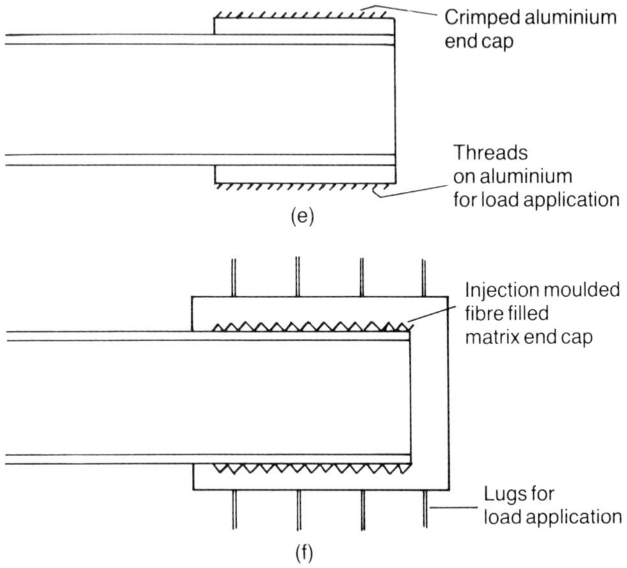

Recently two new joints have been produced; these are

(*a*) the crimp-bonded end fitting developed by Green and Phillips,[6.16] shown in Fig. 6.7(e).
(*b*) the composite end cap fitting developed by Hollaway and Baker[6.17] and shown in Fig. 6.7(f).

These two joints virtually ensure a satisfactory "wet out" of the adherend surfaces in the joint. They are discussed further in section 6.20.

6.7. Butt joints

Butt joints are not generally used in practice because, although they are strong in tension, compression and shear, they are readily fractured under a bending moment situation due to the high peel and cleavage stresses developed. However, this type of joint is used in determining the mechanical properties of structural adhesives in shear, tension and compression. The napkin-ring test[6.19] is often used in this context where the specimen has a circular cross-section.

6.8. Stiffeners

Stiffeners are used when it is required to stiffen essentially flat load-bearing panels. These stiffeners can be any geometric shape, but often "top hat" sections are used; an example of this is shown in Fig. 6.8. These sections can be varied in strength and stiffness by using different configurations. Ideally these top hats will be bonded to the load-bearing plate, rather than bolted, to enable the maximum stiffness of the overall unit to be achieved.

6.9. Flaws in adhesive bonds

Hart-Smith[6.10] has found that, generally, flaws in adhesive joints do not affect the strength of the joint adversely. The reason for this is that the critical stress location in a bonded joint invariably occurs at one end of the overlap and flaws tend to occur in the centre of the joint where the stress levels are low; the flaws consist generally of air bubbles. The problem does not become serious until the bonded joints are large enough to shift the critical location of the stress away from the edge of the joint to a position adjacent to the flaw. In this situation the load redistribution becomes so great that a substantial loss of strength occurs.

6.10. Repairs of adhesive bonds

Joints made from composite adherends can be repaired using adhesive materials. It has been shown, however, that it is vital to dry the laminate thoroughly before repair is commenced.[6.20,6.21] Moisture in the adherend composite during repair will have a detrimental effect upon bond efficiency. As construction repairs normally have to be undertaken in situ, this operation of drying is extremely difficult. A further important step in the adhesion repair is to ensure that thorough cleaning of the surfaces to be bonded is undertaken. It is then important to realize that the surface must not be contaminated.

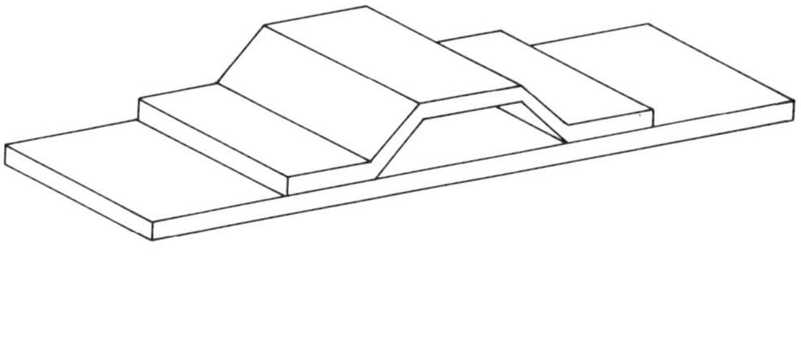

Fig. 6.8. Top hat stiffener

6.11. Fatigue loads

Only limited data are available on the fatigue strength of joints. Renton and Vinson[6.22] and the MIL Handbook[6.23] have given results for glass fibre reinforced epoxy adherends and Sage[6.24] has presented results for carbon fibre reinforced epoxy adherends. Their results indicate that strength values of only about 30% of the static strength are developed at 10^6 cycles. A ductile adhesive is the most satisfactory and the stresses in this bonding agent should be kept within its elastic range; peel stresses should be kept to a minimum.

6.12. Physical effects of bonding

When bonding metal to metal, the adherends require adequate and correct pretreatment of the surfaces if full strength and adequate durability are to be developed. Premature failure of the joint at low loads is almost always due to indequate preparation. The exact treatments depend upon the surfaces to be joined and the adhesive used. Kinlock *et al.*[6.25] have discussed the reasons for such treatment. Similar care is required when joining metal to composites. Reference 6.26 describes treatments for aluminium, stainless steel and titanium. It is usual to only abrade and degrease composite adherends but it is vitally necessary to ensure that all traces of the release agent are removed.

6.13. Analytical analyses for joints

6.13.1. Single lap joint

6.13.1.1. Elastic analysis

The first theoretical analysis for the stress distribution of an adhesive layer between two overlapping metal adherends is usually accredited to Volkersen,[6.27] although previous researchers who investigated soldered joints are recognized as important contributors to this science. The simple Volkersen analysis is often named the "shear lag model", because it considers only the shear deformation in the adhesive caused by differential in-plane straining of the adherends. The mechanism of stress transfer for the shear lag model is given in Fig. 6.9.

The eccentricity of the applied loading in the single lap joint will cause bending of the adherends and a rotation of the overlap region as illustrated in Fig. 6.10. The adhesive must now absorb both adherend stretching and bending strains, giving the complex shear and peel stress distribution shown in Fig. 6.11. Clearly, the Volkersen theory is inadequate to describe the single

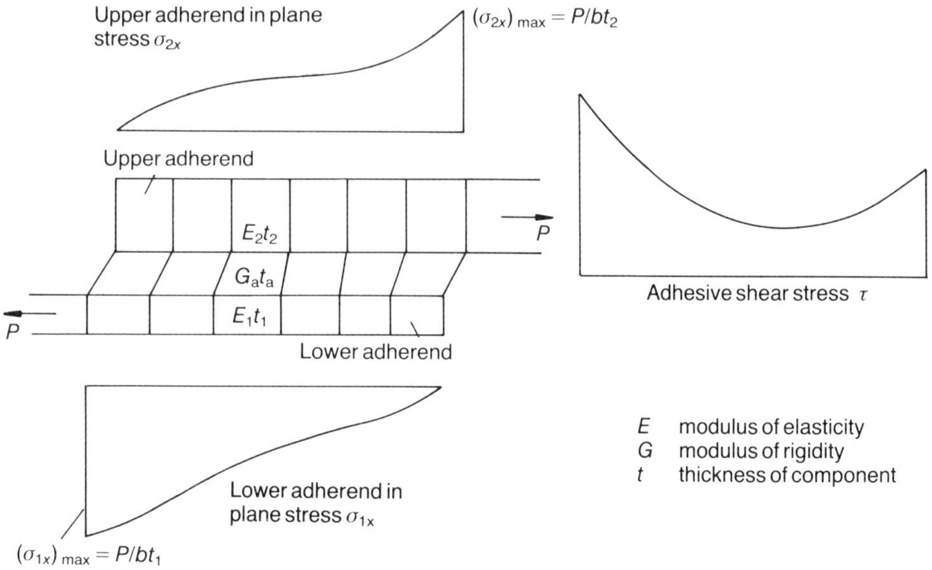

Fig. 6.9. Adherend and adhesive stress distribution for the simple shear lag model

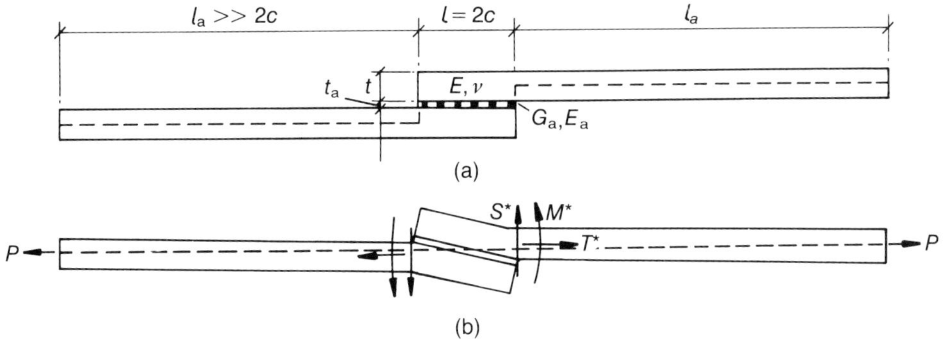

Fig. 6.10. Effect of loading on the single lap joint: (a) unloaded single lap joint showing a variable neutral axis; (b) loaded single lap joint showing an aligned neutral axis and the induced stress resultants to the overlap region

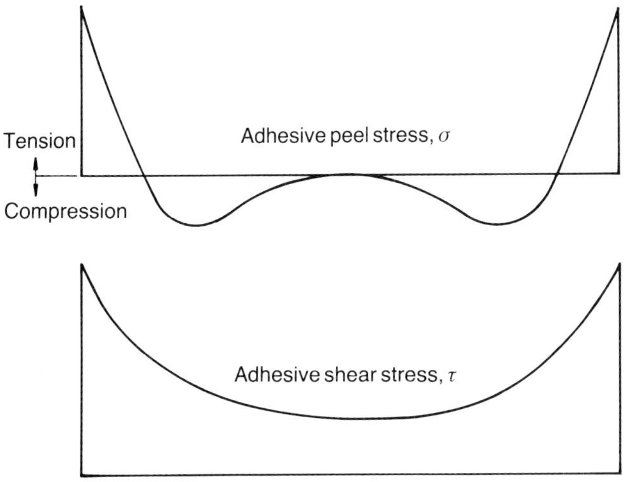

Fig. 6.11. Adhesive stress distributions for the single lap joint

lap joint, but does approximate half of the double lap joint, which, by symmetry of loading, is constrained from rotation.

The mathematical analysis of the single lap joint is dominated by the theory of Goland and Reissner.[6.28] In their work, the adherends were assumed to be identical and isotropic and to obey the classical theory for the infinitesimal bending of a thin elastic beam; accordingly, transverse shear and normal deformations are neglected. In their analysis, the overlap region is isolated and boundary conditions for bending, shear and tension resultants, which are shown as M^*, S^* and T^* respectively in Fig. 6.10, are applied to represent the influence of the adjoining adherends. Expressions for these boundary conditions were obtained from considerations of displacement compatability between the overlap and adherends; because of joint rotation, these were found to be a nonlinear function of the applied load.

Using the standard theory of elasticity, explicit solutions were obtained for the adhesive shear and peel stress distribution in terms of the boundary conditions. The Goland and Reissner analysis was studied for two distinct cases.

(*a*) The adhesive is assumed to be very thin and relatively stiff and is applicable to the bending of wood, or similar low modulus adherends (first criterion).

(*b*) The flexibility of the joint is due mainly to that of the adhesive layer, and is applicable to the bonding of metal and fibre-reinforced adherends (second criterion).

From strain-energy considerations, the following criteria were given for the range of validity of each case

1st criterion $t_a/E_a < t/10E$ or $t_a/G_a < t/10G$

2nd criterion $t/E < t_a/10E$ or $t/G < t_a/10G_a$

The second criterion is appropriate to most bonding problems.

Hart-Smith[6.29] extends the Goland and Reissner work to include the generalization for imbalanced adherends. This work is presented in summary form in references 6.30 and 6.31.

Due to the lack of symmetry, a closed form solution proved impractical and a numerical solution for the governing differential equations and boundary conditions was recommended, although details were not given.

6.13.2. Double lap joint

6.13.2.1. Elastic analysis

The single lap joint is structurally inefficient because of the severe adherend bending strains that are imparted to the glue line by joint rotation. A considerable increase in joint strength is achieved if this undesirable rotation is constrained by adopting the double lap configuration and thus applying load symmetrically. Even so, adherend bending strains are unavoidable due to the eccentric loading path and, as a consequence, the adhesive stress as shown in Fig. 6.12 is developed. It may be seen that the peel stress distribution differs from that previously observed in the single lap joint due to the central adherend being constrained in the transverse direction.

The stress distribution for the double lap joint may also be approximated using the shear lag theory.[6.27,6.32] The adherends may be imbalanced.

Classical analyses for the double lap joint[6.33,6.34,6.35] are intractable to a closed form solution. However, approximate expressions for the peak stress values with a long overlap length are given in references 6.34 and 6.35.

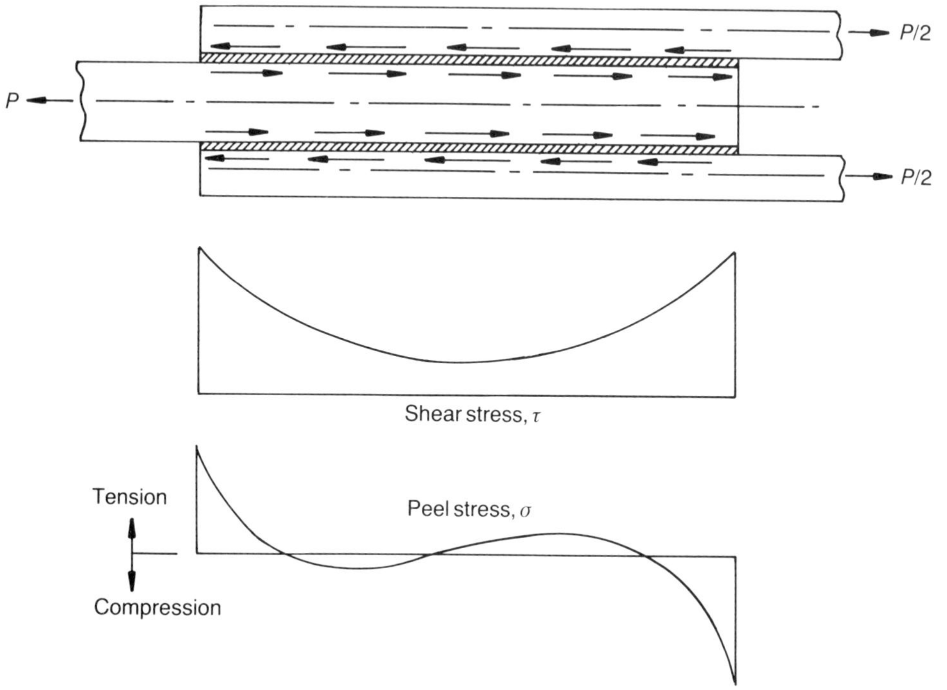

Fig. 6.12. Adhesive stress distributions for the double lap joint

An alternative approach to the analysis of bonded lap joints uses the finite element technique. This method alone is capable of a general analysis, free from assumptions except those implicit in the numerical nature of the analysis.

The linear small-displacement finite element theory has been used[6.36–6.38] to investigate the single lap joint and reference 6.39 has used the same theory to analyse double lap joints. The above mentioned references which investigate the single lap joint are only valid for a low applied load; the large displacement problem of this configuration is analysed in reference 6.40.

The condition of plane strain loading is invariably imposed when analysing the single and double lap joints. This condition reduces the analysis to a two-dimensional problem and either allows the mathematical approach to be tractable or, in the finite element theory, dramatically reduces the computer storage requirements. The tubular lap joint is similarly treated as an axisymmetric problem.

6.13.3. Single and double lap joints

6.13.3.1. Elastic-plastic analysis

For a rigorous treatment of material nonlinearity, the deformation (total strain) theory and flow (incremental) theory are available. While the latter is a general technique, the deformation theory is only applicable to certain structures; it requires that the stress components at all joints in the structure increase proportionally, even during yielding. This would suggest that the deformation theory is only valid for analysing the small displacement problems of the double and tubular lap joints. It would be inaccurate if used to analyse a large displacement problem in a single lap joint.

Grimes *et al.*[6.41] have used the deformation theory of plasticity to analyse the single lap joint as a small displacement problem. The applied load to cause initial adhesive yielding is determined from the Von Mises yielding criterion. Further bending results in plastic straining which is determined by using the deformation theory. Total adhesive strains are given by the summation of

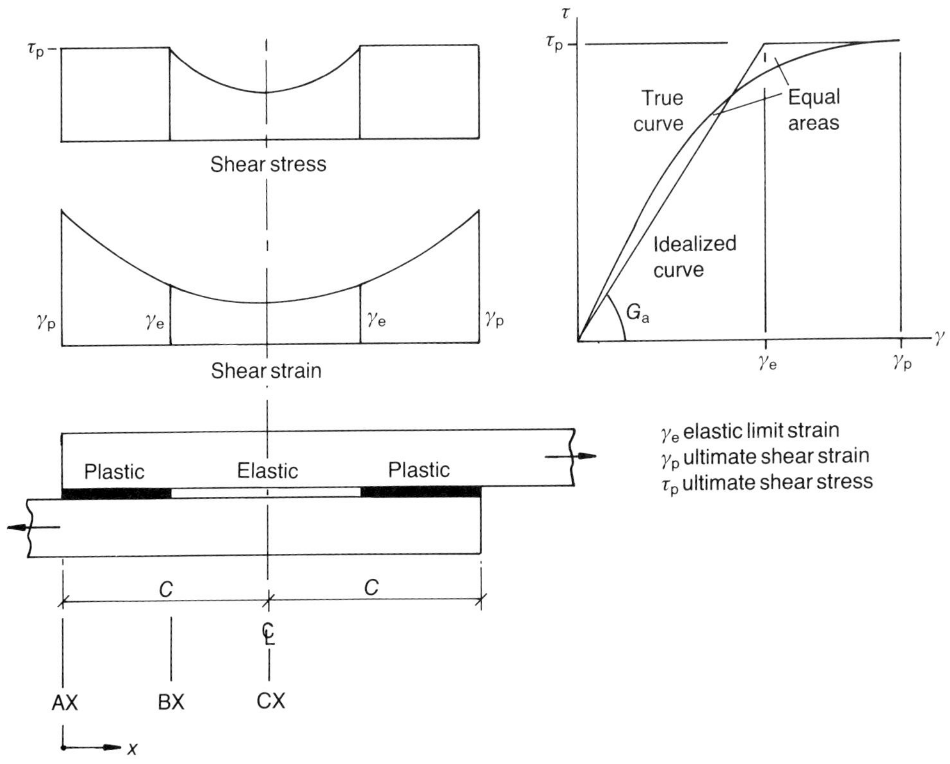

Fig. 6.13. The Hart-Smith elastic-perfectly plastic representation for the single lap joint and the idealized stress–strain curve

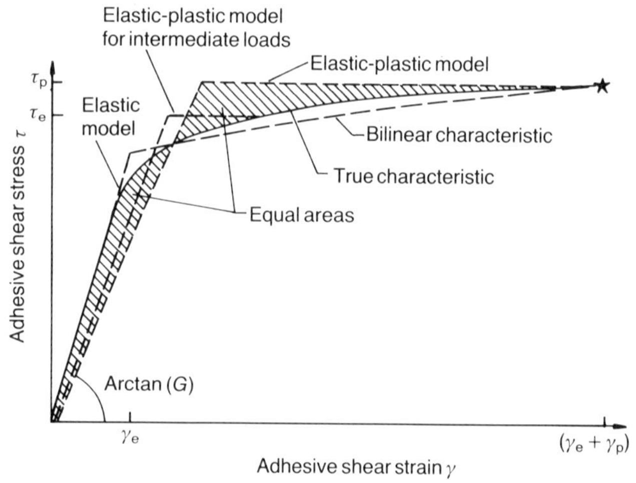

Fig. 6.14. Adhesion shear stress–strain curves and mathematical models (after Hart-Smith,[6.85] by permission of Elsevier Applied Science)

elastic and plastic components, from which two nonlinear governing differential equations are established for adhesive stress distribution. An iterative computer routine is then used to solve these equations.

By comparison Hart Smith[6.42] has chosen a relatively simple approach to adhesive yielding for the single lap joint. This theory does permit large deformations; Fig. 6.13 shows the idealized adhesive shear representation and the adhesive shear stress–strain distribution is discussed in the following section.

6.14. Adhesive characteristics and load transfer

The analyses of the shear strength of adhesively bonded joints are based upon the nonlinear stress–strain curve of the adhesive layer in shear. The reason for this is that modern adhesives, particularly the rubber modified expoxides, have a large plastic strain to failure. The shear characteristics of the adhesives are usually measured by means of the napkin-ring test[6.19] or thick adherend test specimens.[6.14] The adhesive may be characterized by choosing one of the elastic-plastic models suggested by Hart-Smith and shown in Fig. 6.14. The total area under the assumed stress–strain curves for the adhesive is equated to that under their true stress–strain relationships. It will be seen that the majority of the load transfer is within the nonlinear material behaviour and that for the ductile adhesives the contribution of the linearly elastic material behaviour is small. The perfectly elastic model would be used at very low load levels, but for higher loads it would be necessary to apply a large correction factor to the elastic test results. For peak stresses applied between the yield and the failure stress and strain of the adhesive Hart-Smith[6.14] has suggested that model 3 may be used where the maximum intermediate stress–strain value is matched and the elastic portion of the model is selected by equating the elastic and plastic areas on the curve.

Hart-Smith[6.10] assumed that failure of the joint occurs when the adhesive reaches its limiting shear strain. Fig. 6.15(a) shows the adhesive shear stress and shear strains for a double lap joint as the applied load is increased. Fig. 6.15(b) shows the effect of lengthening the overall length of a double lap joint loaded to such a value as to cause plastic deformation in the adhesive regions. A solution is then sought for these equations which satisfies the usual boundary conditions and also ensures continuity of adhesive strains at the elastic-plastic boundary.

The nonlinear theories of Grimes *et al.* and Hart-Smith have been extended to include the double lap joint. The work of Hart-Smith on this subject

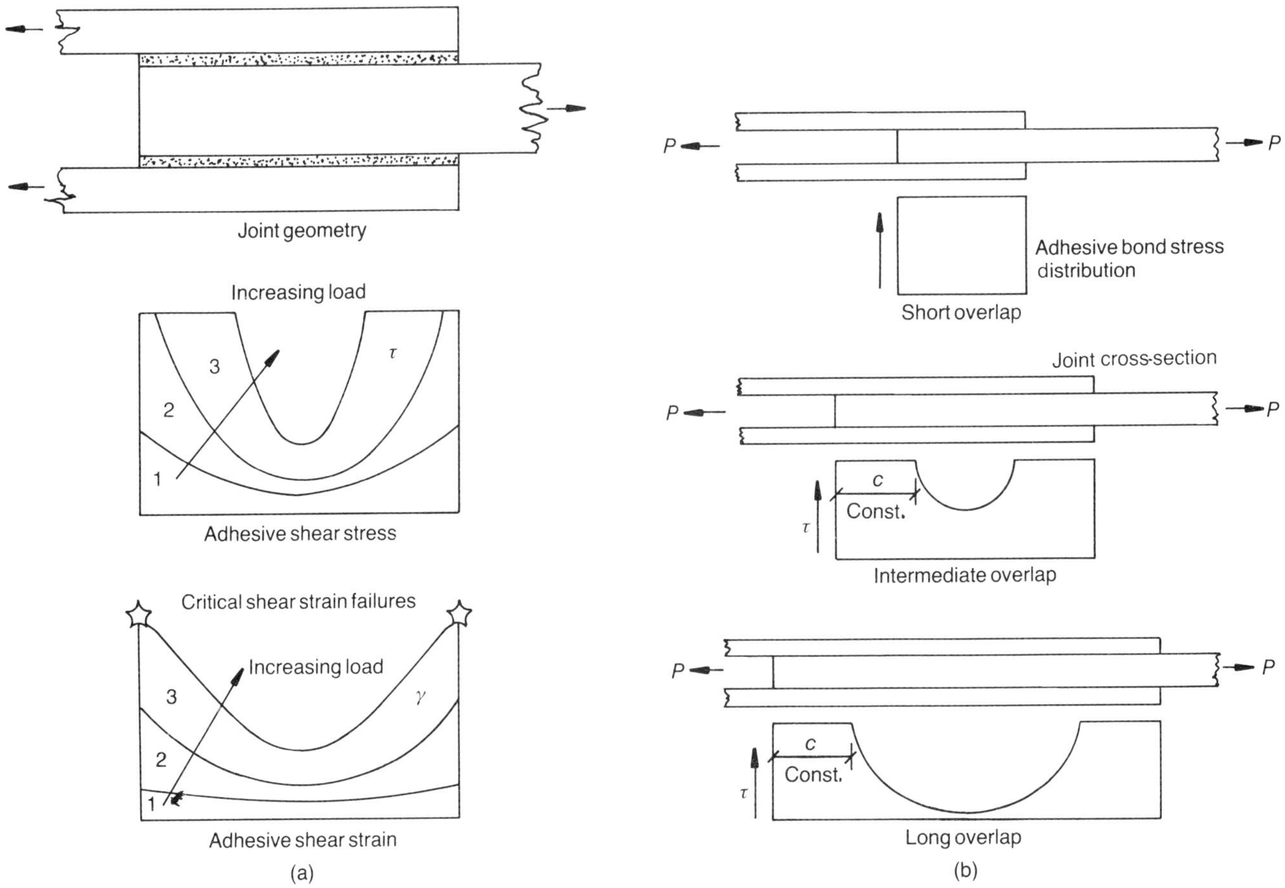

Fig. 6.15. Influence of lap length on adhesive stress distribution: (a) shear stress and shear strain distribution in a double lap joint with increase in load; (b) effect of lengthening overlap length of a double lap joint

is given in reference 6.34, where the shear lag theory first presented by Volkersen[6.27] is adopted to derive the elastic and plastic governing differential equations for adhesive shear stress.

Grant[6.43,6.44] has used an approach similar to Hart-Smith to analyse the double lap joint. The adhesive shear stress–strain curve is expressed as a continuous mathematical function and the Volkersen equations are written in a finite difference form. A numerical solution is then obtained using an iterative computer routine.

6.15. Joint analyses utilizing the finite element analysis

Adams and Peppiatt[6.39] have used the initial stress method advocated by Zienkiewicz[6.45] in an interactive elasto-plastic finite element program to study adhesive yielding. They have recognized that the geometric linear (small displacement) theory is not valid for the single lap joint and therefore confine their attention to the double, bevel and scarf joints. A detailed description of the elasto-plastic theory applied to bonded joints is given by Adams.[6.46] Wright[6.47] has analysed to failure the imbalanced double strap joint using highly orthotropic CFRP adherends. This joint is similar to the conventional double lap joint. Both investigations use the incremental theory and assume the adhesive to be elastic perfectly plastic (see section 6.14).

The study of adhesive yielding in a tubular lap joint has been undertaken by Nagaraja and Alwar.[6.48] The deformation theory has been used with the nonlinear adhesive stress–strain relationship employing a continuous mathematical function.

6.16. Tubular joints

Various end grip configurations for the axial loading of composite tubes have been proposed;[6.49–6.51] in all this work, experimental results were supported by an analytical finite element study.

In 1956 Lubkin and Reissner[6.52] presented the first mathematical analysis to investigate the adhesive stress distribution for a lap joint between two thin-walled cylindrical tubes loaded in axial tension. Their analysis considered only two components of stress: the shear stress to resist axial stretching and the normal stress to resist relative radial displacement of the two tubes. It was assumed that the adherend tubes obeyed the ordinary linear theory of bending and stretching of thin isotropic shells; the adhesive was assumed linear elastic and much more flexible than the adherends. The adhesive shear and normal stress distribution were normalized with respect to the mean adhesive shear stress ($\tau_m = F/2\pi al$) and results were presented for 48 joints having identical adherends. Lubkin and Reissner (Fig. 6.16) showed that a tubular lap joint

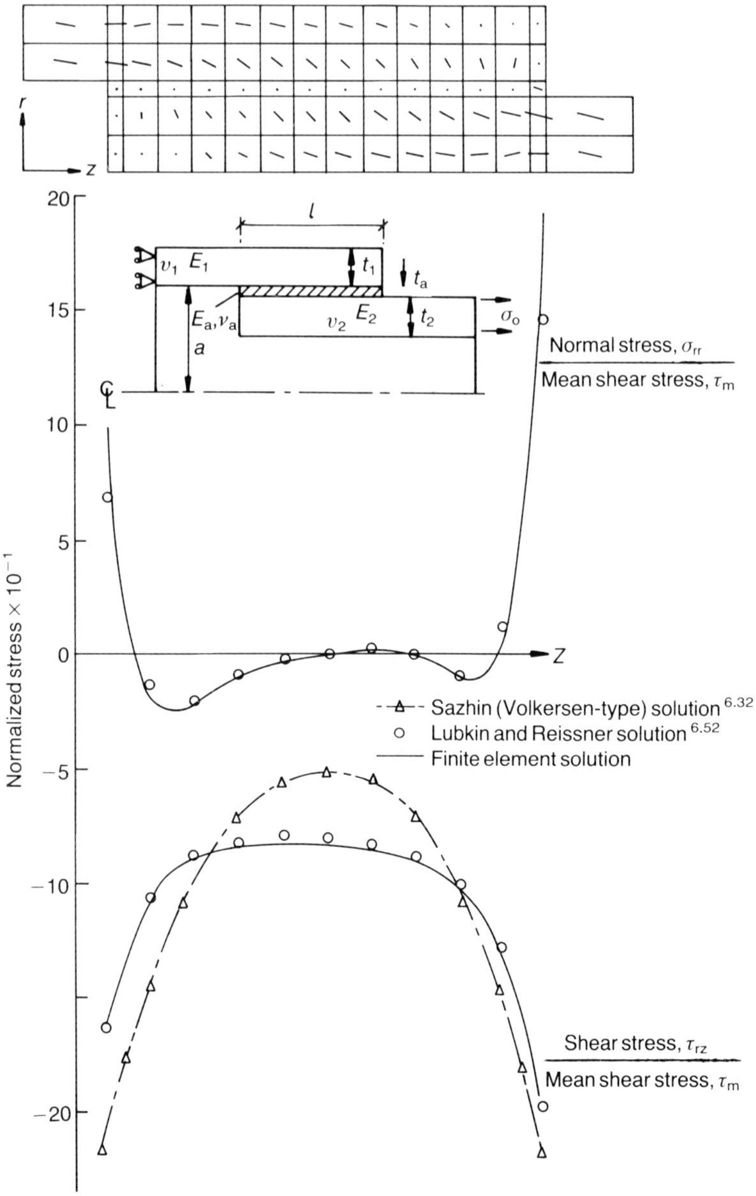

Fig. 6.16. Comparison of classical and finite element solutions

with identical adherends can be uniquely defined by the five independent material and geometric dimensionless parameters

$$\mu,\ E_a/G_a,\ 1/r,\ \beta(=t_a E/E_a t),\ R\ (=t/2a)$$

where for identical adherends

$$\mu_1 = \mu_2 = \mu$$
$$E_1 = E_2 = E$$
$$t_1 = t_2 = t$$

Their analysis requires the solution of three simultaneous differential equations to obtain the normal and shear stress distribution, but a closed-form solution is difficult. Lubkin and Reissner do not give an explicit solution for the stress distribution but instead those for 48 typical joints are cited. However, the dimensionless parameters used to define these joints are not representative for single lap tubular joints with low modulus adherends (viz. aluminium and GRP or GRP and GRP) which require a long overlap length to obtain a high joint efficiency. As no alternative mathematical solution is available to provide the stress distribution for a single lap tubular joint loaded in axial tension, it is necessary to resort to an analytical finite element solution or to apply the analogous single lap joint.

An adhesive bonded tubular lap joint has been analysed more recently using the finite element approach. Nagaraja and Alwar[6.48,6.53] considered visco-elastic adhesive behaviour and nonlinear stress–strain behaviour. Adams and Peppiatt[6.54] used the finite element method to study adhesive tubular lap joints subject to axial and torsional loading; they assumed linear elastic adhesive behaviour. The influence of an adhesive spew fillet was recognized and Adams and Peppiatt investigated the adhesive stress distributions with the classical Lubkin and Reissner solutions.

Figure 6.16 shows the comparison between

(*a*) the stress-analysis solution undertaken by the finite element technique for aluminium adherends
(*b*) the Lubkin and Reissner solutions for a tubular joint defined by the five dimensionless parameters

$$1/r = 10,\ E_a/G_a = 8/3,\ \mu = 0{\cdot}3,\ \beta = 4,\ r = 0{\cdot}05$$

where, for the adherends

$$E_1 = E_2 = E = 69\,\text{kN/mm}^2$$
$$\mu_1 = \mu_2 = \mu = 0{\cdot}3$$
$$t_1 = t_2 = t = 1{\cdot}0\,\text{mm}$$

and for the adhesive

$$E_a = 3\,\text{kN/mm}^2,\ \mu_a = 0{\cdot}33,\ t_a = 0{\cdot}174\,\text{mm}$$
$$l = 10{\cdot}0\,\text{mm}$$
$$a = 10{\cdot}0\,\text{mm}$$

In addition, Fig. 6.16 shows the finite element mesh division with the relative principal stress values and the principal stress directions at the centre of each element.

Axial load is transferred from one adherend to the other by adhesive shear stress. Bending actions are generated within the system because the eccentric loading of the two tubes causes a normal (radial) stress to develop along the interface of the adherends. In addition, radial stresses are developed due to the Poisson's ratio effect from the axial load. From Fig. 6.16 it is clear that the stress distribution is not symmetric; although the adherends are manufactured from the same material, the area stiffnesses are different. Associated with this, the axial stress has a higher value in the inner tube compared with the outer one as the latter has the greater cross-sectional area. Therefore, higher shear strain and shear stress values are developed in the adhesive at the end of the outer tube.

An approximate shear–stress distribution for a bonded tubular joint with dissimilar adherends can be obtained using the elementary analysis of Volkersen,[6.27] which applies to a constrained single lap or a double lap joint. The adhesive shear stress is a result of differential straining of the adherends. A similar analysis by Sazhin[6.32] gives the adhesive shear–stress distribution as

$$\tau_x = (F/b)\,\alpha\left[\left\{\frac{(1-\psi(1-\cosh(\alpha l))\cosh(\alpha x))}{\sinh(\alpha l)}\right\} - \psi\sinh(\alpha x)\right]$$

where the term $\alpha = \sqrt{\beta/\psi}$, the term $\beta = (G_a/t_a)(1/E_1 t_1)$, and the term $\psi = E_2 t_2/(E_1 t_1 + E_2 t_2)$. For the single lap joint, b equals width of joint or for the single lap tubular joint b equals $2\pi a$ approximately, where a is the radius to the centre of the adhesive.

The analysis of Sazhin is superimposed on Fig. 6.16. This analysis fails to predict the true adhesive shear–stress distribution because it can only approximate the tubular joint geometry. However, the analysis may be useful to predict the peak adhesive shear stress in tubular joints with dissimilar adherends.

The classical analysis for the tubular lap joint[6.44] treats the adherends as the isotropic shells. A complex system of governing differential equations and boundary conditions was established but a closed form solution to this has yet to be found.

It has been shown[6.55] that the shear lag theory may approximate the adhesive stress for the tubular lap joint. Reference 6.56 has modified the theory for the double lap joint such that it may approximate the tubular configuration. Reference 6.56 has shown that, with suitable modification, the shear lag assumption reasonably approximates to the tubular configuration. These modifications are as follows.

(*a*) The joint width b now represents the circumference of the tubular lap joint about the midpoint of the glue line.

(*b*) The condition of plane strain loading is ignored.

(*c*) The adherend stiffnesses are modified to account for the curved geometry of the tubes; it may readily be shown that the modified values are given by

$$E_1^* = \frac{a_1 E_1}{a}$$

$$E_2^* = \frac{a_2 E_2}{a}$$

where a_1, a_2 and a are the inner adherend, outer adherend and adhesive centre line radii respectively and E_1 and E_2 are the true adherend stiffnesses.

Furthermore, it should be noted that from a geometric consideration of the tubular lap joint, the inner tube having the lower cross-sectional area will, in all probability, have the least area stiffness. Consequently, yielding will be most evident at the free face from which the inner tube extends.

6.17. Introduction to bolted joints

Initially there was considerable reluctance amongst designers to accept the use of bolted connections for jointing composite materials. At first glance it appeared to be a logical extension to the bolting and riveting of isotropic plates; the latter had been well tested, but designers were reluctant to bolt together anisotropic materials about which they had little knowledge. However, currently there is a wealth of information available on the bolting and riveting of composites and designers should now have much greater confidence in using this method for jointing composite adherends.

This part of the chapter on mechanical joints will discuss mainly double lap configurations, although single lap joints will also be mentioned.

There are advantages and disadvantages in using mechanical fastened joints. The advantages are as follows.

(*a*) There is no specific surface preparation of the components to be joined.
(*b*) The inspection of the joint is relatively easy.

The disadvantages are as follows.

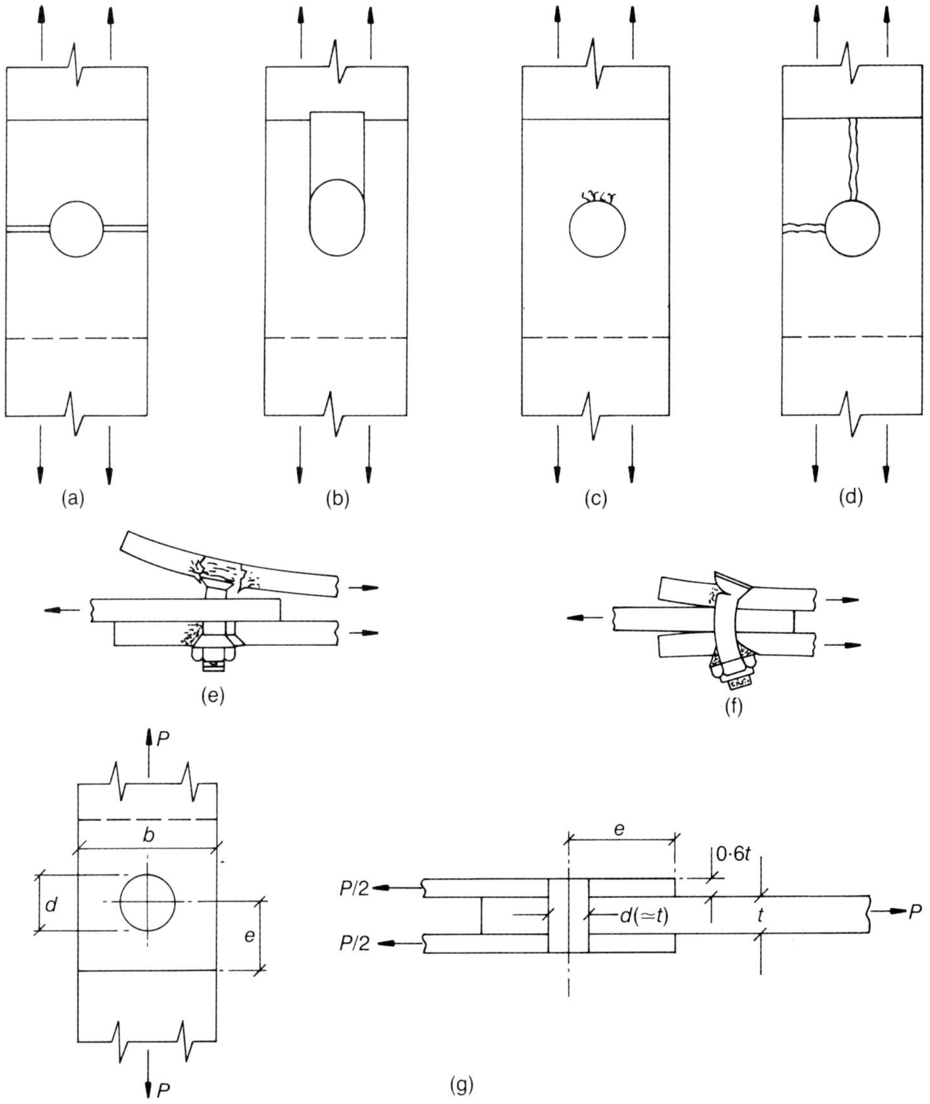

Fig. 6.17. Failure modes and joint notation: (a) tension failure; (b) shear failure; (c) bearing failure; (d) cleavage-tension failure; (e) bolt pulling through laminate; (f) bolt failure; (g) notation used in design of bolts

(*a*) In forming the joint, some of the material of the component parts has to be removed, thus causing higher stress levels and possible stress concentrations in the remaining parts of the material. Consequently composite components can be weakened.

(*b*) Joints tend to be bulky and may incur a weight penalty; however, this is unlikely to cause any problems in the construction industry.

Disadvantage (*a*) could be serious, particularly with anisotropic materials. It will be shown later that the efficiency of mechanical joints is completely dependent upon the anisotropy of the material. It should be noted that whether the jointing system is a bonded or a bolted one stress concentrations will occur.

The choice of fasteners available to the designer is wide. However, because of the unique properties of composite materials, particularly their weakness under certain loading conditions, and the requirement for an efficient joint in terms of strength-to-weight ratio, this choice is necessarily reduced.

The three main systems used by the engineer are: screws; rivets; bolts. The various advantages and disadvantages of these fasteners and their loading characteristics will be discussed later in this chapter. Fig. 6.17 defines the joint notation.

6.18. Failure modes

There are six possible failure modes for composite mechanical joints. These are: tension; bearing; shear; cleavage-tension; pull-out; bolt failure.

Failures within composites are similar to those observed in metals despite the fact that most metals exhibit considerable yielding prior to failure. As yielding in composites does not take place, the mechanism of failure is completely different. However, some form of pseudo-yielding can take place in certain lay-ups in which delamination and partial fibre breakage occur before final failure.

The failure mechanisms are dependent on

- the geometric failure
- the fabrication of the composite and the fibre orientation
- the matrix material.

The geometric factors usually render one of the failure modes predominant. These factors are: the width of the specimen, the distance of the bolt from the free edge running normal to the loading axis and the lay-up patterns of the fibres.

As combined modes can occur during the failure of a joint, very often no single failure mode is clearly defined. In addition bearing-mode damage can generally be identified because the pin damages the composite area adjacent to the loaded half of the hole.

The various failure modes for single fastener joints are shown in Fig. 6.17 and these modes are discussed below.

6.18.1. Tension failure

A tensile failure stress is given by

$$\sigma_G^* = \frac{P_{ult}}{bt} \text{ on the gross area}$$

$$\sigma_N^* = \frac{P_{ult}}{(b - nd)t} \text{ on the net area}$$

As fibre/matrix composites are essentially elastic to failure, stress concentrations will develop and these will tend to give rise to low net failure stresses.

The greater the degree of anisotrophy in the region of the hole, the lower will be this stress value.

6.18.2. Bearing failure

The bearing failure stress is given by

$$\sigma_b^* = \frac{p_{ult}}{ndt}$$

Compressive stresses are developed around the loaded half of the circumference of the bolt or rivet hole. The distinction between bearing failures and tension failures is largely determined by the joint geometry. The most important parameters are the width-to-diameter ratio and the fibre array. The clamping effect from the lateral constraint provided by a tightened bolt is another parameter which will enhance the strength of the joint. This is discussed in greater detail in section 6.19.2.3.

6.18.3. Shear failure

The shear failure stress is given by

$$\sigma_s^* = \frac{p_{ult}}{2et}$$

When the line of action of the load is in the direction of the fibres, the joint shear strength of unidirectional fibre/matrix composites is low compared with the in-plane shear strength.

The difference between the two strengths will give rise to stress concentrations. If fibres are placed at $\pm 45°$ to the line of action, the shear strengths are high and the stress concentrations will therefore be low. Consequently, fibre orientation has considerable influence over the shear failure strengths of joints.

There appears to be little information on the case histories for GRP but Collings[6.57] has shown that for a CFRP composite the shear strength of a $\pm 45°$ fibre orientation is about 100 MN/m^2 while that of a 0° CFRP composite is only 23 MN/m^2. Combining 0° plies with $\pm 45°$ plies increases the shear performance to give strengths greater than either the 0° or the $\pm 45°$ composites alone. Collings has also shown that a reduction in the degree of anisotrophy of a composite, together with a strengthening of potentially weak planes of failure, lead to a decrease in the magnitude of the stress concentration and an increase in shear strength.

6.18.4. Cleavage-tension failure

Cleavage-tension failures are the result of the material being orthotropic and the line of action of the load being parallel to one of the orthogonal axes. To correct this potential failure mode, fibres would be required to be placed in the nonaxial direction.

6.18.5. Pull-out failure

Pull-out failure is usually associated with single shear joints in which rivets and countersunk rivets are employed, although double shear joints are also exposed to this load. As in single and double lap bonded joints, peel forces are set up as the composite components tend to straighten under load, thus imposing an out-of-plane bending situation in the joint. This situation introduces out-of-plane transverse loads and imposes tensile forces in the rivet and through-the-thickness shearing of the rivet head through the laminate thickness.

6.19. Joint strengths of composite materials

The characterization of the strength and stiffness of composite materials is complicated by the large number of variables involved. When considering the strength of load-carrying joints in composite materials, the complexity of the problem is considerably increased because, in addition to the material variables, parameters relating to the type of joint and its geometry must also be considered.

The parameters affecting joint strength can conveniently be divided into three groups. These are: material parameters; fastener parameters; design parameters.

6.19.1. Material parameters

There are several different possibilities for components in fibre/matrix composites. The fibres which could be used in the construction industry cover a range of materials, such as glass, carbon, Kevlar or hybrids of these. The resins would generally by polyester epoxy or phenolic. The material parameters affecting the composite strength will be fibre type, orientation, fibre volume fraction and fibre-surface treatment.

Glass fibre is the cheapest of the fibres mentioned above and in conjunction with polyester forms a high specific strength composite material (GRP). If the fibres are randomly orientated in the composite, the in-plane properties can be considered isotropic, and therefore the composite is not direction sensitive.

However, the fibre volume fraction will need to be considered in strength calculations. If woven fabric is used, the warp direction would normally be parallel to the line of action of the load.

Composites manufactured from an epoxy resin matrix are generally made from unidirectional prepreg, with either carbon or glass fibre, to produce laminates of carbon-fibre reinforced epoxy (CFRP) or glass-fibre reinforced epoxy (GFRP). Generally GRP is 4 mm or more in thickness, whereas CFRP and GFRP are less.

Collings[6.58–6.60] has commented on the high stress concentration factors (SCF) which can arise around holes in composites and suggest that incorporating $\pm 45°$ plies may reduce the SCF. Clayton and Jones[6.61] and Berg[6.62] have noted that surface treatment may influence the SCF. Quinn and Matthews[6.63] have shown that, for pure bearing, placing 90° plies perpendicular to the line of action of the load at or near the surface improves the bearing strength of GFRP and it seems likely that this procedure would also improve the bearing strength of GRP.

Figure 6.18 shows the stacking sequence on the failure load of GFRP and has been constructed from reference 6.63.

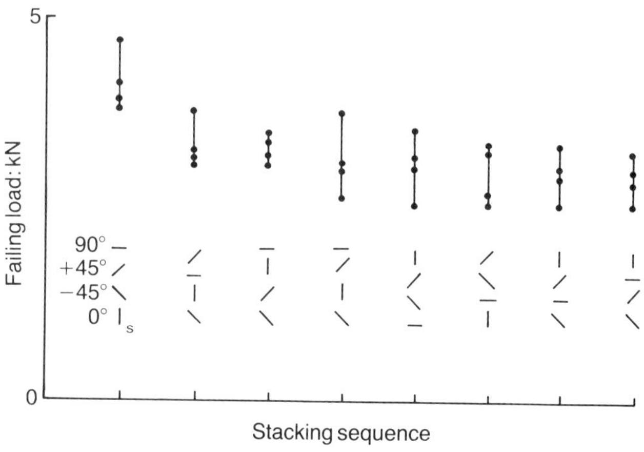

Fig. 6.18. Effect of stacking sequence on failure load for GFRP[6.57] (courtesy of Elsevier Applied Science)

6.19.2. Fastener parameters

The strength of joints is clearly dependent upon the fastener (i.e. whether the joint is held together by screw, bolt or rivet) and upon its parameters. These include: the fastener size; the hole size and tolerance; clamping force; washer size.

Further design parameters include: the joint type (i.e. whether the joint is single lap, single cover, butt etc.); the composite thickness; the geometry of the fastener lay-out; the load direction and rate of loading; the loading type: static or dynamic; the failure criteria.

A complete characterization of the joint behaviour is impossible to undertake because of the very large number of variables. It is generally sufficient to determine as thoroughly as possible the behaviour of a few basic joints in a limited number of materials and systems and to deduce the more important parameters from which the behaviour of other joints and materials can be predicted.

When working with composites it must always be borne in mind that any quantitative information depends upon the material systems being used and this may need to be modified as newer systems are developed.

6.19.2.1. Screws

Screws are the simplest form of fastener and provide a low cost joint where access to the reverse side of the composite is not possible. However, screws are not recommended if the primary role of the joint is structural because both self-tapping screws and tapped-hole screws cause extensive damage to the laminate.

If screw joints are to be used for structural applications, it is possible to cast metal blocks into the composite at the time of manufacture or to mount metal blocks on the composite and then to drill and tap into these in the conventional way.

6.19.2.2. Rivets

Adequate strength may be derived from riveted joints when used with CFRP and GFRP composites for thicknesses up to 3 mm. There are two types of rivet, both of which have varying head size: solid rivet; blind rivet.

The process of forming the rivets involves applying to the composite a lateral pressure which cannot be readily controlled. Consequently, there may be cases in which the riveting operation causes damage to the laminate. The worst damage is caused by blind rivets. In addition the process of forming the rivets will cause varying degrees of lateral clamping directly related to the joint strength; the latter is also increased by the use of washers. Matthews *et al.*[6.64] have suggested that the process of riveting can cause damage to the laminate around the hole, thus influencing the joint strength. The clamping force exerted by the rivet also affects the strength of the joint. If countersunk rivets are to be used to ensure a smooth external surface, they must not be allowed to cut completely through the laminate; this will generally imply a restriction on the minimum thickness of the composite; this minimum thickness will be a function of the load to be transmitted through the joint.

6.19.2.3. Bolts

Bolts offer the greatest strength for mechanically fastened joints. By increasing the bolt-tightening torque, and hence clamping force, the bearing strength is increased. Providing excessive over-tightening of the joint does not take place, no damage will be done to the composite during assembly, but if too high a clamping force is applied, damage may be done by the washers digging into the components being joined. The relationship between the

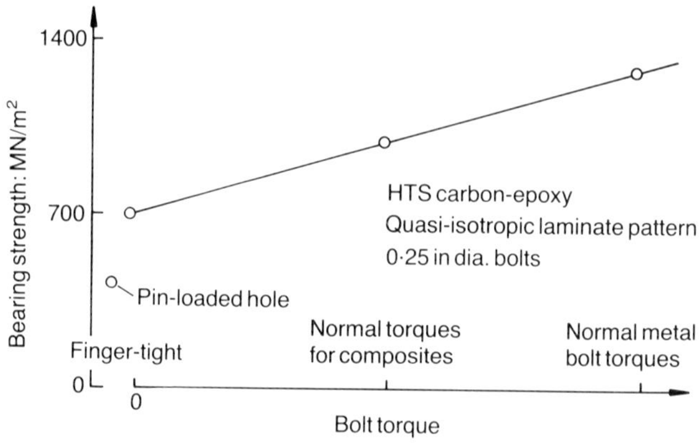

Fig. 6.19. Relationship between bearing strength and bolt torque[6.71] *(courtesy of Elsevier Applied Science)*

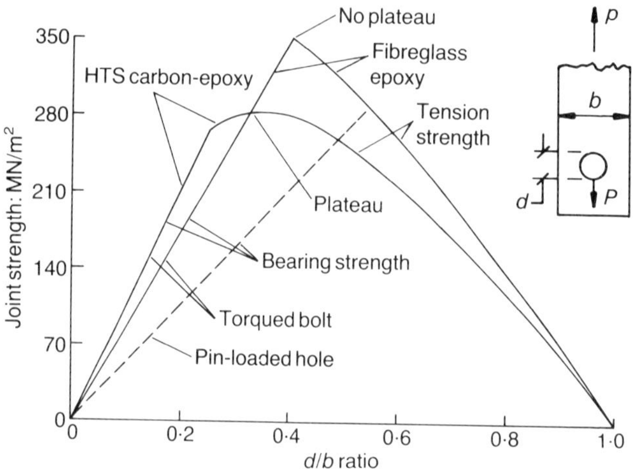

Fig. 6.20. Relation between bearing and tension failures for carbon/epoxy and fibre glass/epoxy laminates as functions of joint geometry[6.81] *(courtesy Elsevier Applied Science)*

bearing strength and the bolt torque has been given by Crews[6.65] and is shown in Fig. 6.19. Godwin *et al.*[6.66] have observed a similar result with glass fibre reinforced polymer laminates. It can be seen that initially almost double the bearing strength can be achieved by torquing bolts to a value which would be used for normal metal bolts. However, the benefits of torquing beyond finger tightness have been questioned because some of the additional tightening will be relieved by creep over the life of the structure. Investigations into the rate of loss of the clamping have shown that the rate of decay is fairly slow;[6.67] it is difficult to control the closing force because of the nature of the riveting procedure. Godwin and Matthews[6.68] have noted that the degree of clamping pressure is determined by the washer size and they have shown that there is an increase in joint strength with carbon fibre reinforced epoxy matrix when washers are used. Kingston-Lee and Rogers[6.69] and Stockdale and Matthews[6.70] have reported that the weaker size can influence the strength of the joint but the hole in the washer should be a close fit to the bolt.

Hart-Smith[6.71] has characterized the strength of single-hole bolted joints as a function of the joint geometry and has produced a relationship between the ratio of the joint strength to the basic composite strength (joint efficiency) and the ratio of bolt diameter to strip width (d/b); this relationship is shown in Fig. 6.20. The values given in Fig. 6.20 are for 6·35 mm diameter bolts in double shear in a quasi-isotropic composite manufactured from high tensile strength carbon fibres and epoxy matrix, but the general form would be relevant for almost any fibrous composite. When the strip width is large, bearing failures are the criterion for failure but, as the bolt becomes large compared with the width of the strip, there is a change in failure mode to that of tension "through the hole" section. Fig. 6.20 shows that the peak strength of the joint occurs either at the tension-failure mode or at an abrupt transition between that and the bearing-failure mode. Hart-Smith has shown that for both cases the failure has a definite value for the ratio of d/b and this is nearer to 1/3 than to the value of 1/4 which is usually associated with bolted joints. He has also stated that bearing failures are more desirable than tensile ones because they are less catastrophic. However, if a bearing failure does occur, it will be associated with a reduced joint strength compared with that for a tensile failure.

In addition Hart-Smith has shown that glass-fibre epoxy composites have a higher ratio of tensile-to-bearing strength than do the carbon epoxy lamintes. Fig. 6.20 shows an abrupt change between the two modes where the maximum strength value lies between ratio *d*/*b* values of 1/3 and 1/4.

6.19.3. Design parameters

The particular function of the joint will dictate whether it is to be manufactured as a single or double lap system. Gibbs and Cox[6.72] give details of examples covering many applications using glass-reinforced plastics. If experimental data are to be used in determining design parameters for joints, care must be taken to ensure that the test specimen configurations are relevant to the particular design being undertaken. Rufolo[6.73] and Kingston-Lee and Rogers[6.74] have shown that, if all design aspects are not taken into account in the analysis, the actual joint could be underdesigned.

A particular difficulty in obtaining design data for manually fabricated composites (and indeed mechanically fabricated composites also) arises from the variation in thickness of the composite. To obtain comparable strength data from such composites, Collings[6.58–6.60] has normalized the results with respect to a standard thickness composite having a fibre volume fraction of 60%. This procedure, however, is not appropriate to low fibre volume fractions.

A number of authors have investigated the effect of end distance (defined in Fig. 6.8(a)). They have all suggested that there is a required minimum ratio value of *e*/*d* to enable the full bearing strength of the joint to be developed. Collings[6.60] has shown that for CFRP bolted joints ± 45° lay-up would require an *e*/*d* ratio of greater than 5, whereas the equivalent value for a quasi-isotropic lay-up would be 3. Rufolo[6.73] has quoted a value of 3 for work undertaken on GRP.

Hodgkinson *et al.*[6.75] have investigated the strength of bolted joints in Kevlar-reinforced plastics and have come to the same general conclusions as other investigators who used GFRP and KFRP. For example they have concluded that the bolt-tightness parameter is important, that the quasi-isotropic composite lay-ups have higher strengths than the orthotropic arrangements and that for most types of joints there appears to be an advantage in designing the composite to include more 0/90° than ±45° lay-ups. In addition, they have shown that double lap joints give higher bearing-failure stresses than single lap joints. Adequate edge distances must be provided and multi-bolt design must ensure adequate distances between bolts so that pitch/diameter ratio (*p*/*d*) exceeds 4 in the case of lines of bolts and lies between 5 and 6 for rows. Bearing stresses for both are similar for single bolt and for multi-bolt arrangements.[6.75]

Values of the ratios *b*/*d* have also been investigated by many research workers. For full bearing strengths to be developed, Collings[6.60] has suggested minimum ratios of 8 and 5 for CFRP composites in which the lay-ups are ±45° and pseudo-isotropic respectively. Matthews *et al.*[6.64] have suggested a ratio of 4 for 0/±45° lay-up in CFRP.

Investigations into the effect of joint strength variation for both layout patterns have shown that it is not possible to use the results of tests on single hole joints to predict the strength of multi-hole arrangements. In addition Pyner and Matthews[6.76] have shown that it is essential to locate accurately the relative positions of the holes in the composite materials to be joined so that high stresses which would eventually lead to premature failure of the composite during assembly can be avoided. Conversely Collings[6.58] using CFRP has shown that the interaction between bolts is insignificant for practical pitches and row spacings. It might therefore be concluded from the work of the latter author that single hole data can be utilized for multi-hole joints provided the group has a tandem configuration with no more than two rows. It has been suggested by Strauss[6.77] that a tandem row is preferable to a parallel row but Rufolo[6.73] has recommended the use of staggered pitch with tandem rows.

With respect to the ratio *d*/*t*, Saddler and Barnard[6.78] have demonstrated that, for pin-loaded holes, in which the data has been derived from various sources and from various composite materials with different lay-ups, the relationship between stress and the ratio of *d*/*t* lies on one curve.

The bearing strength of isotropic composite materials will vary with the direction of the load but Hirst and Matthews[6.79] have shown that CFRP composites are the most sensitive and GRP composites the least sensitive to load direction.

Consideration had been given to placing local reinforcement around the hole area of the composite so that an increase in the bearing strength and stiffness can be effected. Various authors have concluded that for CFRP the extra plies are too expensive, involve a weight penalty and therefore are impracticable. It has been suggested that the area of the composite around the hole should be "softened" to enable the stress concentration factor to be reduced. Methods by which this can be achieved have been introduced by Collings[6.59] and by Clayton and Jones;[6.61] the former has recommended 45° plies in all CFRP lay-ups and the latter have suggested the use of GFRP to soften the CFRP composite.

It can be clearly recognized that the values of *b*/*d*, *e*/*d*, washer pressure and bearing-failure stress are all interdependent. It is difficult, therefore, to give precise values of any of the above parameters with respect to another without knowing the actual lay-up design of the composite, including its stacking sequence and ply directions.

6.20. Jointing and fixing

In the previous sections, methods of analyses and design for the transfer of load from one composite member to another have been presented. Generally, when GRP is used in the form of load-bearing and infill panels the main function of the joint between two composite members is to prevent moisture penetrating to the interior of the units forming the structure; the transfer of load is only of secondary importance. Leggatt[6.80] has identified three basic types of joint

(*a*) the sealed joint
(*b*) the drained joint
(*c*) the overlap joint.

The sealed joint is the simplest one and consists essentially of a sealing material which is compressed between the two adjacent flanges of the GRP units being joined. The compressive force is applied by means of bolts and the sealing material could be either a compressed foam or various forms of sealants and gaskets. A polysulphide sealing compound is an example of an adhesive material often used (section 11.6). The disadvantage of this jointing system is that the compressive force (i.e. the bolts) can only be applied at discrete points and therefore the full compressive value of the force will not be completely effective throughout the length of the joint.

The drained joint is one which will, if penetrated by water, be drained by gravity to the exterior of the building.

The overlap joint is of similar form to that used with the more traditional building materials in which a simple end lap is employed, possibly with convolutions.

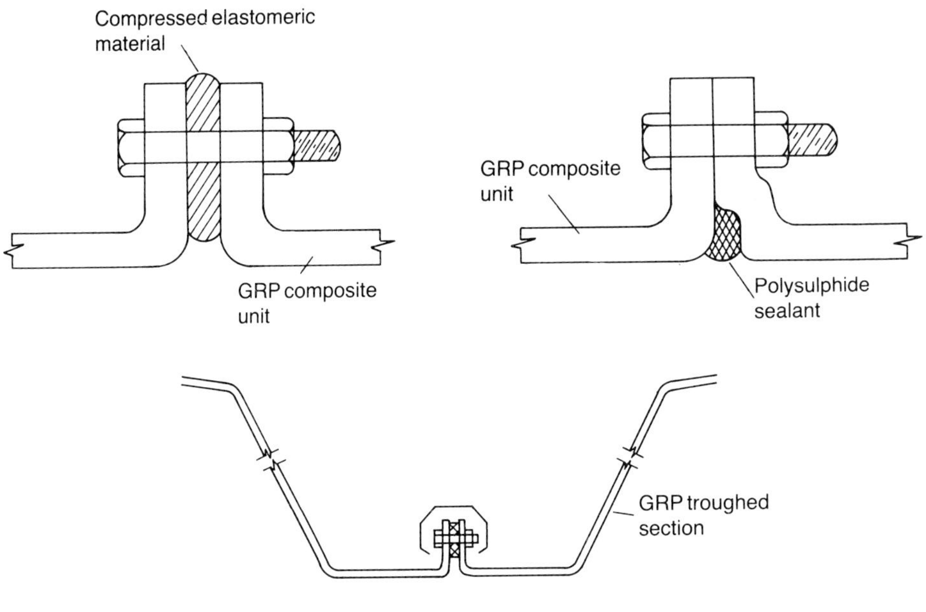

Fig. 6.21. Typical sealed joints

Of the above three joints, the one that is normally used is the sealed joint. Sealants are generally supplied in a two-part pack. When mixed the resulting material cures rapidly to form a flexible but tenacious joint between two surfaces. It should be possible to construct an ideal joint, but this implies that the surfaces to be joined are completely clean and free of moisture. The first condition can be obtained, but it is difficult to achieve the second one. The jointing of structural units is undertaken on site and, however well the surfaces are dried immediately before jointing, it is difficult to achieve a completely moisture-free surface. Consequently, there may be jointing problems and dampness may appear within the structure. Fig. 6.21 shows typical sealed jointing systems for GRP composite units.

6.20.1. Specific bonding applications for composite tube sections

Pultrusion products (the technique of pultrusion is described in section 2.8.2.3) have potential in sectional structural components which could be in the form of plane and space grid systems. In the latter systems, the members are mainly under axial forces, although they could also be subjected to bending moments. The bonded end cap for pultruded tubes, which has been

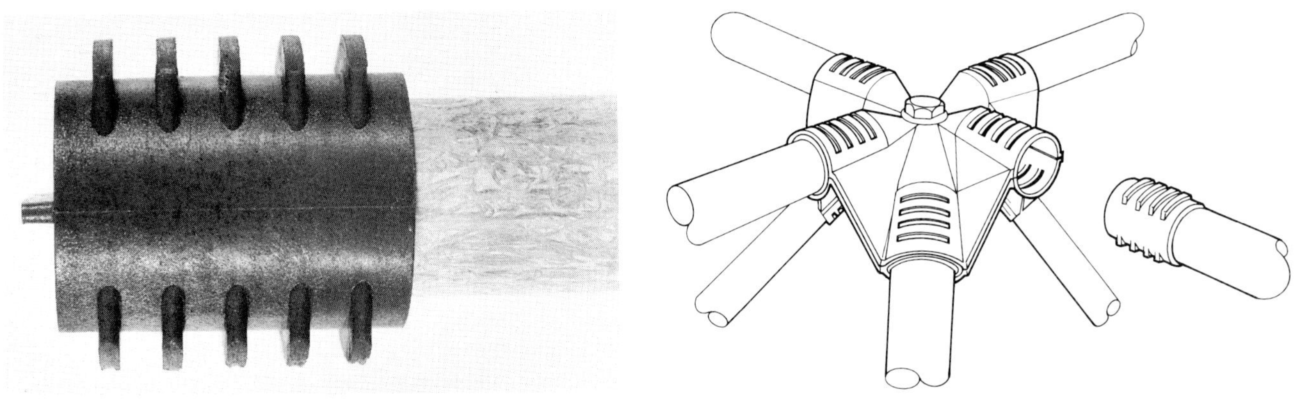

(a) (b)

Fig. 6.22. Bonded end cap for pultruded tube: (a) glass-filled nylon end cap bonded to the pultruded tube; (b) specially formed composite plates forming nodal joints

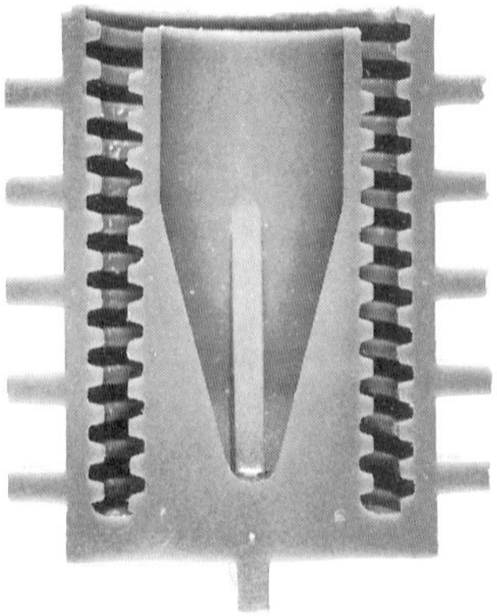

Fig. 6.23. Section through end cap

mentioned in section 6.6, was developed specifically for forming nodal joints in space structures in tubular GRP composites. The system involves bonding onto the pultruded tube a glass-filled nylon end cap, made by the resin-injection process (section 2.7.1), and sandwiching the caps between two specially formed composite plates which form the nodal joints. Fig. 6.22(a) and (b) shows the end cap in position in the pultruded tube and the composite plates in position respectively. The cap consists of an inner core and an outer sleeve which are joined at the base. The outer surface of the inner core and the inner surface of the outer sleeve have a thread moulded into them. Onto the outside surface of the outer sleeve are moulded five pairs of tapered ribs which will mate with slots in the connecting plate when the joint is assembled. The end cap is made in a single injection moulding operation. A cross-section of the cap is also shown in Fig. 6.23, and a detailed discussion has been given by Hollaway and Baker.[6.17]

To form the bond between the tube and end cap, a measured quantity of epoxy adhesive is injected into the space between the inner core and the outer sleeve. The tube is pushed into the space until it reaches the cap's base, thus forcing the adhesive to flow around the threads on either side of the tube and to surface at the top of the cap. The threads are an essential part of the design of the cap as they control the glue line thickness and ensure quality control. The adhesive runs into the threads and forms a chemical bond with the pultruded GRP and a mechanical bond with the nylon.

A nodal joint for skeletal composite sections can also be formed by using the crimp-bonded end fittings[6.16] which have been mentioned in section 6.6. An oversize aluminium sleeve is crimped onto the composite tubular section and in so doing forms a uniform thin glue line. The sleeve may then be threaded to allow the connection of members via a nodal joint, usually aluminium.

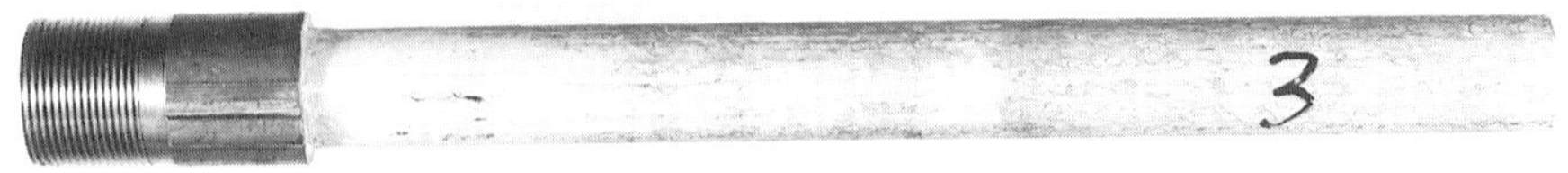

Fig. 6.24. Aluminium sleeve crimped and bonded onto pultruded tube

Fig. 6.25. Eight pultruded members screwed into an aluminium nodal joint

Alternatively, the region outside the crimp overlap may be flattened to accept a bolted type of connector. There are laid-down procedures for the pretreatment of the aluminium and the bond area of the pultruded tube. If the pressure applied during the crimping operation is controlled, the following may be achieved

(*a*) a reasonably uniform thin bond line around most of the crimped joint circumference
(*b*) the transverse crushing of unsupported tubular pultrusions

Figure 6.24 shows an aluminium sleeve crimped and bonded onto a pultruded tube; the end of the aluminium is threaded. Fig. 6.25 shows eight pultruded members which have been screwed into a nodal joint. The latter is made of aluminium block, machined to size and shape and provided with threaded holes in the right location and with the required orientation in order to accommodate the members. A photograph showing a skeletal structure manufactured using pultruded tubes with node joints made from resin injected end caps and special nodal joints is given in chapter 11 (Fig. 11.1).

References

6.1. Matthews F.L. *et al.* The strength of bolted flange joints in GRP pipes. *Proc. 14th Reinforced Plastics Congr. Brighton*, 1975, 143–149.
6.2. Sneddon I.N. The distribution of stress in adhesive bonded joints. In: *Adhesion*, D.D. Eley (ed.). Oxford University Press, Oxford, 1961.
6.3. Koichi M. and Keith R.E. *Fundamentals of selected aspects of deformation characteristics of adhesive bonded joints and metal-adhesive interfaces.* Contract No. DA-01-021-AMC-14693(Z). Redstone Arsenal, USA.
6.4. Benson N.K. The mechanics of adhesive bonded joints. *Applied Mechanics Review*, 1961, **14**, No. 2, 83.
6.5. Kutscha D. and Hofer K.E. *Feasibility of jointing advanced composite flight vehicle structures.* Tech. Report No. AFML-TR-68-391. IIT Research Inst., Ohio, 1969.
6.6. Oplinger D.W. Stress analysis of composite joints. *Advances in jointing technology, Proc. 4th Army Mat. Tech. Conf.* 1975, 405.
6.7. Hart-Smith L.J. *Developments in adhesives*, **2**, 1. Kinloch, A.J. (ed.) Applied Science, London, 1981.
6.8. Matthews F.L. *et al.* A review of the strength of joints in fibre-reinforced plastics; part 2—adhesively bonded joints. *Composites*, 1982, **13**, No. 1, 29.
6.9. Adams R.D. and Wake W.C. *Structural adhesive joints in engineering*, chapters 1 and 2. Elsevier, London, 1984.
6.10. Hart-Smith L.J. *Adhesively bonded joints for fibrous composite structures.* Douglas Paper 7740. McDonnell-Douglas Co., Long Beach, California, 1986.
6.11. Hart-Smith L.J. *Bolted joints in graphite-epoxy composites.* NASA Langley Report NASA CR—144899. Douglas Aircraft Company, 1977.
6.12. Godwin E.W. *et al.* Strength of Multi Bolt Joints in GRP. *Composites*, 1982, **13**, No. 3, July, 268–273.
6.13. Hart-Smith L.J. *Design and analysis of bolted and riveted joints in fibrous composite structures.* Douglas Paper 7739. McDonnell-Douglas Co., Long Beach, California, 1986.
6.14. Hart-Smith L.J. *Differences between adhesive behaviour in test coupons and structural joints.* Douglas Paper 7066 McDonnell-Douglas Co., Long Beach, California, 1981.
6.15. Hollaway L. and Rustum A. Structural Characteristics of Pultruded Glass Fibre Reinforced Polymer Double Layer Skeletal Systems. In: *Proc. Int. Symp. Future for Plastics in Building and Civ. Engng, Liege*, 1984.
6.16. Green A.K. and Phillips L.N. Crimp-bonded end fittings for use on pultruded composite sections. *Composites*, 1982, **13**, No. 3, July, 219–224.
6.17. Hollaway L and Baker S. The development of nodal joints suitable for double-layer skeletal systems made from fibre matrix composite. In: *Proc. Int. Conf. on Space Structures*, 908–913. Nooshin H. (Ed) Elsevier, London, 1984.

6.18. Pickett A. *et al.* Analysis of a crimped and bonded joint for load bearing skeletal members. *Composites*, 1982, **13**, No. 3, July, 257–267.

6.19. DeBryne N.A. *Adhesion and Adhesives*, 91. DeBryne N.A. and Houwink R. (Eds). Elsevier, Amsterdam, 1951.

6.20. Myhre S.H. *et al.* Moisture problems in advanced composite structural repair. *Composites*, **13**, No. 3, 1982, 289–297.

6.21. Hart-Smith L.J. *Design and analysis of bonded repairs for fibrous composite aircraft structures*. Douglas Paper 7133. McDonnell-Douglas Co., 1981.

6.22. Renton W.J. and Vinson J.R. *The analysis and design of composite material bonded joints under static and fatigue loading*. Report AFOSR-TR-72-1627. US Air Force, 1973.

6.23. US Department of Defense. *Military handbook MIL-HD BK-17A*. US Department of Defense, 1971.

6.24. Sage G.N. Pure shear failure of an adhesive with reference to undirectional composite bonded joints. *Proc. Symp. Jointing in Fibre Reinforced Plastics*. IPC, London, 1978.

6.25. Kinlock A.J. *et al.* Durability of adhesive joints. In: *Adhesion science and technology, part B*, 597. Lee L.H. (Ed.) Plenum Press, New York, 1976.

6.26. Althof W. *Strength properties of bonded joints of carbon fibre reinforced composites*. RAE Lib. Trans. No. 1616. Royal Aircraft Establishment, 1971.

6.27. Volkersen O. Die nietkraftoerteilung in ubeansprunchten nietverbindungen mit Konstanten loshonquerschnitten. *Luftfahrtforschung*, 1938, **15**, 41.

6.28. Goland M. and Reissner E. Stresses in cemented joints. *J. Appl. Mech.* 1944, **11**, March, 17.

6.29. Hart-Smith L.J. *Adhesive bonded single lap joints*. NASA CR 1122 36. NASA, 1973.

6.30. Hart-Smith L.J. *Analysis and design of advanced composite bonded joints*. NASA CR 2218. NASA Langley, 1974.

6.31. Hart-Smith L.J. *Stress analysis: a continuum mechanics approach to developments in adhesives*, 1. Kinloch A.J. (Ed.) Applied Science, London 1981.

6.32. Sazhin A.M. Determining the stresses in glued joints between metal plates. *Russ. Engng Jnl*, 1964, **44**, No. 11, 45.

6.33. Volkersen O. Recherches sur la theorie des assemblages colles. *Constr. Metallique*, 1985, **4**, 3.

6.34. Hart-Smith L.J. *Adhesive bonded double lap joints*. NASA CR-1122-35. NASA, USA, 1973.

6.35. Oplinger D.W. Stress analysis of composite joints. Advances in jointing technology. *Proc. 4th Army Mat. Tech. Conf.*, 1975, 405.

6.36. Wooley G.R. and Carver D.R. Stress concentration factors for bonded lap joints. *Int. Aircraft*, 1971, **8**, No. 10, 817.

6.37. Adams R.D. and Peppiatt N.A. Stress analysis of adhesive bonded lap joint. *J. Strain Anal.*, 1974, **9**, No. 3, 185.

6.38. Chun W.W. and Sun C.R. Interfacial stresses and strength of lap joints. *Proc. 21st Conf. on Structures, Structural Dynamics and Materials, Seattle*, 1980, 656.

6.39. Adams R.D. and Peppiatt N.A. Stress analysis of lap joints in fibre reinforced composite materials. In: *Fibre reinforced materials*, 45. Institution of Civil Engineers, London, 1977.

6.40. Cooper P.A. and Sawyer J.W. *A critical examination of stresses in an elastic single lap joint*. TP-1507. NASA, Langley, 1979.

6.41. Grimes G.C. *et al. The development of non-linear analysis methods for bonded joints in advanced filamentary composite structures*. Contract Report No. AFFOL-TR-72-97. Southwest Research Inst., San Antonio, 1972.

6.42. Hart-Smith L.J. *Adhesive bonded single lap joints*. NASA CR 112236. NASA, USA, 1973.

6.43. Grant P. *Strength and stress analysis of bonded joints*. BAe Report SOR(P) 109. British Aerospace, Warton Division, 1976.

6.44. Grant P. Analysis of adhesive stresses in bonded joints. *Proc. Symp. on Jointing in Fibre Reinforced Plastics*, 41. IPC, London, 1978.

6.45. Zienkiewicz O.C. *The finite element method in engineering science*. McGraw-Hill, London, 1971.
6.46. Adams R.D. Stress analysis: a finite element analysis approach. In: *Development in Adhesives* (2nd edn) Kinloch A.J. (ed.) Applied Science, London, 1981.
6.47. Wright M.D. The stress analysis of a butt strap joint in CFRP. *Composites*, 1978, **9**, No. 4, October, 259.
6.48. Nagaraja Y.R. and Alwar R.S. Non-linear stress analysis of an adhesive tubular lap joint. *J. Adhesion*, 1979, **10**, No. 2, 97.
6.49. Steel D.J. The use of carbon fibre composites for axially-loaded tension/compression components. *Proc. Conf. on Designing with Fibre Reinforced Materials*, Paper C 234/77. Institution of Mechanical Engineers, London, 1977.
6.50. Romhi A. *et al.* An elastic and elasto-plastic analysis of a unique joint for pultrusion composites. *Composite Structures*, 1986, **6**, No. 1–3, 197–239.
6.51. Guess T.R. and Haizlip C.B. Jr. End-grip configuration for axial loading of composite tubes. *Exp. Mech.*, 1980 **20**, January 31–36.
6.52. Lubkin J.L. and Reissner E. Stress distribution and design data for adhesive lap joints between circular tubes. *Trans. Am. Soc. Mech. Engrs*, 1956, **78**, August, 1213–1221.
6.53. Alwar R.S. and Nagaraja Y.R. Visco-elastic analysis of an adhesive tubular joint. *J. Adhesion*, 1976, **8**, No. 1, 79–92.
6.54. Adams R.D. and Peppiatt N.A. Stress analysis of adhesive bonded tubular lap joints. *J. Adhesion*, 1977, **9**, No. 1, 1–18.
6.55. Pickett A.K. and Hollaway L. The analysis of elastic-plastic adhesive stress in bonded lap joints in FRP structures. *J. Composite Structures*, 1985, **4**, 135–160.
6.56. Pickett A.K. and Hollaway L. The analysis of elastic adhesive stresses in bonded lap joints in FRP structures. *J. Composite Structures*, 1985, **3**, 55–79.
6.57. Collings T.A. Experimentally determined strength of mechanically fastened joints. *Jointing fibre-reinforced plastics*, chapter 2. Matthews F.L. (Ed.) Elsevier, London, 1987.
6.58. Collings T.A. *The strength of bolted joints in multi-directional CFRP laminates*. RAE Technical Report 75127. Royal Aircraft Establishment, Farnborough, 1975.
6.59. Collings T.A. The use of bolted connections as a means of joining carbon fibre reinforced plastics. In: *Proc. Conf. on Designing with Fibre Reinforced Materials*, paper C 299/77. Institution of Mechanical Engineers, London, 1977.
6.60. Collings T.A. The strength of bolted joints in multi-directional CFRP laminates. *Composites*, 1977, **8**, No. 1, January, 43.
6.61. Clayton G. and Jones D.P. *Further investigations into the effect of counter-sunk holes in a multi-directional CFRP laminate using various fibre/resin combinations*. Report SON (*P*) 137. British Aircraft Corporation, Warton, Preston, 1976.
6.62. Berg K.R. Problems in the design of joints and attachment. *Mechanics of Composite Materials, Proc. 5th Symp. on Naval Structural Mechanics*. 1967.
6.63. Quinn W.J. and Matthews F.L. The effect of stacking sequence on the pin-bearing strength in glass fibre reinforced plastic. *J. Comp. Mater*. 1977, **11**, No. 2, April, 139–145.
6.64. Matthews F.L. *et al.* Bolting and riveting in fibre reinforced plastics. In: *Proc. Reinforced Plastics Congr., Brighton*, 1976.
6.65. Crews J.H. Jr. Bolt bearing fatigue of a graphite epoxy laminate joining of composite materials. *ASTM STP 749*, 131–144. Kedward K.T. (Ed.) American Society for Testing and Materials, 1981.
6.66. Godwin E.W. *et al.* Strength of multi-bolt joints in GRP. *Composites*, 1982, **13**, No. 3, 268–272.
6.67. Shivakumar K.N. and Crews J.H. Jr. *Bolt clamp-up relaxation in a graphite/epoxy laminate*. NASA TM 83268. NASA, USA, 1982.
6.68. Godwin E.W. and Matthews F.L. A review of strength of joints in fibre reinforced plastics, part 1–mechanically fastened joints. *Composites*, 1980, **11**, No. 3, 1155.

6.69. Kingston-Lee D.M. and Rogers K.E. *A Preliminary Evaluation of Proprietary Rivets as Fasteners for Carbon Fibre Laminates*. RAE Technical Memorandum Materials 243. Royal Aircraft Establishment, 1976.
6.70. Stockdale J.H. and Matthews F.L. The effect of damping pressure on bolt bearning loads in glass fibre reinforced plastics. *Composites*, 1976, **7**, No. 1, 34.
6.71. Hart-Smith L.J. Design and empirical analysis of bolted or riveted joints. *Jointing fibre-reinforced plastics*, chapter 6. Matthews F.L. (Ed.) Elsevier, London, 1987.
6.72. Gibbs and Cox Inc. *Marine design manual for fibreglass reinforced plastics*. McGraw-Hill, 1960.
6.73. Rufolo A. *Design manual for joining of glass reinforced plastics*. US Naval Material Laboratory Report, Navship 250-634-1. US Navy, 1961.
6.74. Kingston-Lee D.M. and Rogers K.F. *A preliminary investigation of proprietary self-tapping screws as fasteners for carbon fibre laminates*. RAE Technical Memorandum Materials 200. Royal Aircraft Establishment, 1974.
6.75. Hodgkinson J.M. *et al.* The strength of bolted joints in Kevlar RP. *Proc. Conf. Composite Design for Space Application*, 15–18. ESA SP-243. European Space Agency, Noordwijk, Netherlands, 1985.
6.76. Pyner G.R. and Matthews F.L. *J. Comp. Mater.* 1979, **13**, July, 232.
6.77. Strauss E.L. Mechanical joints in reinforced plastics structures. *Machine Design*, 1960, **32**, March.
6.78. Saddler C.J. and Barnard A.J. *The strength of bolted/riveted joints in CFRP laminates*. Structures Research Note, No. 16. Westland Helicopters, 1977.
6.79. Hirst I.R. and Matthews F.L. The variation of bearing strength with load direction. *Proc. Symp. Jointing in Fibre Reinforced Plastics*. IPC, London, 1978.
6.80. Leggatt A.J. *GRP and buildings*. Butterworths, London, 1984.
6.81. Hart-Smith L.J. Design of adhesively bonded joints. *Joining fibre-reinforced plastics*, Chapter 7. Matthews, F.L. (Ed.) Elsevier, London, 1987.

7 End use performance and time-dependent characteristics

J. R. Crowder (part 1) and C. M. Howard (part 2)

Part 1. End use performance

7.1. Introduction

Composite materials are not new in construction. The ancient Egyptians improved the quality of their bricks by mixing straw with the clay. Horse hair has long been used to increase the impact resistance of plaster. Nowadays steel-reinforced concrete is universally accepted as a major structural material. Polymer composites, a relatively recent addition to this class of materials, have been in use for more than 30 years. Despite this, experience of their long-term structural performance is still somewhat limited in the building environment.

Without such experience, the designer cannot have full confidence in the material, and the purpose of this chapter is to assemble some of the accumulated knowledge of the durability of polymer composites. The factors that affect performance, both in the materials themselves and in the environments in which they are used are discussed. The obtaining of a satisfactory performance through appropriate specification and through control in manufacture is also indicated.

The term "polymer composites" embraces a range of materials, but in this chapter, discussion is restricted to glass fibre reinforced polyesters and epoxies. Reinforced thermoplastics are not used to any significant extent in structural building components; resins reinforced with vegetable fibres are limited in their use, mostly to non-industrial developing countries, and in general are rather susceptible to the weakening action of water. Phenolic/paper/melamine laminates in composites with other board materials are used in panels and facings mainly for their aesthetic appeal and abrasion resistance. In general, the construction industry uses expensive high-performance engineering materials only for very special applications and, for example, carbon fibres or polyimides are hardly used. The laminate systems that are used for structural applications are based mainly on polyesters, with various cross-linking agents; epoxy resins are used to a minor extent with different hardeners, mostly in rather specialized engineering applications. There is a growing interest in phenolic resins for lay-up purposes: good fire and chemical resistance are the main advantages, though they require a separate finishing coat if used for decorative applications. The glass fibre reinforcement most commonly used is in the form of chopped strand mat or chopped rovings. Various modifiers and additives, including colourants, stabilizers and fillers, may be incorporated and influence long-term performance, which can sometimes also be improved with protective coatings.

Laminates can be built up with different layers in various thicknesses and the different component materials of polymer composites are affected differently, and often adversely, by various factors in the environments where they are used. The chemical and physical characteristics of the two principal components, glass and resin, their interaction, the way in which they are laminated together, and their interaction with the environment of use are major factors in determining the durability characteristics of GRP components.

7.2. Environmental factors

In most environments, polymer composites show change with time. In normal indoor ambient conditions, however, the change is slow and is not likely to be of significance. There may be a slow continuation of cure of the resin but this is not likely to be deleterious. The most significant factors in inducing change are

(*a*) elevated temperatures (especially when these fluctuate over a considerable range)
(*b*) moisture, particularly constant wetness
(*c*) adverse chemical environments (acidic, alkaline and organic solvent)
(*d*) natural weathering.

Natural weathering is the principal environmental influence to which structural composites are exposed in construction applications. Moisture plays an important part in this, with warmth and to a lesser extent cold contributing to the effect. Sunlight is very important too, particularly the UV component. Separately, and in combination, these may all contribute to a decline in properties.

7.2.1. Temperature

Temperature is an important factor in the manufacture of GRP composites. The latter should normally be cured (and post-cured) with heat to a state of chemical and physical stability. When curing, the chemical cross-linking process involves polymer chains and active monomers, and has the effect of drawing the chains together, so that a measure of shrinkage is normal. In addition during the process of curing, reaction and combination between the chemically active groups in the constituents reduces further the potential reactivity of the resin system to adverse environments. The further the curing process can be taken in the manufacturing procedure, the greater the degree of dimensional and environmental stability that can be achieved. Although it is desirable to post-cure the composite at elevated temperatures, it sometimes presents practical difficulties for large structural components, and a measure of continuing change in physical properties and dimensions may be expected for a period if a laminate is imperfectly cured. This effect is not usually of much significance unless there is excessive undercure, even if the component reaches a moderately high in-service temperature (e.g. in warm sunlight). In building structures, GRP will not be heated near to the limit of its mechanical performance; this limit is in the region of 200°C for normal resin systems. More deleterious is the effect of constantly fluctuating temperatures. On a microscopic scale, differences in the thermal expansion coefficients of glass and resin may contribute to progressive debonding and weakening of the materials, although the extensibility of the resin system will usually accommodate differential movement. The more resilient the resin, the less the stress-induced weakening of the bond between the major components, although flexible resins are not usually chosen for structural applications because they may not impart adequate strength and stiffness to the composite. On a large scale, cracking and delamination may be induced by

(*a*) the differential movement of the components, or
(*b*) the movement of the substrate on which the component is fixed, or
(*c*) the stresses induced by restrained thermal movement.

With this there will be significant loss of strength, most obviously at points of conjunction, and where fixings are rigid.

Continuous exposure of GRP composites to unusually high temperatures may lead to discoloration of the resin, most noticeably yellowing. Polyesters and epoxies both show this effect, which is likely to be aggravated when flame retardants are incorporated. Apart from this discoloration, the principal practical effect is likely to be an overall hardening of the laminate.

Almost all chemical reactions are speeded up by a rise in temperature and the most obvious effect on composites at higher temperatures is to increase the rate at which either a chemical- or an ultra-violet-initiated moisture reaction proceeds, thereby accelerating rates of deterioration.

7.2.2. Moisture

Moisture is usually a more significant factor than heat in causing deterioration, in that it is likely to have both chemical and physical effects on all the components individually and it may have a severe effect on their interaction.

Cross-linked polyesters are not easily hydrolyzed, even under fairly severe conditions, but they are sufficiently hydrophilic for liquid water to be absorbed, sometimes to quite a significant degree. The consequent plasticizing effect is reflected in measurable change in mechanical properties of the resin, in turn reducing the modulus of rigidity of the composite. Strength properties of wet laminates are commonly found to be of the order of 25 to 30% lower than that of dry specimens[7.1] and wet strength values are sometimes used in design to reflect the possible effects of water and weather. Comparable effects and strength losses of the same order occur with epoxy resins, with some variation between different resin types. The long-term effect seems to be greater than with polyesters, particularly in relation to high-performance applications, with water absorption of as much as 4% occurring and lowering the glass-transition temperature by as much as 80°C.[7.2] Water absorption, both by polyesters and epoxies, also leads to swelling of laminates, usually showing as a small and usually insignificant increase in thickness, though sometimes as warping if the laminate structure is unbalanced; the glass fibre reinforcement has some effect in restraining expansion in the plane of the laminate.

Glass fibres can suffer surface effects from the action of water—their relatively high strength is largely due to the absence of surface imperfections in individual glass fibres; a small amount of solution from the surface of the fibres can produce surface flaws which reduce the overall strength of the laminate. Alkali-resistant glass fibres, intended for use in reinforcing cement products, are said to have better water-resistance than the E-glass normally used for GRP[7.3] and their extra cost may be justified for some situations. Water can also have an effect on surface-active pigments and fillers, and in turn can contribute to changes in bulk or directional properties.

In the short term, water absorption is principally an effect of the surfaces of GRP laminates. However, long-term immersion in water, allowing penetration through the thickness of a laminate may, as well as having the plasticizing effect referred to earlier, also cause the development of water-filled voids or blisters at fibre–resin interfaces. There may be slight solution of the fibre surface, and there is evidence that some organic sizes lead to the development of osmotic pressure which can initiate and aggravate the debonding effect. The blistering effect may occur in unreinforced areas of resin, and can be affected by a number of factors. Quality of materials and of control in the moulding process are most important, notably the choice of gel-coat and laminating resins. There is evidence that polyester resins with higher temperature deflection levels can limit the blistering effect.[7.4] Glass fibres are usually treated with a size to protect them after manufacture, and where there will be constant wetness, a water-resistant size will be advantageous, in minimizing blistering.

Cut edges of GRP laminates, where the ends of fibres are exposed, suffer from a "wicking" action of water along the fibres. There is not a great deal of evidence to show that anything other than a narrow strip at the edge is likely to be seriously affected by water penetration, except in laminates with a high glass content. In the majority of cases, water penetration from the face of a laminate is more significant over most of the area.

The most significant effect of water is on the glass–resin bond. Arguments continue as to the chemical or physical effectiveness of the silane and other treatments that have been used to improve the bond between fibres and resins, but this bond can be weakened by water penetrating from surfaces or edges, causing a fall to as low as 35% of the original strength.[7.5] A rather striking and unexpected feature is that distilled water, probably because its non-ionic characteristics reduce any barrier to its penetration, may cause a more serious reduction in properties than either sea water or tap water. Chlorinated water, used in swimming pools, sometimes has deleterious effects on GRP pool liners, including the bleaching of pigments, as well as the normal bond-weakening action.

Alternating moisture conditions can also be deleterious, by causing differential movement, and progressive attack on the glass–resin bond. Evidence from trials at the Building Research Establishment[7.6] is that wetting and drying on natural exposure is less severe than continuous exposure to water in producing changes at the glass–resin interface. These changes are only partly reversible, and constant wetness has a plasticizing effect on the resin. There seems to be a fair measure of reversibility in the early stages of water absorption within a laminate.

There is a clear need, however, in designing structural laminates for situations of constant, rather than occasional wetness, to select water-resistant resins. Also, alkali-resistant glass fibres can be used to advantage in such situations. Additionally, special provisions should be made, by suitable gel-coats and edge-sealing, to minimize ingress of water to the core of the laminate.

7.2.3. Solution and solvent action

Acidic and alkaline conditions, when they are encountered in the construction situation, in general aggravate the effects of water. They are not, of course, a normal expectancy, and indeed, a properly cured polyester shows good resistance to both. In chemical plant their action may cause some problems, and call for the selection and design of special materials and laminates, possibly incorporating other plastics materials, such as PVC or polypropylene; these would not normally be considered for straightforward structural purposes. Organic solvents also may have more serious adverse effects—ketones and esters in particular can bring about significant attack on cross-linked polyesters and cause laminate breakdown: other solvents have less effect. GRP is liable to damage by strong solvent blends, such as those which are deliberately used in paint strippers. Advice on cleaning and painting of GRP is offered in a Building Research Establishment publication.[7.7] In the most severe cases, which should not be expected in normal construction, chemical or solvent action may lead to serious breakdown of composites; if such action is at all likely, appropriate precautions should be taken.[7.8]

7.2.4. Weather

There are significant differences between climates in different parts of the world, but natural weathering is universally a cause of deterioration of GRP composites. It is worth noting that orientation also has a significant effect. In the northern hemisphere, a GRP panel on a south-facing roof slope will weather less well than one exposed vertically. A panel on the north side of a building may take longer to dry after rain, but will suffer less from the effects

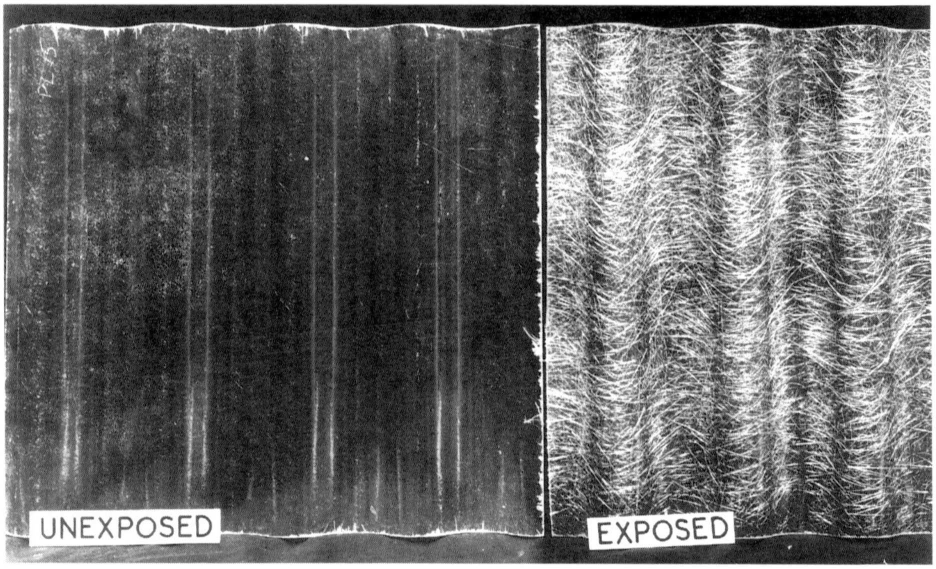

Fig. 7.1. Action of water on GRP increases obviousness of fibres and reduces light transmission

of solar radiation than will one on the south side. The action of weather on GRP in different situations and different climates has been widely investigated.

Depending particularly on the resin component, a major factor is degradation of both polyesters and epoxy resins, brought about by the influence of sunlight. Initially this shows as a discolouration which develops into surface breakdown and erosion. Degradation can also produce a change in the refractive index of the resins used in composites; the consequent mismatch between the refractive indices of glass and resin, which is aggravated by the action of water, is noted as an increased obviousness of glass fibres and loss of light transmission in translucent sheeting (Fig. 7.1).

The UV component in sunlight is largely responsible for this degradation and is powerful enough to initiate the chemical reactions involved; a short wavelength band at 330 nm has most effect on polyesters, but longer wavelengths are also significant, particularly with flame-retardant resins. UV absorbers and stabilizers are used with advantage to mitigate this effect but, in combination, sunlight and water are principally responsible for weathering breakdown. With epoxy resins, which absorb strongly throughout the UV region, exposure to sunlight produces yellowing and surface degradation, though there is little evidence of this seriously affecting the tensile or flexural strengths or moduli. The UV absorbers and the stabilizers lead to little improvement with epoxy resins, though they are sometimes used to improve stability.

Infra-red and visible radiation also play a part in the weathering of the composite, more particularly with coloured and pigmented polyesters which can absorb incident radiation. These radiation bands accelerate rates of degradation by raising the temperature. Wherever chemical reaction plays a part in the breakdown process, then rise in temperature plays a part. A rise in temperature of 10°C approximately doubles reaction rates; resin degradation in consequence proceeds more rapidly in hot climates—two or three times as fast in the tropics as in temperate climates.[7.9] A similar degree of acceleration of the degradation of epoxy resins also occurs in hot climates.[7.10]

In the weathering process, water plays a part as a reagent in some of the chemical degradation reactions, but its most obvious effects are in weakening and breaking the bonds with glass fibres. The photomicrograph (Fig. 7.2) illustrates this breakdown on a moulding where little resin degradation has

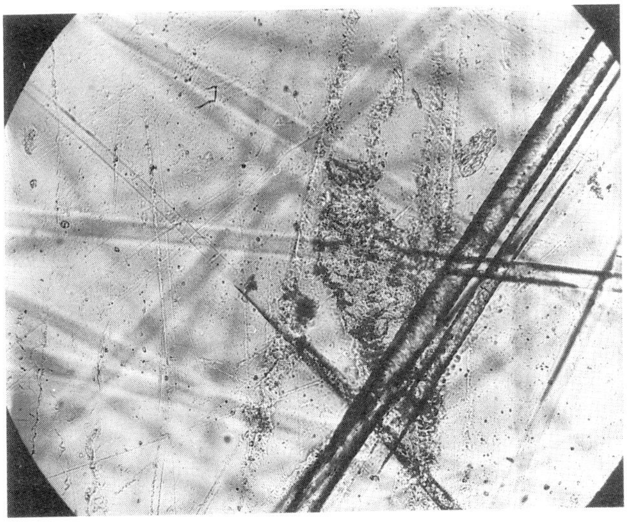

Fig. 7.2. Breakdown of bond between glass fibre and polyester resin in GRP (× 139)

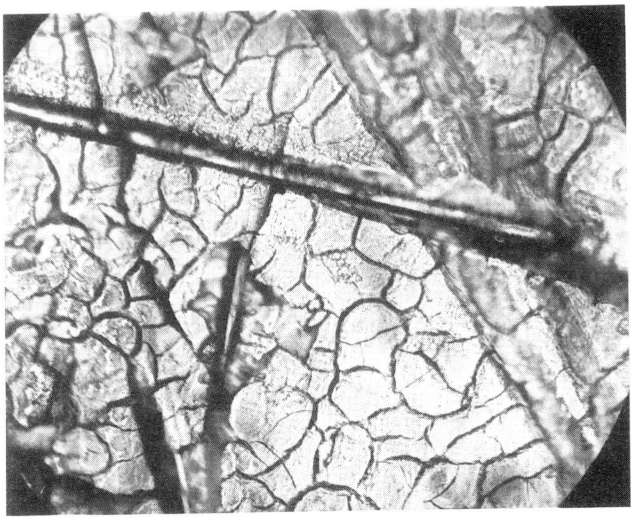

Fig. 7.3. Degradation of flame-retardant resin aggravates the breakdown of bond with glass fibre (× 139)

occurred. Where degradation occurs at the same time, the effect can be much greater (Fig. 7.3). In this instance, severe degradation of a flame-retardant resin has increased the vulnerability of the glass fibre–resin bond. In such cases, from an initial surface effect the process of degradation may develop to have a serious influence on bulk properties, notably strength. With many early resin systems, used in unprotected laminates, such deterioration was not uncommon. Some modern flame-retardant resins, particularly those used to achieve the highest standards of resistance to surface spread of flame in laminates, will still show this effect. The best of currently available modern resins, used correctly in accordance with manufacturers' recommendations, may not show such effects, even in relatively harsh tropical conditions, for periods of many years. The additional protection provided by a gel-coat or surface coating is a valuable additional precaution.

The detrimental influence of weathering also affects the mechanical properties of GRP composites: debonding of the fibres as a result of degradation of the resin reduces the load-bearing capacity. It is not easy to differentiate this effect from the debonding caused by water, but the surface deterioration caused by weathering undoutedly plays a part in strength losses that occur.

There is some difficulty in interpreting the data from mechanical tests carried out on GRP laminates after weathering. Because weathering is largely a surface effect, specimen thickness must have a significant effect on the results, but this is not always considered. In one set of trials,[7.11] 3 mm thick GRP laminates, without the protection of a surface coating or a gel coat, were found to lose between 12% and 20% of their initial flexural strength after 15 years' exposure to natural weathering. It was estimated that, for a 10 mm laminate, the effect of 50 years' weathering would be a reduction of no more than 3% in flexural strength in a temperate climate.[7.11] In a different set of trials,[7.10] 3 mm thick samples made with polyester resins of different reactivity were exposed in the UK and lost between 13% and 32% of their initial flexural strength. These results are compatible with the above findings and with much of the other data available, though no extrapolation was made to a 50 year lifetime.

Tensile strength is a more important design property, but less information is available on how this is affected by weathering. Surface deterioration clearly has a bearing on it: for 3 mm laminates, losses of up to 50% have been reported for high-reactivity resins; for low-reactivity systems, such as might be used for structural components, 8 years' exposure in a wet tropical climate produced losses of no more than 25%,[7.10] and changes in tensile modulus of the order of 10%. Changes in tensile properties were significantly less with duplicate sets of samples exposed in the UK. These changes, both in tropical and temperate climates, were less than that for comparable effects of water; they could certainly be allowed for in designing structural laminates, which would normally be more than 3 mm thick. Epoxy resins, after 8 years in the same tropical exposure trials, showed loss in strength and modulus of the order of 10%. Again, changes on exposure in the temperate UK climate were of a lower order. Amplification of the results of these and other trials is presented by Davis and Sims.[7.12]

The evidence from these and other trials shows that polyester and epoxy resin composites can be designed and produced to be capable of prolonged outdoor exposure in temperate conditions without excessive loss of strength. Even for the tropics, reasonable performance can be expected. The conditions of use, and the deterioration they are likely to cause, must clearly be allowed for in design. This requirement can be accommodated by the application of limit state design principles[7.13] (chapter 4).

It is worth noting that rooflighting sheets are particularly prone to show the effects of weathering because they

(*a*) are often thin
(*b*) lack a protective gel-coat
(*c*) are used in situations of maximum exposure to sunlight and rain
(*d*) are often made with flame-retardant resins.

It might be expected that polymer composites would suffer from an effect akin to stress corrosion in metals when exposed to weathering while under stress, and there is evidence that the combination of adverse environment and simultaneous stress does have a detrimental effect on GRP. It has been shown[7.14] that reinforced polyester samples could sustain for a year no more than 20% of their short-term ultimate stress when exposed in flexure to tropical weathering, though similar epoxy laminates sustained almost 50% of their short-term strength under the same conditions. It is known that reinforced plastics under relatively high stress levels—30% of short-term ultimate stress or higher—may fail within a short time of being exposed to UK weather, but this failure occurs too soon for it to be ascribed to weathering degradation. However, limited term studies at the Building Research Establishment[7.15] found that exposure to UK weather under a tensile load corresponding to 25% of ultimate produced little change in the tensile strength of lightly pigmented GRP. This exposure under tensile loading has now continued for 15 years and, though changes in the materials' properties have not been recorded, it is notable that no sample failure has occurred over this period.

7.3. Materials

All the component materials, and the manner in which they are combined, play an important part in determining the performance of a polymer composite. The contribution of the resin system used is the most important factor in relation to durability.

7.3.1. Resins—polyesters

The three principal types of polyester used in laminating are orthophthalic, isophthalic, and "het-acid" based resins. Orthophthalic types are used as general-purpose resins, isophthalic for better weathering- and chemical-resistance, and het-acid in many instances where flame-retardant characteristics are required. Some gel-coat resins with flame retardant properties, and said to have good weathering characteristics, are based on brominated diols. Styrene is the most commonly used cross-linking agent; a proportion of methyl methacrylate with the styrene has been shown to improve light transmission of laminates, with the considerable bonus also of improving weather resistance. Proper curing of the resins is an important factor in their performance and, in general, the higher the degree of cross-linking, the better is the weathering; flame-retardant resins are often harder to cure than non-flame-retardant ones, and special precautions are needed, especially with gel-coats. The stress which may be imposed on the glass–resin bond by the increase in hardness related to cure is not regarded as being of great significance.

For many structural applications, it is mandatory to achieve some degree of flame retardance and, virtually without exception, the need to use specially compounded polyesters to meet this requirement leads to poorer weathering behaviour.[7.16] It is often possible to design the structure of a laminate to minimize the effect, with the use of a non-flame-retardant gel-coat, but the resulting fire-test behaviour or cost may not always be acceptable. Alternatively, a sprayed paint or lacquer film in place of the gel-coat may give acceptable weathering characteristics. In any case, flame-retardants, which are normally incorporated by resin manufacturers in response to specification, add to the cost of resins and, as a general recommendation, for maximum durability, as well as for cost considerations, the proportion of such additives should kept to the minimum necessary to achieve the requisite fire properties. There is a growing interest in the use of phenolic resins for their flame-resistant characteristics, though they may still require surface protection, mainly to improve their appearance.

Fillers and pigments are also used in resins for a variety of purposes, the former principally to improve mechanical properties and the latter for appearance and protective action. Fillers may also be used to improve flame-retardant characteristics (for example aluminium trihydrate) and to reduce costs but, if not chosen correctly, or too high a proportion is used, they are likely to have an adverse effect on weathering-, water- and chemical-resistance. Thixotropic fillers, mainly finely-divided silicas, are often needed to control gel-coat thickness and flow on complex mould surfaces, but if too much is used, it may be difficult to avoid excessive gel-coat thickness, poor cure or patchiness on weathering. A possible, but limited, advantage in providing a barrier to UV radiation, by the use of fillers, may be counteracted by an increased tendency to water-absorption and sometimes by adverse catalytic effects of pigments on the control of curing of resin.

7.3.2. Resins—epoxies

A range of epoxy resin formulations is also available for laminating, their characteristics depending on the resin constituents and choice of hardeners. Manufacturers are able to advise on appropriate formulations for particular applications, many of which depend on the need for enhanced chemical resistance. Even more than polyesters, epoxies suffer adverse effects on weathering, the evidence being that yellowing and other surface changes usually occur; tensile and flexural strength properties show some decline,[7.10,7.17] and composites made from them show corresponding changes. When fillers and pigments are used, epoxy resins are particularly noted for showing a high rate of chalking, though this is not usually associated with great change in structural properties.

7.3.3. Reinforcement

The glass-fibre reinforcement, which plays a major part in determining the durability of structural laminates, is mostly made from "E-glass", which has a borosilicate composition. It is available in various forms. These forms have been discussed in section 2.3.1. Most common is chopped strand mat, used in the hand lay-up process (section 2.8.1.1) and in preform and sheet-moulding compound (section 2.8.2.2). Such a mat is also effectively built up in the process of spray-up, using glass rovings;[7.18] this allows thickness to be increased over stressed areas, without specific increase in the number of layers. Surfacing tissue (section 2.3.1) of single-fibre filaments may be used to reinforce protective surfaces on laminates. Moulding compounds with chopped glass fibres mixed in are not much used for structural purposes but pultruded sections (section 2.8.2.3), combining continuous glass rovings with mat, are occasionally used for structural building components. Woven rovings and glass fabrics are used to produce laminates with improved strength, which is related to the higher glass content.

Glass fibres themselves are affected by the environment, particularly where moisture is present; hydrolytic action causes surface damage on individual fibres, leading to a disproportionate loss in strength. Alkali-resistant glass fibres, developed for reinforcement of cementitious materials, have better resistance also to the action of water. An appropriate water-resistant size must be used on the glass fibres if the composite is to be used in constant wet conditions. The resin used in lamination provides protection but the major effect of moisture is in weakening the interfacial bond between fibres and resin, and in having a plasticizing effect on the resin. In general, the higher the proportion of resin in a GRP composite, the less is the effect of moisture on the important interface, and a heavily reinforced laminate loses a greater percentage of strength in wet conditions than a resin-rich one.

7.3.4. Laminate structure

The durability of a GRP laminate depends very largely on the quality of the exposed surface. Good performance is most commonly achieved with the use of a surface gel-coat of pigmented resin, which protects the glass fibre reinforcement from the action of moisture and the laminating resin from harmful radiation. The resin used for this is chosen for its good weathering or chemical-resistance properties, and is pigmented to give the required colour.

It is important for uniformity of appearance and behaviour that a gel-coat should be of even thickness; a value of approximately 0·5 mm is usually appropriate. The viscosity of the gel-coat resin is usually an important factor in achieving this, but on a complex surface, it may be necessary to exceed it in places. To avoid the cracking that may be induced by curing shrinkage or flexing, unevenness should be kept to a minimum. For many applications, it is recognized to be good practice to incorporate a glass surfacing tissue which reinforces while maintaining the resin-richness of the surface. This may also help to balance the laminate and is most effective in the case of translucent GRP structural components. Use of polymeric films, particularly of inert fluorinated materials, has been found effective in some cases, notably for translucent rooflights, in providing similar protection, though some polyester films have been found to peel off after weathering. For opaque surfaces, polymeric paints applied after lamination, notably cross-linked polyurethanes, have proved more effective, even where considerable proportions of flame-retardant have had to be incorporated in the laminates.

It will be appreciated that, because the structural performance of a laminate depends on the glass reinforcement in the core, the latter needs to be protected, particularly from moisture ingress, so that it is important to avoid cracking or delamination of the gel-coat. Also, the higher the proportion of resin in a laminate, the less significant is likely to be the effect of moisture but to use too much resin is likely to be uneconomic in terms of strength requirements. Woven rovings or glass-cloth reinforcement may be used more economically in certain circumstances to provide maximum strength for a given thickness, but there will be a lower proportion of resin in the saturated laminate; in general, though, for constructional applications, for which the majority of large components are produced by hand lay-up or spraying process, common practice is to aim to build up a chopped strand mat reinforced laminate with approximately 1/3 glass fibre to 2/3 resin system by weight to attain the optimum. Several layers may be built up together to give the required strength and rigidity, and extra thickness of laminate may be laid up to improve impact resistance. Similar proportions of glass to resin, and similar laminate structure can be achieved by a spray-up process. A laminate with such a moderately low glass-to-resin ratio is more likely to have acceptable weathering properties than one with 50% or more of glass.

The required strength and rigidity may be obtained in a structural component without the use of too much glass and resin by laminating into it structural inserts of metal or timber, including plywood. For indoor situations, where conditions are relatively stable, these can be effective, but in a situation exposed to the weather, differential thermal movement and moisture effects can lead to separation of the different materials. Also, where moisture can reach them, steel and timber components are likely to be degraded, by rusting or rotting respectively. From the point of view of durability, it is preferable to achieve stiffness by incorporating hollow GRP sections, or forming ribs round cellular plastics strips to give a combination of materials with compatible properties. Pultruded GRP may also be used in this application and covered with GRP lay-up; pultruded sections lack resin protection on the outer surfaces, and, if exposed, normally do not weather well.

Sandwich construction can be used to increase the rigidity of polymer composites, notably panels. Mechanically this can be an effective technique, if the core of the sandwich is strong enough to transmit the applied stresses, and if it bonds well to the two skins. Cellular plastics, particularly polyurethanes, are often used for this and can be very effective for relatively unexposed applications. Sandwich components used in situations subject to considerable temperature ranges are, however, particularly prone to blistering and debonding. External structural and cladding panels are a particularly obvious example of this, and a backing of highly efficient cellular plastics insulation can, even in British conditions, produce an annual temperature range of more than 100°C in the outer skin of a dark-coloured GRP panel:[7.19] even the diurnal range can be considerable where the panel is exposed to the south, facing the sun. The consequent differential stresses between inner and outer skins of the panel can lead, not uncommonly, to debonding of the latter, and distortions which are unacceptable from the point of view of appearance, as well as producing excessive stresses at joints and fixing points.

Specifiers generally have quite high expectations of the retention of appearance of GRP composites. Even the situations desribed in the preceding paragraph can, however, induce unevenness of surface, indeed, it is almost impossible to produce perfectly flat surfaces or very smooth curves in GRP. Also, brightly coloured and highly glossy surfaces of GRP panels are especially vulnerable, since even quite small unevenness within the surface, or changes produced by weathering, tend to be very obvious to the eye. More maintenance is likely to be called for in such cases, but it is accepted practice to introduce some form of small-scale texture of a random (e.g. hammered) or ordered (e.g. ribbed) nature, to give a uniform appearance, less susceptible to defects. This should always be borne in mind in including an appearance requirement in the specification for a GRP component.

7.4. Specification

Buildings are expected to have long life, in significantly varying conditions, so that durability is a major requirement for most building components. Composite components must not only perform adequately in such obvious aspects as dimensional fit, strength, rigidity and appearance, but they must continue to retain these characteristics over long periods, in various exposure conditions, and specifications should be drawn up with the aim of achieving this.

Although performance specifications may be drawn up for composite components that state this requirement for long life, it is not possible to ensure, by short-term tests, that the requirement will be met, although accelerated ageing procedures may enable some assessment to be made of long-term performance, in conditions comparable with those of use. It becomes necessary, instead, to assemble in the specification a number of different controls that provide pointers to durability. When these can be applied with appropriate knowledge and expertise, and can be related to the known performance of similar components, then a measure of assurance can usually be achieved.

Few architects or engineers can have detailed knowledge of design, formulation and manufacture of GRP components; in arranging to have such components manufactured, it is usually necessary to call on the services of a specialist consultant, and of an experienced and reputable manufacturer. The former can assist in drawing up appropriate specifications, and ensuring they are adhered to; the latter should be able to ensure that his products meet the specification, and to carry out any necessary detailed design. Both should be able to point to instances of satisfactory performance of GRP components in use as indication of their capability.

In the following section, guidance is offered on meeting the requirement. Other documents which may be consulted provide more detail on some aspects of specification.[7.20–7.24]

7.4.1. Contents of specification

GRP components, when used as structural beams and integral structures, wall and roof panels, domes, canopies, and major architectural features, are generally likely to be purpose-designed, so that a separate specification has to be written for each job. It is possible to incorporate a great deal of detail in such a specification, but the best performance will not normally be obtained by doing so. A simple specification, put into effect by a reliable manufacturer, is just as likely to lead to the desired result.

As this chapter considers the aspects that relate specifically to the durability of GRP components, rather than general structural design features, materials considerations are the most important

- materials: resins (polyesters and epoxies); fillers; colourants; and reinforcement (glass fibres in various forms)
- methods of manufacture and structure of laminate.

These aspects have been discussed in the first part of this chapter. It is also relevant to take into account

- dimensional standards
- appearance
- performance requirements including durability
- control procedures.

7.4.2. Materials

Usually the first essential in specifying the material is that the finished composite component should be able to meet the appropriate standards of fire performance. In general, the taller a building, the better is the fire performance expected of its components, particularly those exposed externally; also, the distance from adjacent buildings and boundaries determines the requirements. Although BS 8200[7.25] relates principally to non-load-bearing structures, it has some relevance to the requirements for load-bearing structural plastics. Although fire regulations are often criticized for requiring too high a standard of performance, and plastics enthusiasts may seek ways to avoid excessive stringency, particularly through "engineering solutions", there is good evidence that UK regulations are among the most enlightened in this respect, and it is still sound practice to use materials that enable the required fire-performance levels to be met. Starting from this point it is usually possible to select a resin system likely to lead to achieving these standards, taking into account any additives, fillers and colourants which may affect the fire properties. Other aspects may then be considered, including the colour and texture of the surface, the laminate structure and the required durability.

7.4.2.1. Resins

Resins can be specified according to various standards. British Standards for polyester resins[7.26] and for epoxy resins[7.27] may be referred to; in the UK there are other resin specifications available, notably those issued by the Ministry of Defence, and worldwide there are a number of other acceptable national standards. Although these standards may be invoked, the specifier must rely heavily on the resin formulator's experience in choosing the most appropriate system to meet his requirements for spread of flame characteristics and durability.

It is a feature of all the flame-retardant materials currently available that they have poorer weathering characteristics than general-purpose types. For some resin systems, the fire properties have been found to diminish with time,[7.28] though in other trials, the opposite was the case.[7.29] Fillers and pigments used in a resin system may also have an effect on the fire behaviour. Because there is some uncertainty in predicting the fire performance of a GRP component from the characteristics of the separate materials used, standard fire tests may be called for on specially moulded samples, meeting the requirements for parts 3–8 of BS 476 *Fire tests on building materials and structures* (see bibliography of chapter 8). For some major projects, it may be justifiable, after a preliminary selection of materials, to make a prototype component which can then be subjected to large-scale fire tests to ensure satisfactory performance.

7.4.2.2. Fillers

A limited proportion of certain mineral fillers, including some based on specially treated calcium carbonate, can improve the mechanical properties of polyesters and epoxies, and laminates made with them. Only a modest proportion—no more than 25%—is recommended in laminating resins, though in decorative surfaces a larger proportion of coarse particles may be used as exposed aggregate to give a decorative effect. It is important that gel-coat resins, which usually incorporate thixotropic agents to give appropriate flow properties for moulding complex shapes, should not be too heavily filled. Uneven texture and performance are likely to result. Aluminium trihydrate and antimony trioxide may also be used as fillers for both laminating and gel-coat resins, to improve flame-retardant properties. It is usually best to rely on the advice of a reputable resin manufacturer in respect of the choice and proportion of filler to be used.

7.4.2.3. Colours and pigments

A great many different pigment combinations are available to produce any particular colour which may be required in a structural component. Certainly not all colour systems will achieve the requisite stability. Light-coloured surfaces absorb less heat than dark ones, and this can give some advantage in minimizing any acceleration of the degradative effects of sunlight.

The use of flame retardants affects the colour retention, as well as other aspects of the durability of a resin system, and in general an exposed gel-coat will retain its appearance better if it contains no flame-retardant additive. A list produced by the British Plastics Federation[7.22] provides a certain amount of help in matching colour and flame-retardance to achieve the best stability, but the specifier again must usually rely heavily on the resin formulator's experience in choosing the most appropriate system.

7.4.2.4. Reinforcements

As retention of the structural properties of a GRP component depends largely on the reinforcement, it is desirable that this should be covered by appropriate specification. There are a number of published British Standards for glass fibre reinforcements;[7.3, 7.18] also a number of American and German standards are available. To include these requirements in a specification gives a measure of control over this aspect of quality and performance.

7.4.2.5. Finished laminate

The quality of a composite can be partially controlled by specifying values against a list of relevant properties, and where possible defining acceptable tolerances. Such aspects as visual appearance, shape and dimensions, colour and texture can be defined nondestructively by reference to measurements or to standard samples. Other important requirements may be stated, but it is less straightforward to ensure that they will be achieved. An assessment can be made of hardness and cure of resin, and nondestructive ultrasonic or acoustic emission tests may help identify major moulding defects. But, apart from these, the limited test procedures that are available for determining strength and environmental resistance are likely to require destructive tests on moulded laminates.

The structure of a laminate is important and, in designing it, it is necessary to aim for a balance between requirements. A gel-coat on the surface determines the appearance, and has a major influence on the weathering performance. It is important to ensure in the moulding process that the glass fibre layers laminated on to the gel-coat are placed before it is completely cured, so that they bond well to it. As far as practicable, a laminate should be built up in one moulding operation, without allowing complete cure to take place between layers until the final lay-up has been completed.

The overall cure of the resin system is important, and though a nondestructive surface-hardness test (Barcol test) may be useful as a quasiquantitative guide, with a laminate built up of different layers, simple sonic techniques such as tapping with a coin may indicate effectiveness of cure. It may, of course, be more appropriate to include the Barcol test in a specification.

As indicated elsewhere, the higher the glass content, the greater is likely to be the susceptibility of a moulding to the influence of water and weathering. Carrying out "ashing" tests to determine this requires the taking of specimens from a moulding, and this is seldom acceptable. Also the value of such tests depends on how effectively the layers of a laminate can be separated and, though such a test may be included to confirm that a specification is being achieved, the results depend to such an extent on the attitude and experience of the person carrying them out, that the best assurance is likely to be satisfactory supervision of workmanship in the fabrication process.

Satisfactory assessment of durability can only be made on the component itself, or on specially fabricated specimens of the material to be used. Resin manufacturers should be expected to have test results available confirming the weathering characteristics of a particular resin system recommended. This would normally have been determined by natural and laboratory weathering test procedures, requiring a longer period of time than could be acceptable in the period of agreeing a contract. It is reasonable in a specification to require evidence of such tests to be provided.

The control procedures written into a specification may include requirements for such properties as water-absorption, change of strength in boiling tests, mechanical properties such as flexural and tensile strength, and behaviour in standard fire tests. These tests are practicable only on specially produced specimens, or with the destruction of a moulded component: a good deal of preliminary investigation is also likely to be necessary to define the limits to be required, except for fire tests, where building regulations may define the limits. The need becomes apparent again for a reputable manufacturer to assist in defining limits and tolerances appropriate to his product. Some confirmatory study is likely to be necessary to ensure that test results can be achieved, and that they have the desired effect. Such requirements can then be agreed for inclusion in the specification, as part of the quality control procedures.

7.5. Quality control

For satisfactory design of polymer composites, it is important that the selection of resin system and reinforcement and the structure of the laminate be given full and critical consideration. It may be necessary to undertake a programme of evaluation to ensure that components so designed and manufactured are capable of giving and maintaining satisfactory performance. It is not easy to define a procedure for GRP which will ensure, by means of a series of tests to be applied routinely during manufacture, that a specification so produced will continue to be met. This has led more than one organization to indicate the requirements principally in terms of good laminating practice.

For example, a document produced by the British Plastics Federation[7.22] proposes requirements for

- good workshop practice
- temperature, relative humidity
- condition of materials (to avoid condensation)
- equipment, especially for weighing and measuring
- preparation of materials and moulds
- application—gel-coat, contact moulding
- cure schedule
- selection and care of materials
- maintenance of records.

These are all necessary to ensure satisfactory products, but for components with structural function it is necessary that a control procedure should aim to achieve more than these limited requirements.

The overall procedure should be established on a statistical basis, so that the frequency of sampling ensures the required confidence levels for performance. The following list includes the most important requirements for a quality control procedure for components manufactured from polymer composites: dimensions, colour, transparency, visual examination, surface condition, nondestructive tests, glass fibre contents and special requirements.

Dimensions: thickness, flatness, overall shape and specified sizes, to be within pre-defined tolerances. Some guidance on the latter is given in the British Standard *Guide to quality control requirements for reinforced plastics mouldings*.[7.30] It may be useful also to check weights both as a means of ensuring appropriate thicknesses, and to minimize variation between nominally identical mouldings.

Colour transparency, texture: comparison with agreed standard samples, which should be large enough to give a fair representation and not be altered by excessive handling.

Visual examination: ensure substantial freedom from cracks, voids, bubbles, blisters, exposed fibres, resin starvation, foreign matter. Dye penetration techniques may be used to detect poor lamination at edges, especially with translucent materials.

Conditions of surface: no tackiness; hardness (Barcol) to be as defined in specification, not showing much variation.

Specialized nondestructive tests: ultrasonic techniques may be used, principally for detecting voids, delamination, inclusions, in polymer composites; acoustic emission techniques may also be used to detect possible cracking or fibre debonding.

Glass fibre content: even when it is possible to obtain a representative sample of the actual component material, or specially moulded test pieces to approximate to its composition and structure, this determination is meaningful only if some separation of layers can be achieved, or if the principal requirement is to determine the weight of glass reinforcement per unit area. "Residue on ignition" is a more commonly determined property, and can serve a useful purpose in a quality control procedure, particularly if allowance can be made for incombustible filler content.

Special requirements: normally a specification will include a requirement for one or more strength properties and relevant elastic moduli: tensile, compressive, shear strengths are important in relation to structural design; impact and bolt-bearing strengths may be required for particular components; chemical-resistance, water-absorption and resistance to natural and laboratory weathering tests; fire properties. All of these may be relevant to the performance of a component and should be determined on prototype materials and samples, but they are not usually as effective as quality control procedures, although for strict control, it may be appropriate to include them in a specification.

Though the above requirements can be incorporated in a defined quality control procedure, they may not be entirely effective in maintaining the required quality levels. It cannot be too strongly emphasized that proper design, and choice of materials, coupled with skill and care in workmanship, remain the most effective means of ensuring satisfactory manufacture of composite components, most especially in relation to long-term performance.

Part 2. Time-dependent characteristics

7.6. Introduction

A polymer-based composite material is a combination of two phases: a relatively stiff and essentially elastic fibre network; and a flexible viscoelastic polymer matrix. The mechanical response of the composite material is a combination of these two distinct behaviour states and, therefore, will be time-dependent to a lesser or greater extent. Hence, in the design of structural and semi-structural engineering components which are to be fabricated from reinforced polymer composite materials, it is necessary to consider the time-dependent characteristics of the material to be used.

Often the governing design criterion, for the type of polymer composite materials used in the construction industry, is that of deflection, resulting in engineering components being stressed well below the short-term or long-term ultimate strength of the material.[7.31] Therefore, this section will concentrate on aspects concerned with the long-term deformational characteristics of these materials, with particular emphasis on the determination and interpretation of material property data and its use in the design procedure.

However, before discussing in detail the time-dependent effects of polymers and polymer-based composite materials, it is necessary to define some of the terms used throughout this section. *Creep* is the time-dependent strain change exhibited by the material when subjected to a constant applied stress. *Creep-recovery* is the time-dependent recovery of creep strain on removal of the creep load. *Creep-rupture* is the time-dependent failure of the material during the creep process. *Stress-relaxation* is the time-dependent stress change exhibited by the material when subjected to a constant applied strain. *Creep compliance* is the creep strain divided by the applied stress.

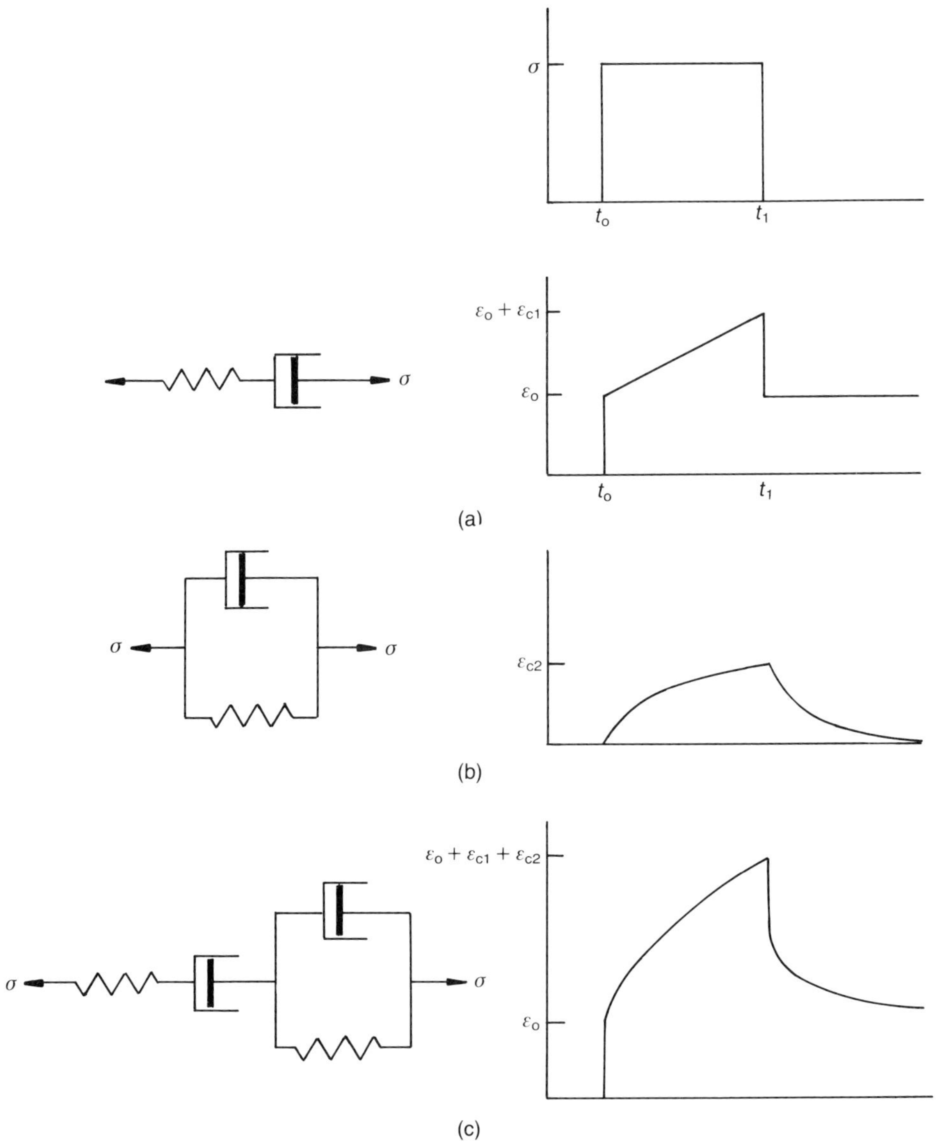

Fig. 7.4. Maxwell fluid element model: (a) Maxwell fluid; (b) Kelvin solid; (c) standard linear solid

7.7. Creep of polymers

The behaviour of polymer materials is described as being viscoelastic. The term implies a behaviour state which incorporates both elastic (solid) and viscous (fluid) components. Certainly this appears to provide a qualitative description of the observed mechanical response of polymer materials when subjected to long-term load. On the application of load the material deforms instantaneously, as an elastic solid, but if the load is maintained the material continues to deform to some extent, in a similar manner to a viscous fluid. In fact early models were developed, in the attempt to quantify this behaviour, which used combinations of elastic spring and viscous dashpot elements.

The simplest of these mechanical analogues is referred to as the "Standard Linear Solid". The model consists of Maxwell fluid element, a linear spring in series with a linear dashpot, shown in Fig. 7.4(a), connected in series with a Kelvin solid element, a linear spring in parallel with a linear dashpot, shown in Fig. 7.4(b). The creep and creep-recovery behaviour of the model describes, in a qualitative manner, the creep behaviour exhibited by many polymers. Fig. 7.4(c) shows the response of the model to a creep and creep-recovery cycle. At the instant that the load is applied the Maxwell spring deforms elastically. The model then continues to deform, at a decreasing rate with time of loading, until at long times of loading a steady state creep rate is established. On removal of the load the elastic deformation due to the Maxwell spring is recovered, instantaneously, followed by the gradual recovery of the creep deformation accumulated by the Kelvin element. A non-recoverable or plastic component of displacement results due to the Maxwell dashpot in the model. Although this simple model describes the phenomenon of viscoelasticity displayed by polymer materials, in reality the time-dependent behaviour of polymers is more complex than can be quantitatively described by this elementary form.

At low applied stress levels many polymers behave in a linear viscoelastic manner.[7.32] In other words, the strain response of the material, at any time of loading, is a linear function of the constant applied stress. This relationship is described by equation (7.1). In addition, linearity implies that the behaviour of the material when subjected to changes in the applied stress is equivalent to the superposition of the individual stress step changes. This is demonstrated for the case of creep and creep-recovery in Fig. 7.5.

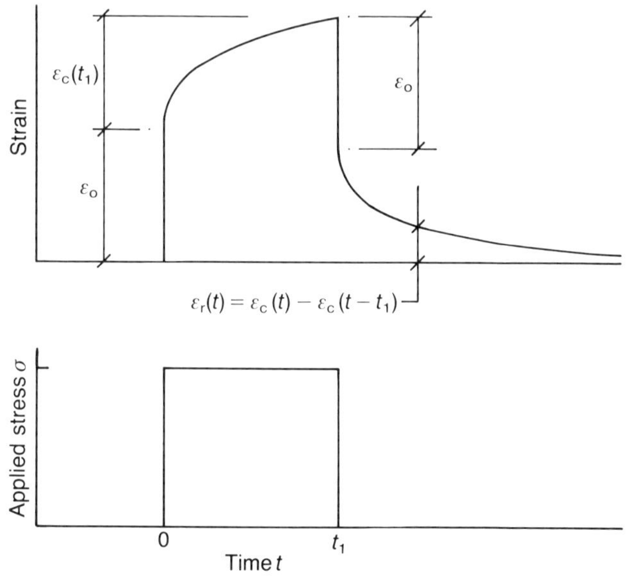

Fig. 7.5. Linear viscoelastic behaviour

$$\varepsilon = f(t)\sigma \tag{7.1}$$

The concept of linear superposition can be extended further to the case of a number of stress steps or for a continuous stress history. This is known as the linear superposition principle and is described in its general form of the hereditary integral, and for the case of a continuously applied stress history it is given by equation (7.2).

$$\varepsilon = \int_0^{t_1} f(t - \tau)\,\frac{d\sigma}{d\tau}\,d\tau \tag{7.2}$$

The mechanical spring/dashpot model is of the linear form and, therefore in theory could be used to quantify the behaviour of polymers under low applied stress. In fact this can be achieved by adding together a number of solid elements to form chains. In this case each component models a particular part of the viscoelastic response. However, to model the behaviour of real materials adequately, very long chain models may be required. This would result in complicated equations if the input was anything but a constant applied stress or strain.

However, the concept of linearity is only applicable at low applied stresses because at higher levels the viscoelastic behaviour of polymers can become highly nonlinear.[7.33] The effect the nonlinear stress dependence on the creep behaviour can be seen by examination of equation (7.3). Here, the creep compliance is both a function of stress and time. Similar superposition rules have been developed to deal with nonlinearity under certain restrictive conditions.[7.34] The simplest form is that of magnitude nonlinearity, where the time and stress functions of the creep compliance term are separable (equation (7.4)). In this case the superposition integral is simply a modified linear integral (equation (7.5)).

$$\varepsilon = f(t, \sigma) \tag{7.3}$$

$$\varepsilon = f(t)\,g(\sigma) \tag{7.4}$$

$$\varepsilon = \int_0^{t_1} f(t - \tau)\,\frac{dg(\sigma)}{d\tau}\,d\tau \tag{7.5}$$

7.8. Creep of reinforced polymers

The extent to which the stiffness and strength characteristics of a composite material are affected by the time-dependent mechanical properties of the matrix can generally be related to the properties of load being carried by the polymer. This, in turn, is dependent on

(*a*) the type of polymer and its stress history
(*b*) the direction of alignment, type and volume fraction of fibre reinforcement
(*c*) the nature of the applied loading
(*d*) the temperature and moisture environment to which the element is exposed.

An insight into the effects of these parameters on the general properties of composite materials can be gained through the examination of the behaviour of simple micromechanical models. Simple relationships between volume fraction of fibre, elastic modulus and stress, which assume strain compatibility within the laminate (as is discussed in section 2.10) allow an estimate to be made of the extent to which the matrix phase contributes to the overall stiffness of the material. Provided that this contribution is small, the effect of the time-dependent changes in the properties of the matrix material on the

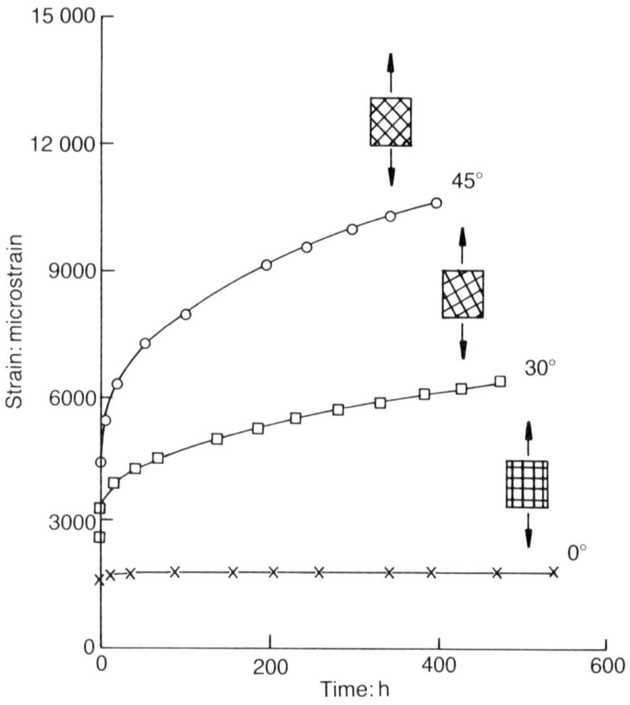

Fig. 7.6. Typical creep behaviour of a glass bidirectional woven roving fibre-reinforced polyester laminate

overall composite is also likely to be small. A good example of this would be the case of a laminate which is reinforced by unidirectional continuous fibres. If loaded in the fibre direction, the polymer matrix will be under low applied strain and hence will contribute very little to the stiffness of the laminate. Therefore, the creep of the composite material in this situation will be negligible. However, if the same composite is now loaded perpendicular to the fibre axis, in the unreinforced direction, the component of load taken by the polymer matrix is significant, and in this case the creep of the laminate could be important. Needless to say, this latter situation can easily be avoided by placing fibres in the directions of the envisaged loading. Indeed, this type of simple analysis is used at the material design stage to ensure that the engineering component will have sufficient stiffness and strength in the directions of loading. However, the types of fibre reinforced polymer used in the construction industry are often reinforced with discontinuous randomly orientated or with continuous bidirectional woven glass fibres. In these types of laminate, the contribution of the polymer matrix towards the overall stiffness of the element becomes far more significant. This can result in the materials having time-dependent mechanical properties which may need to be evaluated to establish the long-term performance of the component.

Figure 7.6 shows a typical creep strain response of a glass bidirectional woven roving fibre reinforced polyester laminate with different directions of loading in relation to the orthogonal fibre axes. As would be expected, the behaviour of the laminate when loaded in the fibre direction is virtually time-independent. However, when loaded in directions other than those of the fibre axes significant creep occurs. The worst case is that when the fibre network has the least influence on the stiffness of the composite, at 45° to the two fibre axes.

Many attempts have been made to establish the observed viscoelastic characteristics of general laminated composite materials by consideration of the time-dependent behaviour of the polymer matrix together with the theoretical micromechanical material models used in materials design. The analysis methods can be divided into two broad classifications: lamina models based on the analysis of the behaviour of fibres within a viscoelastic matrix; and laminate models which are considered to be a combination of a number of homogeneous orthotropic viscoelastic laminae. Foye[7.35] used a finite element analysis to model the behaviour of a simple unidirectional lamina and then constructed a laminate model based on classical lamination theory. An alternative approach was adopted by Brinson and his co-workers.[7.36] Here the laminate analysis was based on the experimentally derived viscoelastic properties of the unidirectional lamina on which the laminate was constructed.

Difficulty arises, however, when considering the application of these viscoelastic material models to the types of laminate and types of service environments under consideration here. Generally, in the development of a realistic model, it is necessary to simplify the problem by making the assumption that the material microstructure remains intact. It is, therefore, assumed that the creep of the composite will be due solely to the viscoelastic flow of the polymer matrix. At higher-than-ambient temperatures near to the glass transition temperature (T_g) of the polymer this assumption may be valid but in the general working temperature range for construction materials (-10°C to $+40$°C) the primary mechanism of creep is more complex. On account of the high density of geometric discontinuities in this type of fibrous composite, such as fibre ends and voids, and the brittle nature of the polyester and epoxy resin systems used, it is likely that the majority of the observed creep, stress-rupture and stress-relaxation at ambient temperatures is due to the time-dependent growth of fibre/matrix debonds and resin cracks.[7.31] This damage growth mechanism of deformation often results in nonlinear viscoelastic behaviour even when the material is lightly loaded.[7.37] Clearly, with such a complex mechanism behind the observed creep response of these materials, there is necessity for the direct measurement of long-term material properties if a reliable estimate of the in-service performance is required.

7.9. Measurement techniques and procedures

The basic test techniques used in the characterization of the time-dependent behaviour of polymers and polymer-based composites are relatively straight forward procedures. Two primary test modes are normally adopted which represent the two extremes of response: the creep test, and the stress-relaxation test. The creep test is the most popular due to the inherent simplicity of dead-load testing and the relative ease of displacement measurement. Generally, the loading actions involved are either uniaxial tension or flexure.

A problem arises, however, due to the long-term nature of the necessary tests. Long-duration testing, by definition, can be expensive both in terms of time and cost. The data are normally required in order to consider the life-time performance of the structural component and therefore, in theory, it would be necessary to conduct the characterization test for that period of time. There is little difficulty involved in test durations up to a period of a few days or even weeks. However, the life-time of components used in the construction industry is often of the order of decades. It is, therefore, necessary to be able to predict, with some confidence, the long-time behaviour of a material from relatively short time test data.

The extrapolation of creep test data is desirable as it can significantly reduce the cost of the test programme as well as facilitate the long-term characterization of a material. However, such extension of data must be done on a rational basis if any confidence is to be placed on the resultant data. The simplest and most straight forward method is to predict the long-term data by fitting a mathematical equation to the available short-term test data. The extrapolation is then simply based on the assumption that the continuous creep process

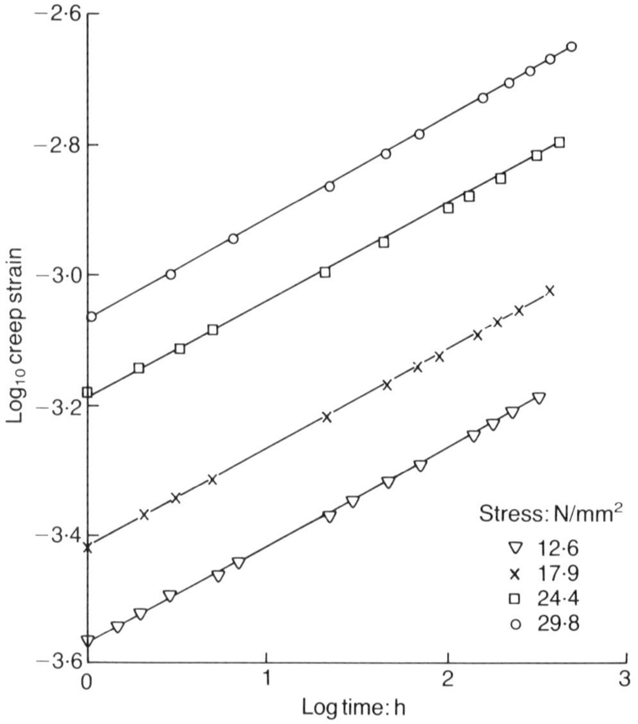

Fig. 7.7 (left). Creep data for a glass fibre chopped strand mat reinforced polyester laminate

Fig. 7.8 (below). Extrapolation procedure using TTSP

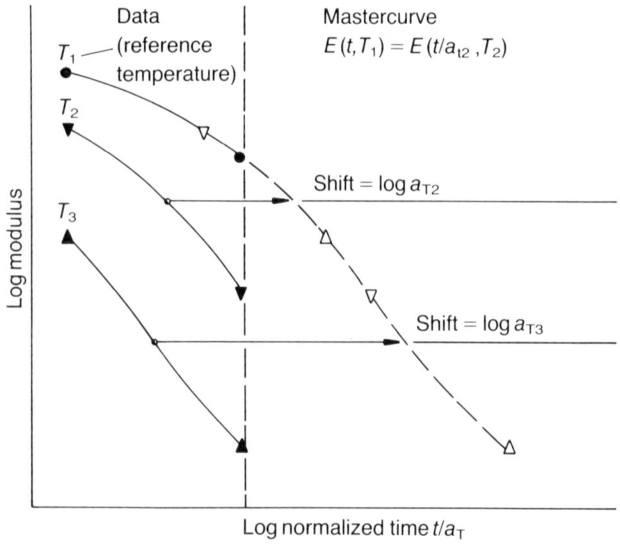

will obey the same empirical function at the extended times. It has been found that the creep of many polymers can be modelled by a power-law function, identified by the linear nature of the data when plotted on logarithmic axes. Fig. 7.7 shows creep data for a glass fibre chopped strand mat reinforced polyester laminate which illustrates approximate power-law behaviour over a large time scale. The method might be used to yield a relatively confident extrapolation of the original data up to twice the test duration. But it would be difficult to justify, with any confidence, further extrapolation on this basis.

Extrapolation of test data for a period in excess of twice the test duration requires a more rational basis for the choice of equation. Ideally the form of the equation should be dictated by the mechanism of creep deformation. However, as discussed previously, a rigorous model of the complex viscoelastic damage growth does not exist. It is clear, therefore, that any procedure must be justified on an empirical basis.

One such technique, which has been used successfully to extrapolate short time creep data for many polymer composites, utilizes the effect of increased temperature on the creep behaviour. A family of creep curves, for different test temperatures, can be modelled by a single creep equation. The effect of higher temperatures is to simply cause the creep process to accelerate. This phenomenon is known as the time temperature superposition principle (TTSP).[7.38] The effect of changing temperature is the same as applying a multiplication factor to the time scale. The test procedure involves the measurement of creep data at a range of different test temperatures but at the same stress level. The resultant creep compliance data is then plotted against a logarithmic time scale. Provided there is time–temperature correspondence, a master curve can be constructed by horizontally shifting the higher temperature curves along the log time axis, as is demonstrated in Fig. 7.8. The amount of shift necessary to achieve the superposition of the data yields the reduced time factor for each test temperature. A similar relationship has been found to exist between applied stress and time for some polymers and polymer-based composite materials.[7.39, 7.40] The time stress superposition principle (TSSP) is based on the nonlinear response of the creep compliance. If all or part of nonlinearity of the stress-dependence has the effect of accelerating the creep response, this property can be used to extrapolate the creep data at lower stress levels. The method has been used, for a glass fibre chopped strand mat reinforced polyester laminate, to enable the extrapolation of the creep data to 1 year from an original test programme duration of 2 weeks (Fig. 7.9).

A useful method of presenting both creep and stress-relaxation test data for engineering applications is the isochronous plot (Fig. 7.10). A number of isochrones are projected up from the time axis through a family of creep

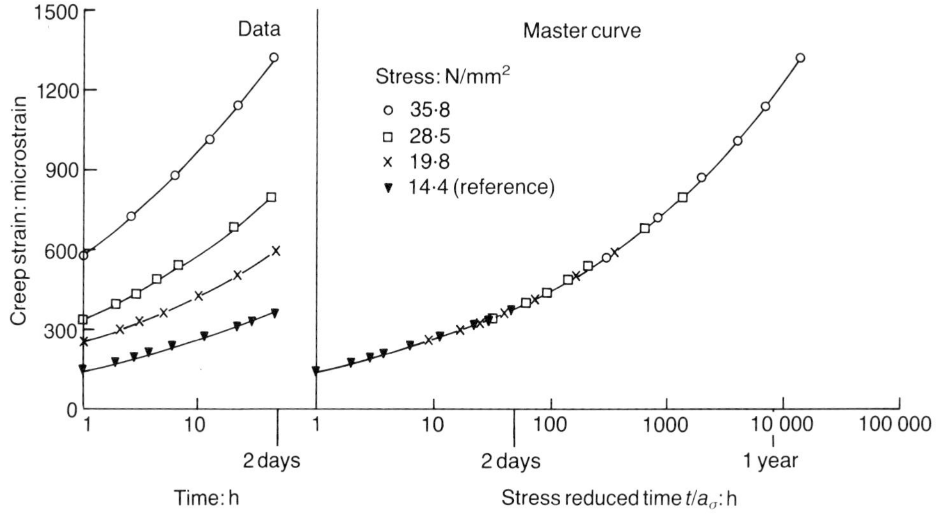

Fig. 7.9. Extrapolation of creep data using TSSP for a chopped strand mat glass fibre reinforced polyester laminate

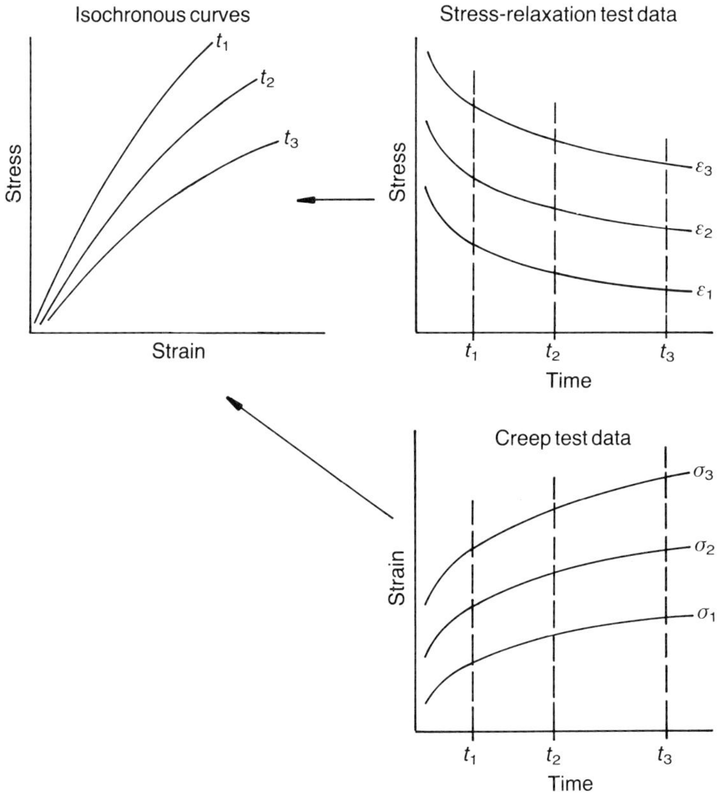

Fig. 7.10. Isochronous curves from creep or stress-relaxation test data

curves, or stress-relaxation curves, and the resultant isochronous data is plotted on stress–strain axes. The result is a family of stress–strain curves for different times of loading. This allows quick examination of the relative magnitude of the creep strain in relation to the elastic strain as a function of stress and time. Fig. 7.11 shows the isochronous stress creep strain data, plotted from creep curves, for a chopped strand mat glass fibre reinforced polyester laminate with a fibre volume fraction of 15%. It can be seen that at low applied stress the material behaves in a linear viscoelastic manner and that the magnitudes of the creep strains are relatively small even at long times of loading. However, at higher applied stress levels the behaviour can be seen to be nonlinearly viscoelastic and the relative magnitudes of the creep strains are far more significant.

Another important consideration when designing with polymers is to know to what extent the creep strain is recoverable on removal of stress and what length of time is required to achieve the creep-recovery that does take place. A useful technique of data presentation can be used which allows ease of examination of creep-recovery data and also comparison between materials. Turner[7.41] shows that creep-recovery data are more easily evaluated if normalized with respect to the maximum creep strain and plotted against a reduced time, or normalized time, scale. It was shown that provided the creep compliance, for the material under question, is of the form where the time and stress dependence are separable, as in equation (7.4), the normalized recovery data, from a number of tests conducted at different applied stress levels, should be coincident. This then indicates that the time taken to achieve creep-recovery is the same regardless of the magnitude of the creep stress. The exercise has been conducted on the creep-recovery data for a chopped strand mat glass fibre reinforced polyester laminate and is shown in Fig. 7.12. It can be seen that the data for different applied creep stress levels is coincident at

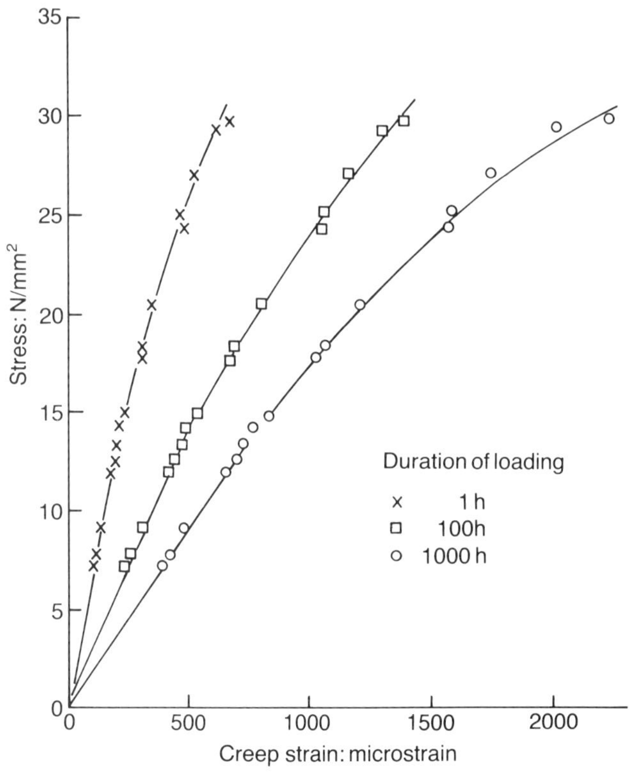

Fig. 7.11 (left). Isochronous stress/creep–strain curves for a chopped strand mat glass fibre reinforced polyester laminate generated from creep data

Fig. 7.12 (below). Normalized creep recovery data for a chopped strand mat glass fibre reinforced polyester laminate

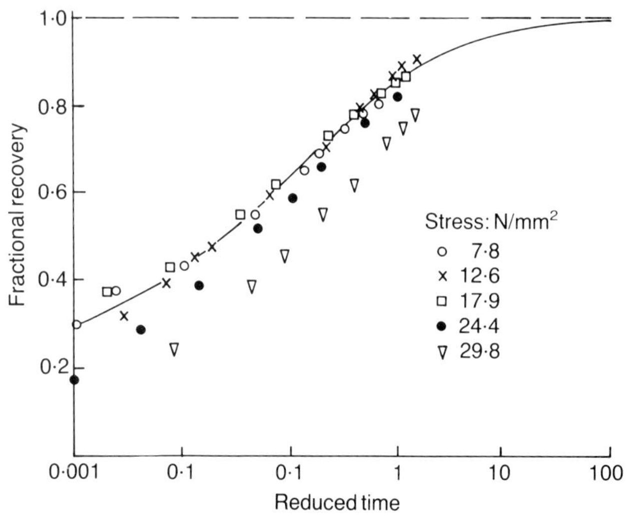

moderate to low stresses but for higher applied stresses the data indicates that recovery is delayed. In fact, it was found that at higher applied stress levels the stress and time functions of the creep compliance equation were not separable thus causing a change in character of the recovery data.[7.37]

7.10. Analysis and design

In many situations where fibre reinforced polymers are used in structural and semi-structural applications, the proportion of fibre in the loaded directions is high in order to yield sufficient element stiffness. An example of this is the use of pultruded glass fibre reinforced polyester sections for support structures in the chemical industry. In the type of structural section used, the unidirectional fibre volume fraction can be as high as 60%. Clearly, the behaviour of these predominantly uniaxially stressed members will be virtually time-independent. In this situation elastic analysis and design procedures are appropriate and are used. However, as previously discussed, it is not always possible to place a sufficient quantity of fibre in all the directions of loading, such that time-independence of behaviour can be achieved. In addition, materials are often relatively lightly reinforced and, hence, may have significantly time-dependent mechanical properties. In these situations it may be necessary to consider the time-dependent nature of the material behaviour in the analysis and design procedure.

At present there is no unified approach to the inclusion of time-dependent material properties in the analysis and design of fibre reinforced polymers. However, the pseudo-elastic method of analysis and design, which has been proposed as a method for the design of plastic panels[7.42] has emerged as the most appropriate approach to date. The method involves the substitution of isochronous material properties in place of the corresponding elastic constants in conventional elastic design equations.

Clearly, in a method of this type, there are two distinct situations which govern the choice of the type of data to be used in the substitution. If the component to be designed is a part of a structure which is statically determinate then the loads applied to that element will not be affected by the magnitude of the deformation of that element. In this case the applied load can be considered as a constant. The direct substitution of isochronous values, which have been generated from creep data, into the deflection equation will yield a true estimate of the creep response of the structure. An example of this type of structure can be seen in the stressed skin of a long cylindrical pressure vessel. Here, the predominant applied stress is the hoop tensile stress. The direct substitution of the isochronous modulus and Poisson's ratio into the elastic equation then gives an estimate of the deformation behaviour of the structure.

However, in the situation where a component is subjected to a constant applied displacement, such as in a structure which can be described as being kinematically determinate, the displacement of the element is unaffected by any change in the load carried. In this case, the analysis of the structure should be based on the substitution of isochronous data generated from stress-relaxation test data, as the element strain will remain constant. This situation may arise in the analysis of bolted structures fabricated from viscoelastic materials. The through-the-thickness properties of fibre reinforced laminates can be highly time-dependent. In a bolted plate, significant bolt strains may be applied to the composite material. In such a situation the applied strains will remain relatively constant but the viscoelastic behaviour of the polymer material will cause stress relaxation to take place.

It is clear, therefore, that the appropriate test data should be used for each case being considered. However, there are many situations where a structure will neither behave in a statically or kinematically determinate manner. In these situations, where the stress carried by the component is both a function of the geometry of the structural configuration and the stress–strain–time characteristics of all the elements of the structure, the material may exhibit the combined behaviour of both creep and stress-relaxation. In fact this form of behaviour might be best described as being creep-induced stress relaxation. Studies have shown that this form of behaviour can be modelled adequately by the use of an incremental or time-stepping numerical analysis procedure.[7.43] However, provided that the magnitude of the viscoelastic response of the material can be considered as relatively small compared to the elastic response, adequate estimates of both deformation and stress-relaxation, or stress-intensification, can be made using the pseudo-elastic method.[7.44] In fact, in many cases an estimate of the long-term stress and maximum deformation is all that is required to enable an adequate and yet economic design to be achieved.

This section has attempted to address some of the major factors which influence the time-dependent effects of polymers and polymer-based composite materials when used to fabricate engineering components. At present, the use of reinforced polymers in the construction industry is hindered by the lack of confidence in the long-term durability of these lightweight materials. However, the limited number of long-term investigations into the problem of durability has shown that, provided adequate care is taken in the material design, fabrication and finishing, and the appropriate material testing and long-term structural analysis and design methods are employed, fibre reinforced polymer composite materials can exhibit excellent durability characteristics.

References

7.1. American Society of Civil Engineers. *Structural plastics design manual.* ASCE Report No. 63, 1970. American Society of Civil Engineers, New York, 1970.

7.2. Wright W.W. *Composites*, 1981, **12**, 201.

7.3. British Standards Institution. *E glass fibre chopped strand mat for the reinforcement of polyester resin systems.* BS3496: 1973. British Standards Institution, London, 1973.

7.4. Tipping G. Water resistant systems for the marine industry. *Proc. British Plastics Fed. Reinforced Plastics Congr., Nottingham*, 1986, Paper 32.

7.5. Holmes M. and Just D.J. *GRP in structural engineering*, 227. Applied Science Publishers, London, 1983.

7.6. Building Research Establishment. *Internal records.* Building Research Establishment, Garston, 1963.

7.7. Whiteley P. and Gardiner D. *Painting plastics.* Information paper IP11/79. Building Research Establishment, Garston, 1979.

7.8. Oliver P.C. Chemical resistance. In: *Glass reinforced plastics*, Parkyn B. (ed.), chapter 17. Butterworth, London, 1970.

7.9. Building Research Establishment. *Durability and application of plastics.* BRE Digest 69. Building Research Establishment, Garston, 1977.

7.10. Ministry of Defence and British Plastics Federation. *Weathering of glass reinforced polyester and epoxy resin systems.* Ministry of Defence/British Plastics Federation, PERME, Waltham Abbey, 1982.

7.11. Scholz D. *Kunststoffe*, 1978, **68**, 556.

7.12. Davis A. and Sims D. Weathering of polymers. In: *Composites*, chapter 12. Applied Science, London, 1983.

7.13. Head P.R. and Templeman R.B. The application of limit state design principles to fibre reinforced plastics. *Proc. British Plastics Fed. Reinforced Plastics Congr., Nottingham, 1986.*

7.14. Barrett D. and Steel D.J. *MEXE Technical Note 4/67.* Ministry of Defence (Army Dept.), Christchurch, 1967.

7.15. Norris J.F. *et al. Composites*, 1976, **7**, 165.

7.16. Forrest J.W. and Crowder J.R. The weathering of plastics and rubbers. *Proc. PRI Symp., London*, 1976, Paper E7.

7.17. Scott K.W. and Matthan J. Weathering. In: *Glass reinforced plastics*, Parkyn B. (ed.), chapter 16. Butterworth, London, 1970.

7.18. British Standards Institution. *Glass fibre rovings for reinforcement of polyester and epoxy resin systems*. BS3691: 1969. British Standards Institution, London, 1969.

7.19. Probert C. *Surface temperatures of glass reinforced plastics panels*. BRE Internal Note. Building Research Establishment, Garston, 1976.

7.20. National Building Specification Services. *GRP cladding features*. National Building Specification, Part G61. National Building Specification Services, London, 1983.

7.21. Parvin K. Specifications. In: *Glass reinforced plastics*, Parkyn B. (ed.), chapter 11. Butterworth, London, 1970.

7.22. British Plastics Federation, Reinforced Plastics Group. *Guidance notes on the construction of glass-reinforced plastics cladding panels*. British Plastics Federation, London, 1981.

7.23. American Society for Testing and Materials. *Glass fibre reinforced polyester plastic panels*. ASTM Specification D3841-86. American Society for Testing Materials, Philadelphia, 1986.

7.24. Hollaway L. *Glass reinforced plastics in construction*. Surrey University Press, Glasgow, 1978.

7.25. British Standards Institution. *Design of non-loadbearing external vertical enclosures of buildings*. BS8200: 1985. British Standards Institution, London, 1985.

7.26. British Standards Institution. *Unsaturated polyester resin systems for low-pressure fibre reinforced plastics*. BS3532: 1962. British Standards Institution, London, 1962.

7.27. British Standards Institution. *Epoxide resin systems for glass fibre reinforced plastics*. BS3534, now withdrawn. British Standards Institution, London.

7.28. Parkyn B. *Brit. Plast.* 1959, **32**, No. 1, 29.

7.29. Malhotra H.L. and Crowder J.R. *Weathering properties of fire-retardant plastics rooflights*. Fire Note No. 11. HMSO, London, 1969.

7.30. British Standards Institution. *Guide to quality control requirements for reinforced plastics mouldings*. BS4549, part 1: 1970. British Standards Institution, London, 1970.

7.31. Johnson A.F. *Engineering design properties of GRP*. British Plastics Federation, London, 1979.

7.32. Ferry J.D. *Viscoelastic properties of polymers* (2 edn). Wiley, New York, 1980.

7.33. Lockett F.J. Assessment of linearity and characterisation of nonlinear behaviour. Mechanics of viscoelastic media and bodies. In: *Proc. IUTAM Symp., Gottenburg*, 1974, 1–25.

7.34. Schapery R.A. On the characterization of nonlinear viscoelastic materials. *Polymer engineering science*, 1969, **9**, July, 295–310.

7.35. Foye R.L. Creep analysis of laminates. *Composites reliability*. ASTM STP580, 381–395. American Society for Testing Materials, Las Vegas, 1975.

7.36. Tuttle M.E. and Brinson H.F. Prediction of the long-term creep compliance of general composite laminates. *Experimental Mechanics*, 1986, **26**, 1, 89–102.

7.37. Howard C.M. and Hollaway L. The characterisation of the nonlinear viscoelastic properties of a randomly orientated fibre/matrix composite. *Composites*, 1987, **18**, 4, 317–323.

7.38. Aklonis J.J. and MacKnight W.J. *Introduction to polymer viscoelasticity* (2 edn), 36–56. Wiley, New York, 1983.

7.39. Cessna Jnr L.C. Stress-time superposition for creep data for polypropylene and coupled glass reinforced polypropylene. *Polymer Engineering Science*, 1971, **13**, May, 211–219.

7.40. Franceschini A. *et al.* On feasibility of accelerated creep measurements in some polymeric materials. *Proc. Joint US/Italy Symp. on Composite Materials, Capri*, 1981.

7.41. Turner S. Creep in glassy polymers. In *Physics of glassy polymers* (Haward R.H. ed.), 223–279. Applied Science, London, 1973.
7.42. Johnson A.F. and Sims G. *Design procedures for plastics panels*. National Physical Laboratory, Teddington, 1987.
7.43. Howard C.M. Creep and stress-relaxation in structural composites. In: *Proc. Int. Conf. on Composite Materials, Milan*, 1988, **1**, 63–71.
7.44. Lockett F.J. and Cousins R.R. Validity of the pseudo-elastic method of stress and deformation analysis for plastics. *Plastics and Rubber: Materials and Applications*, 1977, November, 145–151.

8 Fire properties of polymers and composites

D. B. S. Berry

8.1. Introduction

Fire is a complex subject. One of the reasons is that, in the context that is being considered, it is not a deliberate occurrence but the result of a number of unfortunate coincidences. No two fires are exactly the same. The materials involved are different, the essential oxygen supply to support the fire is different. Even the positioning of the flammable materials with respect to each other can have a drastic effect on the rate of progress of a fire and its intensity. There is, of course, at the same time, a fear of and a fascination with fire. Onlookers will stand and watch a fire—oblivious of the danger they are in, young children are drawn to a fire or flame until they finally burn themselves.

When considering fire in buildings the designer has to realize the huge variety of situations which can occur. He must take account of the fact that the use to which the building is put, and hence its contents, will have a more significant effect on the fire risk than the materials of construction.

An added complication is that the use of flammable materials in buildings is governed by regulations and statute. These regulations are designed to make the occurrence of a fire less likely and also to limit the danger posed by a fire if and when it occurs. They can not, however, cover every eventuality but will lay down general guidelines. These guidelines will tend to limit or prohibit the use of flammable materials and try to specify safe ways of using materials with some flammability. The ultimate in safety would be to specify the use of materials which did not burn at all. However, many materials which are considered non-combustible will, in fact, burn under favourable conditions. Steel wool is an example; if a flame is applied then it will rapidly oxidize and flame. However, if regulations limited materials to those considered to be non-combustible then everyone's lives would be far less comfortable and convenient than they are now. For instance, there would be no padded furniture and most decorative surfaces would be banned. Therefore, when fire regulations are complied with there is still some risk. The regulations only seek to minimize the risk.

The designer's responsibilities go further than complying with regulations for regulations only consider health and safety. The designer must also consider the financial consequences of a fire. A good design or application will not only consider ways of reducing the danger of death or injury but will also consider limiting the spread and scale of damage to property if a fire occurs.

8.2. Plastics and fire

Polymeric materials are organic in that they are principally formed from carbon, hydrogen and nitrogen; hence are all flammable to various degrees. In spite of this they have found an increasing market in the building and construction industry but their penetration into this market has been slowed by their relatively poor fire performance. The plastics industry has, therefore, expended considerable efforts in trying to improve the fire properties of polymers and reduce their flammability (mainly, some feel, with little real effect). Such improvements as have been achieved are marginal and are concerned almost exclusively with reducing only slightly the probability of ignition if a flame is applied. They have not appreciated that ignition is only one very small part of a fire incident and have been perhaps unable to improve the materials' performance when a fire has started. The building industry, on the other hand, has concentrated on using plastics materials in a way which limits the possible dangers of their inherent flammability. Two examples of this are PVC (polyvinyl chloride) gutters and polythene pipes. PVC is by its chemical make-up one of the less flammable plastics, because it includes chlorine in its chemical composition. In addition to this its position on the eaves of a house means that its potential for causing harm is minimal. With plastic water pipes the contents of the pipe assists in preventing ignition of the pipe, plastic pipes normally melt releasing the pipe's contents; in melting they normally sag and are then much more difficult to burn.

Plastic foams—like expanded polystyrene—have also found applications. Although these tend to burn well because of their open structure—despite adding chemical fire retardants—they have been widely used in the building industry because of their very desirable thermal insulation properties. They are, however, always enclosed to protect them from flames, for instance within the cavity of a wall or covered by aluminium foil to make insulated wall linings for commercial buildings.

One success which has been achieved by polymer producers is the development of glass-reinforced plastics (GRP) with adequate fire properties to satisfy regulations and requirements. These materials have been used and make significant and impressive applications, particularly in the external cladding of a number of prestige buildings. Among the earliest of these was the Mondial House international telephone exchange on the banks of the River Thames (Fig. 5.4).

In order to design a panel or component in reinforced plastics for successful application, the designer must be familiar with three aspects, namely

(*a*) the building regulations
(*b*) the tests to fulfil the requirements of the building regulations
(*c*) how these tests are applicable to the various stages in the development of the fire and the test objective.

8.3. The UK Building Regulations

Since 1975, the UK Building Regulations[8.1] have been couched in much more general and bold terms than previously. The Regulations set out the "Requirements" and in Approved Documents explain and suggest some ways of meeting these Requirements. The Requirements cover internal fire spread (surfaces), internal fire spread (structure) and external fire spread. The regulation is printed in full in Fig. 8.1.

The internal fire spread (surfaces) requirement calls upon the designer to make sure that flames do not spread over the surfaces of walls and ceilings and when they are ignited they do not give off sufficient heat to fuel the fire significantly.

The internal fire spread (structure) requirement lays down that the stability will be maintained for a reasonable period. That means collapse will not occur too quickly. It also calls for the structure to be divided into compartments by fire-resisting walls so as to limit the spread of fire once it has started.

The external fire spread requirement states that the roof and walls shall offer an adequate resistance to the spread of fire. Usually this concerns spread of flame across the surface.

These requirements state the bold objectives of the Building Regulations but the actual interpretation of "adequate" and "reasonable" are set out in

Internal fire spread (surfaces)
B2. In order to inhibit the spread of fire within the building, surfaces of materials used on walls and ceilings—

(*a*) shall offer adequate resistance to the spread of flame over their surfaces; and (*b*) shall have, if ignited, a rate of heat release which is reasonable in the circumstances.

Internal fire spread (structure)
B3. — (1) The building shall be so constructed that, in the event of fire, its stability will be maintained for a reasonable period.

(2) The building, or the building as extended, shall be sub-divided into compartments where this is necessary to inhibit the spread of fire within the building.

(3) Concealed spaces in the structure or fabric of the building, or the building as extended, shall be sealed and sub-divided where this is necessary to inhibit the unseen spread of fire and smoke.

(4) A wall common to two or more buildings shall offer adequate resistance to the spread of fire and smoke.

(5) For the purposes of sub-paragraph (4) a house in a terrace and a semi-detached house are each to be treated as being a separate building.

External fire spread
B4. — (1) The external walls of the building shall offer adequate resistance to the spread of fire over the walls and from one building to another, having regard to the height, use and position of the building.

(2) The roof of the building shall offer adequate resistance to the spread of fire over the roof and from one building to another, having regard to the use and position of the building.

Fig. 8.1. Building Regulations—the Requirements

the accompanying *Approved Document B—Fire Spread.*[8.2] This suggests a number of ways in which the objectives of the regulations can be achieved but from the point of view of the designer in plastics materials his product is likely to have to undergo specified tests and to reach certain criteria for compliance. The tests are principally those set out in BS 476.[8.3] Some plastics manufacturers believe that the tests were designed to prevent the use of plastics materials in buildings. This is not the case. Some of the tests actually predate the commercial use of plastics materials. The tests are designed to simulate as closely as possible in the laboratory some of the conditions which can occur during a fire. In the description of the tests given below those conditions will be described as an introduction to the test procedure.

8.4. Fire tests

These requirements are usually complied with by compliance with the various parts of BS 476, *Fire tests of building materials and structures.*[8.3] There are several parts to the standard, each dealing with a different aspect of resistance and reaction to fire. For the sake of completeness each part of BS 476 will be discussed briefly to give the outline of the type of properties which it examines.

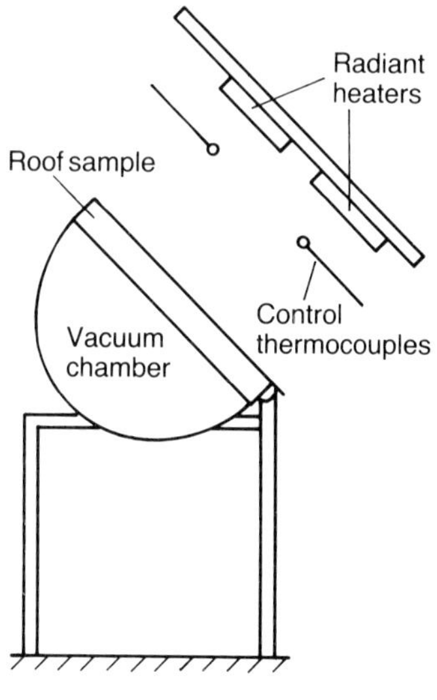

Fig. 8.2. Roof test apparatus

BS 476. part 3: *External fire exposure roof test*[8.4]

This test is to determine the performance of a roof and is meant to simulate the effects on a roof of a nearby building catching fire. This will, of course, heat the roof by irradiation and could either cause it spontaneously to catch fire or to give off flammable gases. In addition to this, sparks or flaming brands could be deposited on the roof as a source of ignition.

This situation is simulated by irradiating the roof sample approximately 1 m square by radiant gas heated panels (Fig. 8.2). These are set parallel to the roof at a distance of approximately 600 mm. During the test a small gas flame is applied to the surface of the sample in order to simulate flaming brands and ignite any flammable gas which is being given off. Roofs can be tested either horizontal (F) or sloping (S) and this determines the first letter of the classification. The second letter is determined by the time when flame penetration of the sample occurs and the third letter is determined by the distance flames spread across the surface of the specimens.

BS 476. part 4: *Non-combustibility test for materials*[8.5]

This specification includes a definition of non-combustibility and determines whether materials with or without coatings used in construction meet

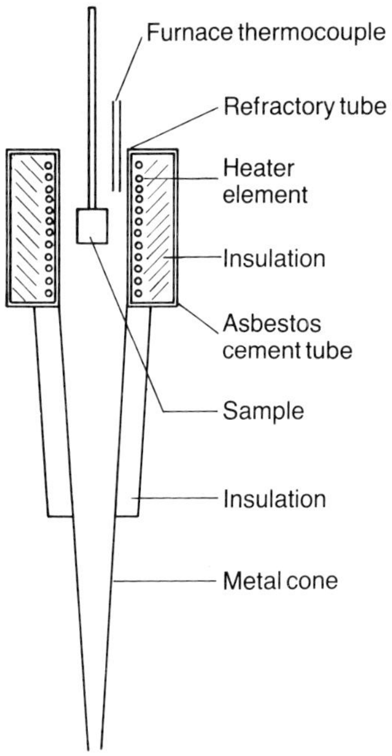

Fig. 8.3. Non-combustibility test

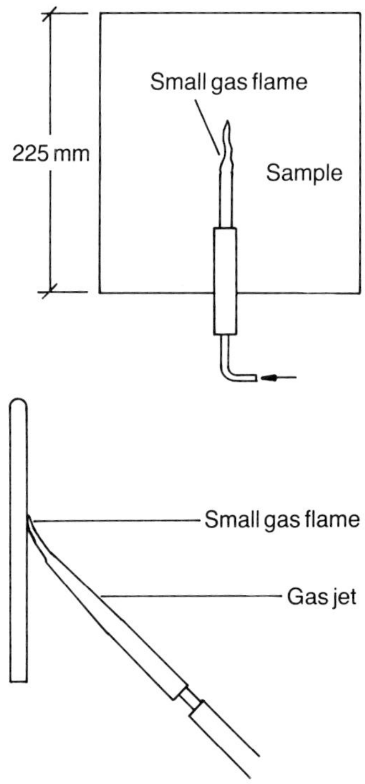

Fig. 8.4. Ignitability test

the definition. A cuboid sample with a thermocouple in its centre is lowered into a furnace at 750°C. The sample shall not flame for more than 10 s or increase the temperature of the furnace or sample thermocouple above the initial furnace temperature by more than 50°C. In practice, a sample with more than 4% organic content is very unlikely to meet these conditions and so a "non-combustible" commercially viable plastic or plastic composite is unknown at the time of writing (Fig. 8.3).

BS 476. part 5: *Method of test for ignitability*[8.6]

This test is solely meant to determine how easily a material will ignite when subjected to a small flame. In the test the flame is approximately 10 mm long and is applied to the middle of a vertical specimen. The fire performance of a material cannot be assessed on this test alone. The practice of polymer manufacturers proudly quoting the result of this test (resistance to ignition by small flaming, thereby implying that this defines adequate fire performance) is to be depreciated. This test refers only to the very start of a fire and not to the material performance when the fire catches hold.

The test rig is shown in Fig. 8.4. It consists of a gas jet which applies a controlled flame approximately 10 mm long to the centre of a vertically held sample sheet 225 mm square, with the flame impingeing at an angle of 45°.

BS 476. part 6: *Method of test for fire propagation for products*[8.7]

This test takes account of the combined effects of ignition characteristics, the rate of release of heat and the thermal properties of the products. It is specifically aimed at determining the likelihood of a material to accelerate or contribute to the increasing intensity of a fire. It measures the heat which a sample gives out when subjected to radiant heat and gas jets from electric heaters with a row of gas jets. The sample is 228 mm square and is one wall of the fire propagation box, as can be seen in Fig. 8.5. The gas jets are ignited at the start of the test and the electric heaters are switched on at 2 min 45 s at 1·8 kW and then altered to 1·5 kW at 5 min. The test lasts 20 min. The progress of the test is monitored by the temperature indicated by thermocouples set in the chimney. Two tests are carried out, one with a non-combustible sample in place and the second with the test sample. Typical time-versus-temperature graphs are shown in Fig. 8.6. For a good result the area between the calibration graph and the sample graph should be small. In fact, two indices are calculated, one for the first 3 min and the second for the whole test. These are respectively i_1 which should be not greater than 6 and I which should be not greater than 12.[8.2] In practice if the sample burns to any significant extent then heat will be generated and the sample will not comply with the requirements.

BS 476. part 7: *Surface spread of flame test for materials*

This test determines the tendency of materials to support spread of flame across their surfaces. It was originally developed to assess whether the spread of flame along the walls of a corridor with a fire at one end was faster than a person could run to escape. It consists of a 1 m^2 radiant panel which gives out approximately 300 kW. A sample which is 900 mm long by 225 mm high is placed with its short edge approximately 225 mm from the face of the radiant panel, at right angles to it. A small flame approximately 75 mm to 100 mm long plays on the end of the panel nearest the furnace for 1 min. The progress of any flame along the sample is monitored and for a Class 1 result flame must not travel more than 165 mm within the 100 min test duration. A Class 1 result is the best and results up to Class 3 are permitted. These limits can be seen in Table 8.1.

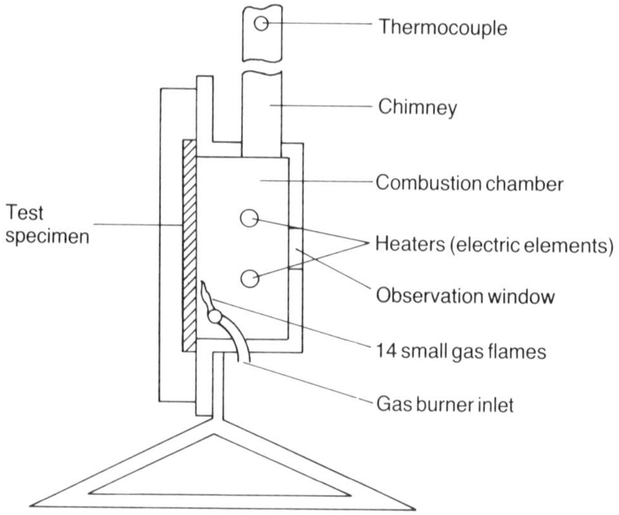

Fig. 8.5. Fire propagation furnace

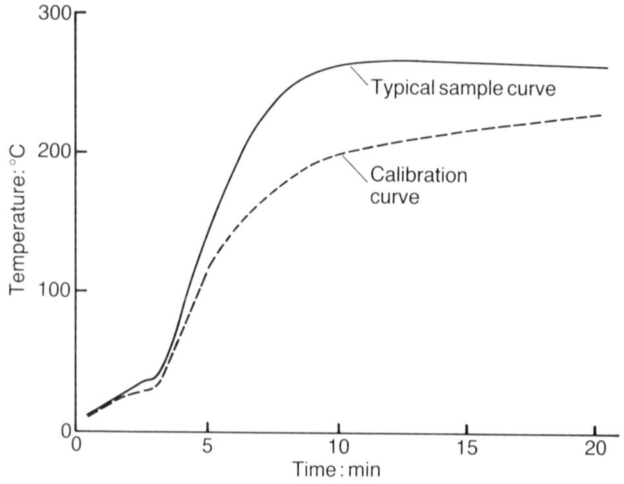

Fig. 8.6. Fire propagation test curves

BS 476. part 8: *Test methods and criteria for the fire resistance of elements of building construction*

The tests described previously are called reaction to fire tests and are designed to show how a material reacts when subjected to fire conditions, i.e. does it catch fire, does it spread flame or does it give off heat to feed the fire? The fire resistance test, on the other hand, is designed to show how a construction will hold back fire. On one side of the large sample, the test both simulates the temperature rise that would occur in a normal fire and gauges the time for which the floor or partition would hold back the fire before it is penetrated or collapses. It is very unusual for a material on its own to be subjected to this test. Usually the material has to be built into a wall or floor with other materials in order to be suitable for the test. The sample is 3 m square and forms the outside wall of a large furnace. The temperature inside the furnace is raised to follow the curve shown in Fig. 8.7. During the test the temperature of the face not exposed to heat is recorded by means of fixed and roving thermocouples. Any gaps that appear are measured. Hot gas and flame penetration or collapse of the sample determine the time to failure, as does an average increase in temperature of 140°C or any individual thermocouple indicating a temperature increase greater than 180°C. The reason for the temperature requirement is to make sure that combustible materials on the

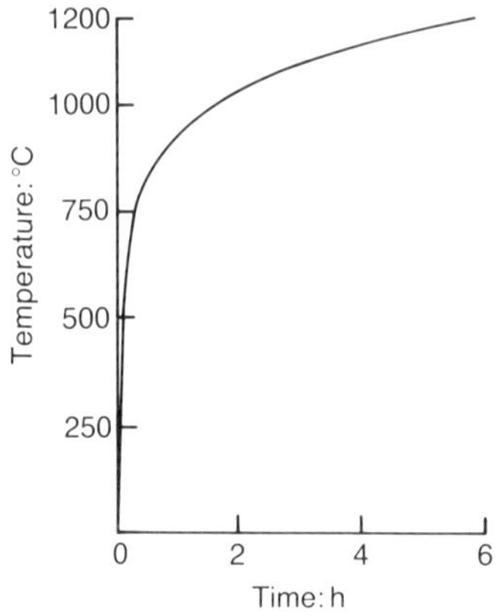

Fig. 8.7. Fire resistance curve

Table 8.1. Flame spread classification

Classification*	Flame spread at 1·5 min		Final flame spread	
	Limit: mm	Tolerance for one specimen in sample: mm	Limit: mm	Tolerance for one specimen in sample: mm
Class 1	165	25	165	25
Class 2	215	25	455	45
Class 3	265	25	710	75
Class 4		Exceeding Class 3 Limits		

* The flame spread on any specimen of the sample shall not exceed the limit assigned for the class with the proviso that for one specimen only in the sample the flame may exceed this limit by the tolerance shown.

side of the wall or floor not exposed to the fire will not catch fire and so effectively breach the fire break. Results are given to the nearest minute and structures are usually classified in terms of time to failure designated to the nearest 30 minutes. Thus it is possible to have a half-hour partition or a half-hour panel, a one hour panel, etc.

8.5. Test results and regulations

The UK Building Regulations take the test results and specify what grade of fire performance is necessary for a particular application, specifying, for instance, a half-hour fire barrier or fire door, or where a material with a Class 1 surface spread of flame classification shall be used. In some applications a non-combustible material complying with BS 476: part 4[8.5] will be required. As an alternative to this, a material with a Class 0 surface can be used. A Class 0 classification is peculiar to Building Regulations, as it is not a British Standard classification. It designates a material with a Class 1 surface spread of flame classification combined with indices of performance in the fire propagation test of i_1 not greater than 6 and of I not greater than 12. Class 0 is the rating which designers should be aiming for in designing panels or products for use in the building and construction industry.

8.6. The development of a fire

As indicated earlier in this chapter, the different flammability properties of materials are important at different stages in a fire's development. A fire progresses through five stages and the different BS 476 tests apply to the different stages. The stages are conveniently defined as: ignition; development; spread; full burning; decay.

In the ignition stage a small flame or hot radiant material comes into contact with a flammable material and causes it to ignite. The BS 476: part 3 ignitability[8.4] test obviously refers to this ignition stage, where the fire starts by the material catching fire then the fire develops.

As the intensity of the fire increases the rate of increase depends, among other things, on the rate of temperature rise. This in turn depends on the amount of heat given out by the materials which are burning in this stage. It is the fire propagation test indices which define the performance of a particular material. In the spread stage the area of the fire can increase by flame spreading along burning surfaces; the surface spread of flame test, which measures the rate of progress of a flame along a wall or ceiling or other cladding is the applicable test. The fire can also be very quickly spread by a flashover. This is where a material some way from the fire has been heated to give off flammable gases and these spread towards the flames and ignite. It can also be where materials reach their flash point temperature and spontaneously ignite. A spectacular case of flashover was seen by television viewers watching the disastrous Bradford football ground fire (11 May 1985), when a man's hair and the back of his coat caught fire even though he was 100 m from the flames. There is no test for flashover despite intense activity to develop one.

In the fire propagation test, the surface spread of flame test and the non-combustibility test, the samples are subjected to intense radiation from a radiant source. This is entirely in harmony with real fire conditions where radiation and temperatures are very high. The fire performance of materials at these high temperatures is far different from their performance at room temperature: they are much more likely to ignite. Therefore, if a fire test is to be really informative, it is essential that it mirrors all these conditions. This is why the spectacular tests of blow lamps playing on materials without ignition are eschewed by fire professionals. It also explains why the oxygen index test is falling out of favour.

8.7. Fire tests and GRP

Glass-reinforced polyester is commonly used in composites for buildings; these can be compounded relatively easily by the manufacturers, either by the use of heavy loadings with fillers or by the addition of flame retardants, to produce a Class 0 resin. Manufacturers sell Class 1 or Class 0 resins, but it must be understood that this classification relies on testing a composite panel with a specific glass/resin ratio and a specific form of glass in the make-up. If, in the design of a panel, a designer decides to deviate significantly from the tested laminate construction, then he could well discover that the performance of his product does not meet the Class 1 or the Class 0 classification. Therefore, a close inspection of the test certificate should be made.

It is, of course, common to carry out fire tests on samples taken from actual production. The fire performance can be severely reduced if the sample has not been properly cured. It is, therefore, preferable to cure the sample well and not to test for at least 14 days after the sample manufacture. Green laminates can give anomalously bad results.

8.7.1. Weathering

One drawback with the Class 1 and Class 0 resins is that their weathering characteristics are usually poor. They are especially sensitive to shortcomings in the curing procedure and if not properly cured, will be liable to chalking within a few months. In order to obviate these problems many designers use a gelcoat or flowcoat of a material with a lesser fire performance, either a Class 2 gelcoat resin or an acrylic or other type of surface coating which may not be Class 1. Because the gelcoat is so thin its effect is minimal and the finished panel can still retain its Class 1 surface spread of flame classification.

Composites can also be used to produce panels which will achieve 0·5 h fire resistance but in this case the GRP is only used as the surface. The fire resistance and the thermal-insulation properties are provided by a glass fibre or mineral fibre core between the GRP faces. This technique was successfully used in the walls of the walkway at Terminal 2, Heathrow Airport, which had a Class 0 surface and a 0·5 h fire resistance when the outside face was tested.

8.7.2. Improving the fire properties of resins

The method used by manufacturers to improve the fire performance of polyester resins is by mixing additives into the resins: for instance, in the addition of borax, which froths up during the fire, forming an intumescent coating. Alternatively halogenated waxes are added, which inhibit flaming. These are almost always used with a synergist, for instance antimony oxide. Antimony trihydrite is often used as a heat-absorbing additive because this undergoes an endothermic reaction in a fire, cooling the flame. Other ways of flame-retarding the polyester resins are adding mineral fillers or modifying the chemistry, forming a het-acid resin.

Resin manufacturers are well versed in these modifications and are only too willing to supply resins containing a suitable fire retardant system.

8.7.3. Tests to simulate service conditions

The shapes of composites for buildings, their applications and attitudes in use and their combination with other materials, sometimes raises doubts as to the fire performance of a particular installation or product.

Sometimes a client will be unaware of the various types of plastics materials and needs to have the fire properties of GRP demonstrated. In these cases *ad hoc* tests which simulate as closely as possible the conditions of use will be useful. Many tests have been performed on cladding panels where a 12 kW gas flame (equivalent to 60 central heating boilers) is applied to the bottom of the panel and the progress of the flaming combustion up the cladding pane is noted. Usually in cases such as this the flame travels approximately 1·5 m and stops. This is true even when a non-fire-retarded gelcoat has been applied over a fire-retarded GRP core.

In a test carried out by a British local authority, three small huts were constructed with three different roofs. The first had a polymethyl methacrylate (PMM) roof, the second a glass-reinforced plastics roof and the third a standard close-board roof with bitumen felt over. Identical fires were started in each hut. The PMM roof lasted just 3 min, the bitumen felt roof 15 min or so and the GRP roof 30 min without collapse.

8.7.4. Designing GRP products for fire performance

When designing GRP products for fire performance, the designer should decide what properties are required in order to comply with regulations. If this means that a Class 0 or Class 1 performance is required then he will need to specify a suitable resin and gelcoat. The resin will have to be a Class 1 resin but the gelcoat does not have to be. The designer should specify a range of gelcoat thicknesses that will not adversely affect the fire properties. If the material is to be bonded to a substrate then it must be tested on that substrate as the change in the temperature produced by the substrate could adversely affect the fire performance. The designer will also need to specify the glass/resin ratio, the form of glass reinforcement and the laminate thickness. Finally a thorough cure cycle should be demanded.

When designing for a fire resistance rating, the engineering aspects of the construction assume great importance. A laminate with good fire properties should be specified for both faces. The infill between the faces will have to have a thermal conductivity such that the maximum temperature on the face not exposed to the fire will be lower than the specified maximum for the period required. It will also be necessary to maintain stability throughout the fire tests and one essential requirement is that the faces are tied together. Often the panels will need to be stiffened by added metal members and fixings.

8.8. Unreinforced polymers and fire

Unreinforced thermoplastics and thermosetting resins have also been used extensively in the building and construction industry. These polymers do not have the same standard of fire properties as the glass-reinforced plastics. Being organic, they tend to ignite easily and to melt at relatively low temperatures. This means that they cannot be classified under the normal tests of BS 476. The exceptions to this general statement are polycarbonate and PVC, which can be classified by the surface spread of flame test, although certain difficulties are encountered with polycarbonates when they melt in the fire propagation test.

Unreinforced polymers are, therefore, subjected to different tests in order to enable them to be used in construction and the products in which they are used are severely restricted. They are extensively used in applications where their fire properties are not important or in small quantities where they do not pose a hazard. In spite of this, PVC corrugted roof sheet has proved popular and has gained a wide acceptance for use as roofing sheets in lean-to's or car ports and special provisions have been made in the Building Regulations to cover this. The material does, however, need to be tested in accordance with BS 2782: method 508A[8.10] which is a small flame ignition test similar to BS 476: part 5.[8.6] Tests to simulate service conditions have shown that in a fire the sheet tends to melt and fall away without spreading the fire or contributing significantly to its development. Plastics foams are widely used with a small amount of fire retardancy. These are always enclosed by masonry or some other fire barrier to prevent their ignition.

8.9. Smoke and toxicity

One aspect of the fire properties of plastics which has not hitherto been controlled by regulation is the generation of smoke and toxic gases when materials burn. The view was generally taken that most deaths were as a result of burning or oxygen depletion and production of carbon monoxide and carbon dioxide during combustion. There are, however, signs that this will assume increasing importance in the future, for the subject has already been raised in the current round of discussions to amend the UK Building Regulations.

One of the reasons why these properties have not been regulated may be that there are no widely accepted tests. Tests are, however, being developed with which to ascertain the generation of smoke and toxic gases and while these can no doubt be improved, recent fires which have occurred in Britain are generating pressure from the public for a change. The main difficulty is that fire conditions can vary over a very wide range of temperature and oxygen availability. This naturally affects which toxic gases are generated and how much smoke is produced. Therefore, a simple small-scale test is often not fully representative of the wide range of toxicity which can be produced.

Two tests are commonly being used. These have been developed by the Ministry of Defence and are Naval Engineering Standard 711 and NES 713. This latter is entitled *The determination of the toxicity index of products of combustion from small specimens of material.* In the test, the specimen is burnt by a specially adapted bunsen burner flame with an external oxygen source in an airtight 1 m^3 box. The atmosphere produced by this combustion is analysed subsequently for 14 different gases. They are: carbon dioxide (CO_2); carbon monoxide (CO); formaldehyde (HCHO); nitrogen oxides ($NO + NO_2$); hydrogen cyanide (HCN); acrylonitrile (CH_2CHCN); phosgene ($COCl_2$); sulphur dioxide (SO_2); hydrogen sulphide (H_2S); hydrogen chloride (HCl); ammonia (NH_3); hydrogen fluoride (HF); hydrogen bromide (HBr); phenol (C_6H_5OH). This is not intended as a complete list of all possible gases that can be found in products of combustion, but it does represent those most commonly produced in quantity, on which toxicity data can therefore be based.

The most interesting aspect of this specification is the way in which the results are calculated. A toxicity index is determined which is the sum of the toxicity factors of the 14 selected gases. The toxicity factor for each gas is derived from the calculated quantity of each gas that would be produced when 100 g of material is burnt in air in a volume of 1 m^3. The resulting concen-

Table 8.2. The concentration of gas in parts per million considered fatal to humans for a 30 minute exposure time

Gas	Concentration: ppm
Carbon dioxide	100 000
Carbon monoxide	4 000
Hydrogen sulphide	750
Ammonia	750
Formaldehyde	500
Hydrogen chloride	500
Acrylonitrile	400
Sulphur dioxide	400
Nitrogen oxides	250
Phenol	250
Hydrogen cyanide	150
Hydrogen bromide	150
Hydrogen fluoride	100
Phosgene	25

tration is expressed as a factor of the concentration fatal to a man after 30 min exposure. An index of 1 for a given volume would, on average, bring about death in 30 min. Thus the toxicity factor is weighted against gases which are highly toxic, as is shown by Table 8.2. The concentrations of each gas are divided by the number in the Table. Relatively innocuous gases like carbon dioxide are divided by a large number and only have a small contribution to the total toxicity, while more poisonous gases like phosgene are divided by a small factor and thus more significantly affect the final toxicity index. Table 8.2 shows that phosgene gas is the most toxic and carbon dioxide is the least.

8.9.1. Smoke

NES 711 is entitled *Determination of the smoke index of the products of combustion from small specimens of materials.* A 75 mm square specimen is supported vertically and exposed to 2·5 W/cm^2 of irradiance from an electrical heater and then a row of gas flames along its bottom edge. The smoke generated collects in the approximately 900 mm × 900 mm × 600 mm combustion chamber and the smoke density is monitored by a vertical light beam. A mixing fan is used to disperse the smoke throughout the chamber.

The graph of light transmittance against time is then used to calculate a smoke index. This is based on the time taken to reach 70%, 40%, 10% and minimum transmittance. The test lasts 20 min or until minimum transmittance has been reached. The test is weighted against those materials producing smoke early in the test, so the faster the smoke is produced, the higher the smoke index. The same apparatus is used to test to North American specifications (e.g. those of NFPA and ASTM), British specifications (e.g. BS 6401) and aircraft specifications (e.g. Airbus ATS 1000). Although the apparatus is the same, the test procedures vary slightly, as do the expression of results.

8.10. Fire engineering and Building Regulations

Although the tests and requirements to produce a GRP with the best possible fire performance have already been described, there are other considerations to be taken into account for safe use of the material. Among these is the use of the materials in situations where they pose less hazard in the event of a fire.

A common method of achieving the above aim is to limit the size of the panel or component. For instance if an external cladding is more than 15 m high then a Class 0 surface is required. Below this, a lesser performnce is acceptable. The reasoning behind this can be appreciated; a small fire starting at the bottom of cladding will easily spread upwards if the flame spread characteristics are poor and in a multi-storey building this could soon involve many floors.

Another approach is to position the material so that it reduces any possible hazard. One of the requirements of the UK Building Regulations is designed to prevent fire spread from one property to another. They require that any cladding etc. within 1 m of the boundary shall be Class 0. Designers can therefore use material with poorer fire properties if they are kept away from the boundary.

From these examples it will be clear that any designer in GRP should be well conversant with the Building Regulations in order to make the safest and the most efficient and economical use of his material.

8.11. Changing regulations

In 1985, new Building Regulations were introduced in the UK, the object of which was to give the designer more freedom and not to tie him down to rigid rules. The Approved Documents which expand and explain the essential requirements of the regulations give suggested solutions to the problems of safe design but are not meant to be slavishly followed. The examples usually follow the dictates of the previous regulations because they are tried and tested over many years.

It is, however, open to the designer to pursue a different solution. These can be based on test data to foreign standards or on non-standard tests, perhaps large-scale tests simulating actual fire conditions. It will be the designer's responsibility to convince the building control officer of the validity of his design. If necessary he can appeal for a "Determination" from the Department of the Environment.

Regulations are not static; in 1986 a review of the regulations was initiated and proposals were made by the Department of the Environment and circulated for comment. The proposals with respect to fire were to reduce the fire resistance requirements of certain partitions, those proposals apparently being based on a report issued by the Fire Research Station entitled *Fire safety in buildings*.[8.11] These proposals were not well received. There was, of course, opposition from manufacturers of fire resistant materials but in addition to this there seemed to be a widespread reaction to any suggestion of reducing fire safety. In view of this reaction these proposals are to receive lengthy consideration and changes are not expected in the near future.

The review document also suggested that smoke and toxicity should also be introduced into the regulations and invited comments.

A more dramatic effect on the use of materials in building will be the result of the completion of the internal European market in 1992. The objective is to reduce all barriers to trade within the European Community. In addition to removing tariff barriers (import taxes) and customs regulations the aim is to remove the technical barriers to trade. Principally these are the result of differing product standards in different countries.

As a first step a *Construction products directive*[8.12] has been adopted. This directive is designed to bring the building laws and regulations in the different countries closer together. The document sets out seven "essential requirements" with which products must comply. The essential requirements are similar to those embodied in the UK Building Regulations. One of the essential requirements is safety in case of fire. The document defines the requirement as follows.

> "The construction works must be designed and built in such a way that in the event of an outbreak of fire in the works:
>
> (*a*) occupants can leave the works or be evacuated unharmed
> (*b*) the safety of rescue teams is ensured
> (*c*) the fire cannot spread to neighbouring works or parts of them
> (*d*) the generation and spread of fire and smoke within the works themselves is limited."

Therefore, the requirement covers life safety and the spread of fire.

Products will be deemed to comply with the requirements by two routes, either by achieving a "CE" mark or by achieving a "European Technical Approval".

The CE mark will be obtained by certified compliance testing in accordance with a European Standard (EN). There are a few European Standards now but many more will be needed within the next few years. The Committee Européen Normalization is the organization mandated to produce these standards. Their main executive arms will be the national standards-making bodies. The British Standards Institution (BSI) is responsible for the administration of the development of the fire standards. It is generally accepted that new fire standards acceptable to all the EEC countries (and the European Free

European Community CE Communauté Européenne

Fig. 8.8. CE mark

Trade Area, EFTA, countries) could not be developed by 1992 since different countries have widely different approaches to fire and fire safety. A proposal that has been widely accepted is the interim adoption of the "three test option". The three tests would be the British surface spread of flame test, the German "Brandschact" test and the French "epiradiateur" test. This is of course a political decision and it is difficult to see how the three tests can encompass the facets of flame spread, heat release, ignition and combustibility in the same way as BS 476: parts 7, 6, 5 and 11 or 4 respectively.

With minor differences the fire resistance test is the same in many European countries and so harmonization of this test presents few difficulties. The relevant ISO standard will probably form the basis of the new specification.

When successful tests have been completed by an accredited laboratory, the manufacturer will be able to mark his product with the CE mark as shown in Fig. 8.8. CE marked products will be offered for sale freely throughout the community. The most significant aspect of this harmonization is that when a European standard is published then within 6 months all conflicting national standards must be withdrawn. It therefore appears that designers and manufacturers will be faced with different tests and their results to be interpreted in the design of buildings and in the selection of materials and products.

With innovative products for which there are no standards the manufacturer will be able to apply for a European Technical Approval which will probably be similar to an *Agrément* certificate.

How plastics materials will fare under these new fire tests is impossible to predict; there may be a necessity to improve the fire properties of the products or the requirements may be relaxed. However the standards change, the designer will be faced with the necessity to be more knowledgeable about the tests and their interpretation if he is to design safely.

8.12. Conclusions

The objective of national building regulations is to safeguard health and life and, as such, these regulations are one of the predominant factors governing the use of materials in building.

The organic nature of plastics materials means that the fire performance of plastics products must be carefully considered because organic materials are likely to burn. This generally means that plastics materials are used in small quantities or in non-hazardous applications or are encased in some way to exclude flame and oxygen.

Reinforced plastics have been widely used because the attempts to improve their properties to BS 476 standards has been successful. Plastics will continue to be widely used in buildings but their performance will be thoroughly scrutinized if smoke and toxicity become the subject of regulation.

References

8.1. *The Building Regulations.* HMSO, London, 1988.

8.2. *The Building Regulations, Approved Document B—Fire Spread.* HMSO, London, 1988.

8.3. British Standards Institution. *Fire tests on building materials and structures.* BS 476. British Standards Institution, London.

8.4. British Standards Institution. *External fire exposure roof test.* BS 476: part 3. British Standards Institution, London, 1975.

8.5. British Standards Institution. *Noncombustibility test for materials.* BS 476: part 4. British Standards Institution, London, 1970.

8.6. British Standards Institution. *Method of test for ignitability*. BS 476: part 5. British Standards Institution, London, 1979.
8.7. British Standards Institution. *Method of test for fire propagation for products*. BS 476: part 6. British Standards Institution, London, 1981.
8.8. British Standards Institution. *Surface spread of flame test for materials*. BS 476: part 7. British Standards Institution, London, 1987.
8.9. British Standards Institution. *Test methods and criteria for the fire resistance of elements of building construction*. BS 476: part 8. British Standards Institution, London, 1972.
8.10. British Standards Institution. *Methods of testing plastics*. BS 2782. British Standards Institution, London.
8.11. Malhotra H.L. *Fire safety in buildings*. Building Research Establishment, Borehamwood, 1986.
8.12. Commission of the European Communities. *Proposal for a Council Directive on the approximation of the laws, regulations and administrative provisions of the member states relating to construction products*. Commission of the European Communities, Brussels, 1987.

Bibliography

British Standards

British Standards Institution.
Guide to fire test methods for building materials and elements of construction. PD6520: 1988. British Standards Institution, London, 1988. (Gives a valuable review of BS test methods with commentary.)

British Standards Institution.
External fire exposure roof test. BS 476: part 3: 1958. British Standards Institution, London, 1958. (This standard has been withdrawn but is still cited in UK Building Regulations. The test method is very similar to the 1975 version but the classification method is different.)

British Standards Institution.
External fire exposure roof test. BS 476: part 3: 1975. British Standards Institution, London, 1975. (Designed to measure the ability of a representative section of a roof, rooflight, dome light or similar components to resist penetration by fire, when its external surface is exposed to heat radiation and flame. Also measures the extent of surface ignition.)

British Standards Institution.
Non-combustibility test for materials. BS 476: part 4: 1970 (1984). British Standards Institution, London, 1984. (Includes definition of non-combustibility. Determines whether materials, with or without coatings, used in construction or finishing of buildings meet the definition.)

British Standards Institution.
Method of test for ignitability. BS 476: part 5: 1979. British Standards Institution, London, 1979. (Measurement of ignitability characteristics of essentially flat and rigid building materials and composites when subjected to a small flame.)

British Standards Institution.
Method of test for fire propagation for products. BS 476: part 6: 1981. British Standards Institution, London, 1981. (A method to determine the fire performance of products used as internal linings in buildings. Takes account of the combined effect of factors such as the ignition characteristics, the amount and rate of heat release and thermal properties of products in relation to their ability to accelerate the rate of fire growth.)

British Standards Institution.
Method for classification of the surface spread of flame of products. BS 476: part 7: 1987, British Standards Institution, London, 1987. (Method for measuring the lateral spread of flame along the surface of a specimen in the vertical position, together with a classification system based on rate and extent of flame.)

British Standards Institution.
Test methods and criteria for the fire resistance of elements of building construction. BS 476: part 8: 1972. British Standards Institution, London, 1972. (Methods of test and criteria for determining fire resistance of walls and partitions, floors, flat roofs, columns, beams, suspended ceilings protecting steel beams, door and shutter assemblies, glazing and ceiling membranes. Replaced by parts 20–23 of *BS 476* but remains current as it is cited in legislation.)

British Standards Institution.
Guide to the principles and application of fire testing. BS 476: part 10: 1983. British Standards Institution, London, 1983. (Describes the general principles and application of methods in this series of standards for fire testing of building products, components and elements of construction.)

British Standards Institution.
Method for assessing the heat emission from building materials. BS 476: part 11: 1982. British Standards Institution, London, 1982. (Describes a method for assessing the heat emission from building materials when inserted into a furnace at a temperature of 750°C.)

British Standards Institution.
Method for determination of the fire resistance of elements of construction (general principles). BS 476: part 20: 1987. British Standards Institution, London, 1987. Also ISO 834. (General details of test conditions, specimens, apparatus and criteria for fire resistance testing. Appendices give general guidance information, method for residual loadbearing capacity and operating instructions. Replaces *BS 476: part 8: 1972* which remains current as it is cited in legislation.)

British Standards Institution.
Methods for determination of the fire resistance of loadbearing elements of construction. BS 476: part 21: 1987. British Standards Institution, London, 1987. (Detailed test requirements for beams, columns, floors and flat roofs and walls. Replaces *BS 476: part 8: 1972* which remains current as it is cited in legislation.)

British Standards Institution.
Methods for determination of the fire resistance of non-loadbearing elements of construction. BS 476: part 22: 1987. British Standards Institution, London, 1987. (Detailed test requirements for partitions, fully insulated, partially insulated and uninsulated doorsets and shutter assemblies, ceiling membranes and glazed elements. Replaces *BS 476: part 8: 1972* which remains current as it is cited in legislation.)

British Standards Institution.
Methods for determination of the contribution of components to the fire resistance of a structure. BS 476: part 23: 1987. British Standards Institution, London, 1987. (Detailed test requirements for the contribution made by suspended ceilings to beams and by intumescent seals to doorsets and shutter assemblies. Replaces *BS 476: part 8: 1972* which remains current as it is cited in legislation.)

British Standards Institution.
Method for determination of the fire resistance of ventilation ducts. BS 476: part 24: 1987. British Standards Institution, London, 1987, Also ISO 6944.

British Standards Institution.
Methods for measuring smoke penetration through doorsets and shutter assemblies. BS 476: part 31. British Standards Institution, London.

British Standards Institution.
Method of measurement under ambient temperature conditions. BS 476: section 31.1: 1986. Also ISO 5925/1, British Standards Institution, London, 1986. (This method measures air leakage of door and shutter assemblies under ambient temperature conditions as being representative of smoke penetration occurring from the early states or at a remote position from a fire occurring in a building.)

British Standards Institution.
Methods of testing plastics. BS 2782. British Standards Institution, London.

British Standards Institution.
Introduction. BS 2782: part 0: 1982. British Standards Institution, London, 1982. Also ISO 291, ISO 3205. (Essential information for testing plastics to be read in conjunction with separately published methods that are listed together with relationship with the 1970 methods. Appendices are identical to ISO 291 for standard atmospheres and ISO 3205 for preferred test temperatures.)

British Standards Institution.
Flammability of a test piece 550 mm × 35 mm of thin polyvinyl chloride sheeting (laboratory method). BS 2782: part 1: method 1400: 1980 (1987). British Standards Institution, London, 1987. (Replaces method 508C of *BS 2782: 1970*.)

British Standards Institution.
Flammability of a small, inclined test piece exposed to an alcohol flame (laboratory method). BS 2782: part 1: method 140E: 1982. British Standards Institution, London, 1982. (Replaces method 508D of *BS 2782: 1970*.)

British Standards Institution.
Determination of flammability by oxygen index. BS 2782: part 1: method 141: 1986. British Standards Institution, London, 1986. Also ISO 4589. (Gives arbitrary methods for testing the ignitability of plastics materials in the form of small bars or panels of thin film or sheet, as indicated by the oxygen concentration necessary for ignition by a small flame of test specimens at 23°C.)

British Standards Institution.
Laboratory method of test for assessment of the horizontal burning characteristics of specimens no larger than 150 mm × 50 mm × 13 mm (nominal) of cellular plastics and cellular rubber materials when subjected to a small flame. BS 4735: 1974 (1984). British Standards Institution, London, 1984. (Describes a small scale laboratory procedure for use in controlling manufacturing processes to ensure consistency of production. This method should not be used to assess potential fire hazard in use.)

UK Building Regulations

Mandatory rules for means of escape. HMSO, London. (Applied to buildings (including dwelling houses and flats) of more than 3 storeys.)

Cone calorimeter

Babrauskas. Development of the cone calorimeter—a bench scale heat release rate apparatus based on oxygen consumption. *Fire and Materials*, 1984, **8**, No. 2, 81.
Huggett. Estimation of rate and of heat release by means of oxygen consumption measurements. *Fire and Materials*, 1980, **4**, No. 2, 61.
Parker J. Calculation of the heat release rates by oxygen consumption for various applications. *Fire and Sciences*, 1984, **2**, 380.

Other publications

Hollaway L. (ed.) *The use of plastics for loadbearing and infil panels*. Manning Rapley, 1976. (Chapter 9 describes a fire test on building panels to simulate extreme fire conditions. Chapter 12 describes how plastics can be used under the constraints of the German Building Regulations and DIN Specification 4102.)
Smitz J. V. (ed.). *Testing of polymers*. **2**, Interscience, 1966. (Chapter 11 describes American Society for Testing and Materials specifications for building materials, ASTM E84 *Tunnel test* and ASTM E162 *Surface spread of flame test*.)
Titow V.W. and Lanham B.J. *Reinforced thermoplastics*. Applied Science, London, 1975. (Chapter 9 discusses flammability test methods and ratings (British and American) and the improvement of performance by using additives.)
Modern plastics encylopedia, 658 to 663. McGraw-Hill, New York, 1986/7. (Gives a comprehensive list of flame retardants for use in plastics.)
Legatt A.J. *GRP and buildings*. Butterworths, London, 1984. (Chapter 7 gives suggestions for 'real fire' situation testing, a short section on fire tests and hints on fire design.)

9 Structural elastomeric bearings and resilient seatings

P. Grootenhuis

9.1. Introduction

There are a number of trends in the development of building sites and in construction which require isolation from ground-borne vibration. Often the remaining vacant sites in an inner city are the undesirable ones, in that the site is close to or over some railway tunnels. The trend to install double windows in order to keep out noise from road traffic can lead to interference and annoyance being created by a low frequency rumble from an underground train. The expectancy of the prospective occupants of a new building of a quiet environment is much greater than hitherto. The most critical period from a vibration and noise isolation aspect can be during the letting or selling time when the building is empty. A sudden disturbance from a train, unseen and unexpected, can have a most off-putting effect on a client. It will inevitably reduce the value of the property.

The noise one can sometimes hear in a building over a railway tunnel is caused by structure-borne vibration transmitted up into the building from the foundations. A separation or vibration isolation in the structure can reduce these transmitted vibrations sufficiently to bring the disturbance below the general background noise level. The isolation of structures on rubber bearings has been in practice for more than 25 years. The technique has been pioneered in the UK where there are many buildings, mainly in London, supported solely on rubber blocks. One of the largest complexes, blocks of flats in Ebury Street in Westminster, has a total mass of 24,000 tons. The technique has been introduced into Sweden, the USA and in the Far East.

The design considerations for supporting structures on rubber bearings will be discussed but only as applied for isolation from ground-borne vibration. The extension of the technique for protecting a building against damage from earthquakes is a separate matter and will not be referred to; for a detailed discussion see reference 9.1.

9.2. Trends in development

There are now very few underdeveloped or vacant sites in the major cities. Where there are any, the reason is that there is often a railway or underground in the near vicinity. However, the pressures are such that prestige buildings have been erected on the most unfavourable sites. One of the earliest examples is the Barbican Redevelopment in the City of London, where the four tracks of railway (two British Rail tracks into Moorgate Station and two for the London Transport District and Circle lines) pass right through the site. In the redevelopment, the railway tracks were realigned and covered over and the opportunity was taken to isolate the tracks on the world's first major floating track slab.[9.2,9.3] A description of the floating track slab is given later in this chapter. Similar situations have been experienced in Birmingham and Glasgow, where concert halls and cultural centres have been sited either over or close to main line railways.

Another typical example where a hitherto undesirable site has been profitably used is the development of the Wellington Hospital in London. The

Fig. 9.1. The Wellington Hospital, London, mounted on resilient bearings

directors of a shipping company which operate cruise liners diversified its activities by setting up a private hospital. The developer found a site with a railway cutting with tracks for British Rail (BR) diesel trains from Marylebone Station. However, it was a difficult construction site because under the road were the two tunnels for the Metropolitan Line trains out of Baker Street Station, and further underground were the twin tunnels of the Jubilee Line. Nevertheless, the hospital was built on this apparently totally unsuitable site. A general view is shown in Fig. 9.1 and a part section in Fig. 9.2, showing the proximity of the four underground tunnels and the BR tracks passing through the basement, boxed in a heavy concrete surround about 1 m thick. The entire building above the basement is supported on resilient bearings with an isolation natural frequency of about 8 Hz. An air void on plenum space was inserted to attenuate the airborne noise path. The operating theatres are at ground floor level and one of the specialist operations frequently performed is for eye surgery. No disturbance from any of the railways can be either heard or detected in the building above the basement. The shipping company has sold out to a private medical care company which, having found the hospital to be most successful, has recently built a second one further up the railway line. This second hospital has also been mounted on resilient bearings with an isolation natural frequency of 9 Hz and the trains underneath cause no problems.

A further example of a major inner city development which will extend into the next century is the new British Library. The early proposal to build the British Library in Bloomsbury, close to the British Museum, had to be abandoned because of objections to the demolition of buildings with preservation orders. A large and substantially vacant site, the old coal yard beside St. Pancras Station, was located by the planners. This site is traversed by the twin Victoria Line and the twin Northern Line tunnels, with the District Line under the Euston Road along the southern boundary and a BR main line tunnel at the northern end. This was not a very promising location for quiet reading rooms. The basements will be extended to a depth below some of the underground railway tunnels, which are to be encased in the concrete structure. However, so far no antivibration and structure-borne noise control measures have been incorporated in the design.

9.3. Trends in construction

When some vibrational energy has entered a structure, it is transmitted and dispersed by compression, flexural and shear waves. The response of any particular part of the structure will depend on the proximity to the input and even more on the local dynamic characteristics. Obviously, when a floor slab or a wall has a resonant (natural) frequency close to a predominant input frequency, a magnification of the response will occur. The trend to construct buildings as continuous structures without any expansion or movement joints is helping the transmission of structure-borne noise. The further trend of light-weight construction and the use of materials with low damping means that the response to even a low level input can be quite strong. The older construction of brick piers with beams on plates and lintels had many more discontinuities and greater damping.

A recent tendancy to erect more steel-framed buildings with welded joints is also providing less damping. Similarly, prestressed concrete construction can offer hardly any damping for low level structural vibrations because the shrinkage cracks found normally in in situ concrete have been eliminated. The modern forms of construction, either in steel or with prestressed beams, has enabled designers to introduce very large span floors with a first natural frequency less than 5 Hz. The response of such large span floors can be quite severe. To counteract the undersirable features of these trends, more damping can be introduced by incorporating resilient rubber seatings—to be discussed later on.

The magnification at a structural resonance is called the Q-factor, which is a non-dimensional ratio of the response at a resonant frequency to the motion

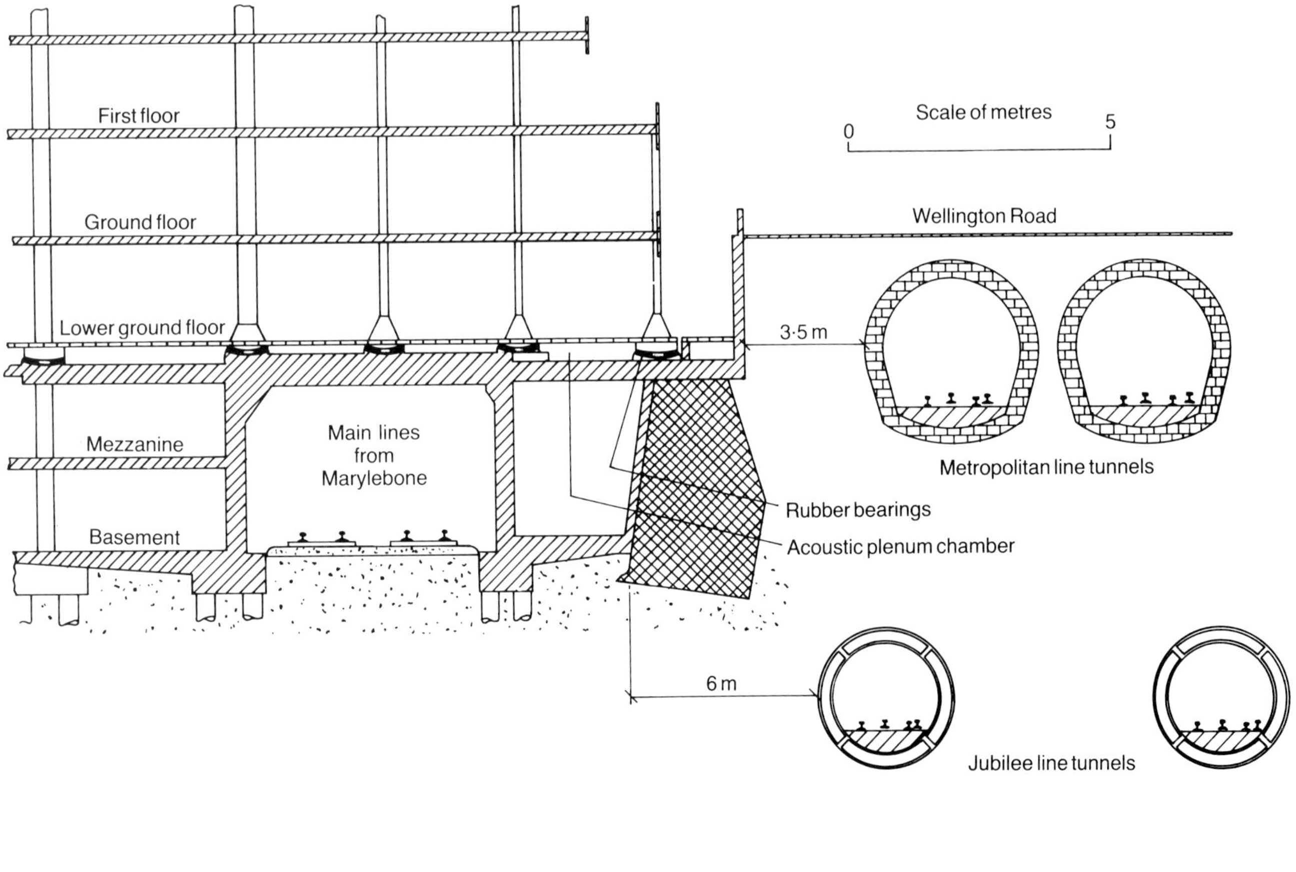

Fig. 9.2. Part section of the Wellington Hospital showing the British Rail tracks and the London Transport tunnels

Table 9.1. Typical damping properties of engineering materials for small alternating stresses. The values can vary considerably with heat treatment, temperature, frequency and with the method of construction (friction in joints etc.)

Material	Loss factor material only	Q-value for built-up structures
Aluminium alloy	0·0001–0·002	50–1000
Alloy steel	0·0002–0·005	
Mild steel	0·0005–0·01	
Titanium	0·0001–0·002	
Cast iron	0·02 –0·04	25–50
Structural timber	0·005 –0·008	125–200
Concrete		
reinforced	0·01 –0·02	50–100
prestressed	0·003 –0·01	100–300
Plastics—rubber	0·01 –2	0–100

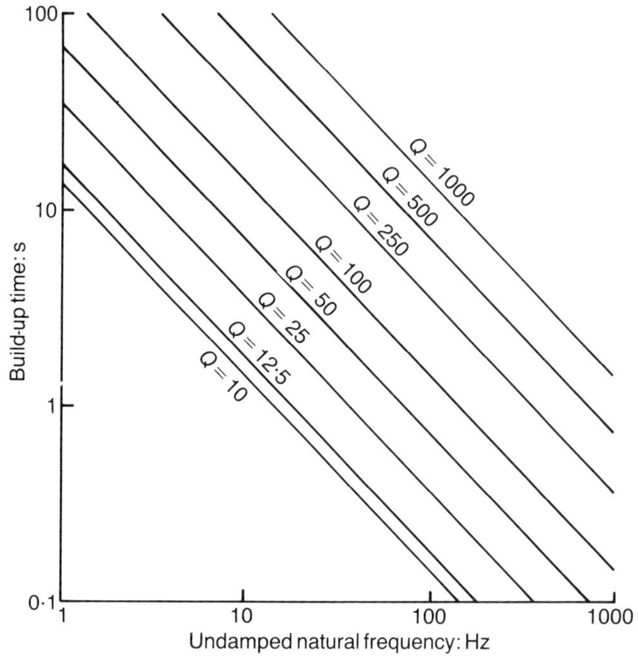

Fig. 9.3. Resonance build-up time of a one degree-of-freedom system for different values of the Q-factor

input. There are several ways of denoting the degree of damping in a material and in a structure. Some equivalences will be given later. Typical values of the Q-factor are given in Table 9.1. The difference between in situ and prestressed concrete is quite large and as the use of prestressed construction often permits a reduction in mass per unit length, the dynamic response of such a structure can be many times greater than that of a more conventional concrete construction. It is therefore helpful for reducing the dynamic response to add as much as is practicable, or "loose" in situ concrete infall on a floor made up of prestressed beams, or on a steel-framed structure.

The full response at a resonant frequency of a structure will take time to build up. It is argued sometimes that a full state of structural resonance cannot be built up by the passing of an underground train because there is insufficient time. This is unfortunately not true. The build-up time for a simple structure as a function of the natural frequency is given in Fig. 9.3. Note that both scales are logarithmic. The different curves are for values of the Q-factor, denoting the level of damping, ranging from 10 to 1000. A value of 10 implies quite heavy damping and a value of 1000 is for a free-standing, all welded metal structure. For a system with zero damping (Q-factor of infinity) the response at resonance is infinite but it will take an infinite time to reach.

A typical floor natural frequency is 20 Hz and with a Q-factor of 12·5 the time taken to reach the maximum response at resonance is about 1 s (Fig. 9.3). This is much shorter than the time taken by a train to pass a point. A train of length 60 m, travelling at a speed of 60 km/h, will take 3·6 s to pass a point. Where a tunnel passes underneath a building the time will be even longer and can be up to 10 s or longer. A full state of resonance can then be built up quite readily.

9.4. Human sensitivity to vibration

Human perception of structure-borne vibration is not easily quantified because of a wide variation in sensitivity and because of an emotional element. No single scale for vibration levels has been devised so far, as has now been established for air-borne noise for which an NR-value can be specified as a design level. A detailed review of perception and comfort levels will not be given. Work is still in progress to establish accurately the threshold values for people standing, sitting and lying down, as reported by Parsons and Griffin.[9.4]

The degree of discomfort and the possible startle effect by vibration and shock depends upon duration, the number of occurrences and whether it occurs during the day or at night time. It is now generally accepted that the vibration intensity can best be described by the peak velocity. The threshold for perception of a person standing is at a peak velocity slightly greater than 0·1 mm/s in the frequency range 5–100 Hz. A considerable amount of information relevant to permissible levels of vibration in buildings has been assembled in the German DIN 4150 Standard (1970 and 1975). A draft revision was prepared in 1986 which has not yet been adopted. The principal ISO document on human perception is 2631 and a draft addendum with recommended levels for different classes of buildings was issued in 1981 as DOC No. 81/78512. This addendum has been translated into BS 6472 (1982) *Guide to evaluation of human exposure to vibration in buildings (1–80 Hz)*. This document is also undergoing revision. A simplified summary of typical recommended maximum vibration levels, expressed as peak velocity, for different classes of buildings is given in Table 9.2. A further simple guide-line[9.4] is that for domestic situations a vibration level must be regarded as unacceptable when it is twice the threshold value, i.e. about 0·2 mm/s. A much higher level can be tolerated for intermittent vibrations from such temporary operations as pile driving.

9.5. Vibration damage of structures

There is even less agreement internationally on the levels of vibration and shock which will result in structural damage. This is largely on account of the differences between North America and Europe in house construction and in the ages of buildings. The timber-framed buildings in North America can

Table 9.2. ISO proposal for maximum vibration velocity levels in the frequency range 5–80 Hz for human comfort in buildings

Construction area	Time*	Guide values			
		Continuous vibration and those occurring repeatedly with interruptions		Rarely occurring vibrations	
		KB	V: mm/s	KB	V: mm/s
1. Residential, weekend house, etc.	Day Night	0·2(0·15) 0·15	0·3(0·2) 0·2	4·0 0·15	5·7 0·2
2. Urban residential and commercial	Day Night	0·3(0·2) 0·2	0·4(0·3) 0·3	8·0 0·2	11·5 0·3
3. Commercial, including offices	Day Night	0·4 0·3	0·55 0·4	12·0 0·3	17·0 0·4
4. Industrial	Day Night	0·6 0·4	0·85 0·55	12·0 0·4	17·0 0·55
5. Special area according to type of usage and proportion of housing	Day Night	0·1–0·6 0·1–0·4	0·2–0·85 0·2–0·55	4–12 0·15–0·4	5·5–17·0 0·2–0·55

* Day = 0600 h–2200 h; night = 2200 h–0600 h.

Table 9.3. Vibration damage levels

Type of building	Peak particle velocity below which damage would not be expected: mm/s
Ancient buildings with architectural finishes in a poor state of repair	2–5
Generally sound buildings	
architectural damage	5–8
light structural damage	12–15
severe damage	30

withstand considerably higher levels than the older brick and stone buildings in Europe. The type of excitation also has a marked effect, whether steady state, intermittent or shock. For a summary discussion see reference 9.5.

Different criteria for assessing structural damage are in existence in a number of countries; one of the more detailed is the German DIN 4150—Part 3, 1986. There is also an ISO discussion document, ISO/DIS 4866 (1986) *Measurement and evaluation of vibration effects on buildings*. Account is taken of soil conditions and different forms of excitation for several classes of building. A review has been written by CIRIA on the damaging effects of piling operations.

The susceptibility of a structure to damage by vibration or shock is normally expressed in terms of the peak particle velocity in mm/s, measured either at the foundation or at midspan of a floor higher up in the building. There is currently much discussion on whether to include a 5% damage criteria. It would appear that the velocity for a certain degree of damage is independent of frequency for the range normally associated with ground-borne vibration and shock. Some typical values of peak particle velocity to cause different degrees of damage are given in Table 9.3, but these figures should not be taken as mandatory. All these levels are well above the human sensitivity threshold.

The effect of vibration on the load-carrying capacity of foundations is even less well established. There may be an increase in differential settlement of the foundations which can cause structural damage. Under certain conditions a liquefaction of the soil can be caused by continuous vibration of quite low amplitudes.

9.6. Basic facts of isolation

Static forces and moments (couples) are transmitted through a resilient material unchanged; it is only by alterations in geometry that local pressures and stresses can be reduced. When dynamic forces, moments and motions are being transmitted then it follows from Newton's second law that

(*a*) to isolate a dynamic force, it must be permitted to move a mass
(*b*) to isolate an oscillating moment, it must be permitted to move an inertia
(*c*) to be isolated from an alternating motion, a mass or an inertia must form part of the system.

A structural transmission path, and especially when a polymeric bearing forms part of it, can be modelled by a complex stiffness, $k(1 + i\eta)$, where k is the elastic stiffness, η is the loss factor (the ratio of the quadrature or loss stiffness to the elastic stiffness) and $i = \sqrt{-1}$. The theory of vibration isolation and of transmission is well established, including an allowance for structural damping. The transmissibility ratio T between the transmitted force

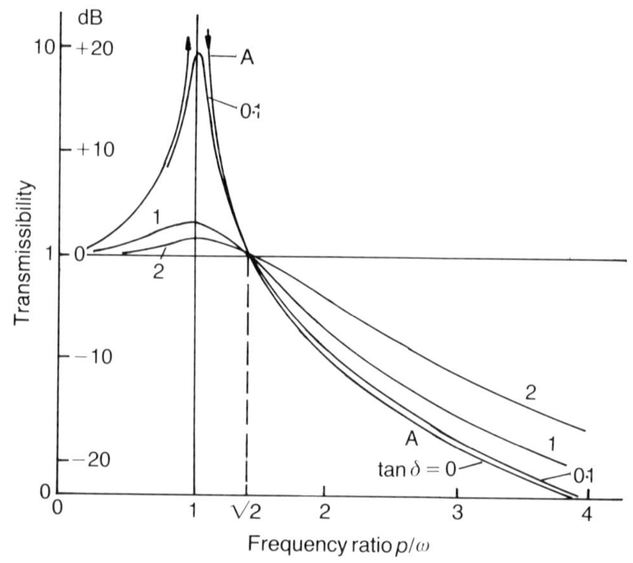

Fig. 9.4. Transmissibility frequency curves for different values of damping

and the input force or between the response to the input motion for a single path, can be obtained from the same equation

$$T = \left[\frac{1 + \eta^2}{(1 - p^2/\omega^2)^2 + \eta^2}\right]^{1/2} \tag{9.1}$$

where p/ω is the frequency ratio, with p the input frequency and ω the resonant undamped frequency of the transmission path including the mass m or the inertia I of the supported load. The natural frequency for motion in translation is $\omega = (k/m)^{1/2}$ and for rotation $\omega = (ka^2/I)^{1/2}$, where a is the moment arm. The transmissibility ratio has been plotted to an abscissa of the frequency ratio in Fig. 9.4 for several values of the loss factor η. At resonance, i.e. at $p/\omega = 1$, a magnification occurs known as the Q of the system and from equation (1)

$$T = \left(1 + \frac{1}{\eta^2}\right)^{1/2} = Q \tag{9.2}$$

For normal values of structural damping, the loss factor is much less than unity and approximately

$$Q = \frac{1}{\eta} \tag{9.3}$$

The scale for the transmissibility ratio in Fig. 9.4 is either logarithmic or in dB relative to a ratio of 1. A very large transmission occurs near resonance, for a frequency ratio = 1, and damping can have a beneficial effect in reducing the peak value. However, all the curves cross over at a frequency ratio of $\sqrt{2}$ and at the transmissibility ratio of 1. The effect of damping at a frequency ratio greater than $\sqrt{2}$ is not beneficial. This can lead to a compromise, often necessary in engineering, to have some damping in the resilient bearings to reduce the effects of resonance but not excessively to affect the isolation at higher frequencies. The effect of damping well above resonance is not very severe; the decrease in isolation is 3 dB for $\eta = 1$.

The transmissibility curves given in Fig. 9.4 are for a single degree-of-freedom system. The structure above the resilient bearings will have several resonant modes of vibration and these will reduce the isolation effectiveness at the resonant frequencies. A strong, lightly damped mode may well produce a peak above the transmissibility ratio of 1, even though the frequency ratio is well above $\sqrt{2}$.

The dynamic characteristics of the structure below the resilient bearings can also have an effect on the overall performance. A deep-piled foundation can be the main receiver of ground-borne vibration and it is unlikely that strong resonances are excited because of the large amount of ground surrounding the piles. There may be resonant frequencies associated with the ground beams between pile caps although the excitation is often not very strong. An exception would be where there are strong Rayleigh waves from a source of vibration at the surface and special precautions should then be taken.

A raft foundation or strip footings have a different dynamic response. This type of foundation can be regarded as a rigid body resting on an elastic half space. Vibration of the rigid body introduces compression and shear waves into the half space which are radiated away, in the absence of a strongly reflecting sub-layer. The natural frequency of a mass on an elastic half space can be predicted with reasonable confidence for a uniform soil with a known value for the shear modulus, Poisson's ratio and the density.[9.6,9.7] A footing of finite length will have six degrees of freedom each with a different natural frequency. There are the vertical and two horizontal motions and three rotational oscillations. These effects of the rotation about the vertical axis may not be very important. Some of the motions may be coupled.

The response at resonance is dependent largely upon the ratio of the mass, or of the inertia of the foundation block, to that of the bowl of soil immediately underneath the base. As an example, taken from reference 9.7, the non-dimensional response for vertical motion is shown in Fig. 9.5 of a base with a square and a rectangular area. The family of curves is for different values of the mass ratio b and have been plotted as a function of a frequency

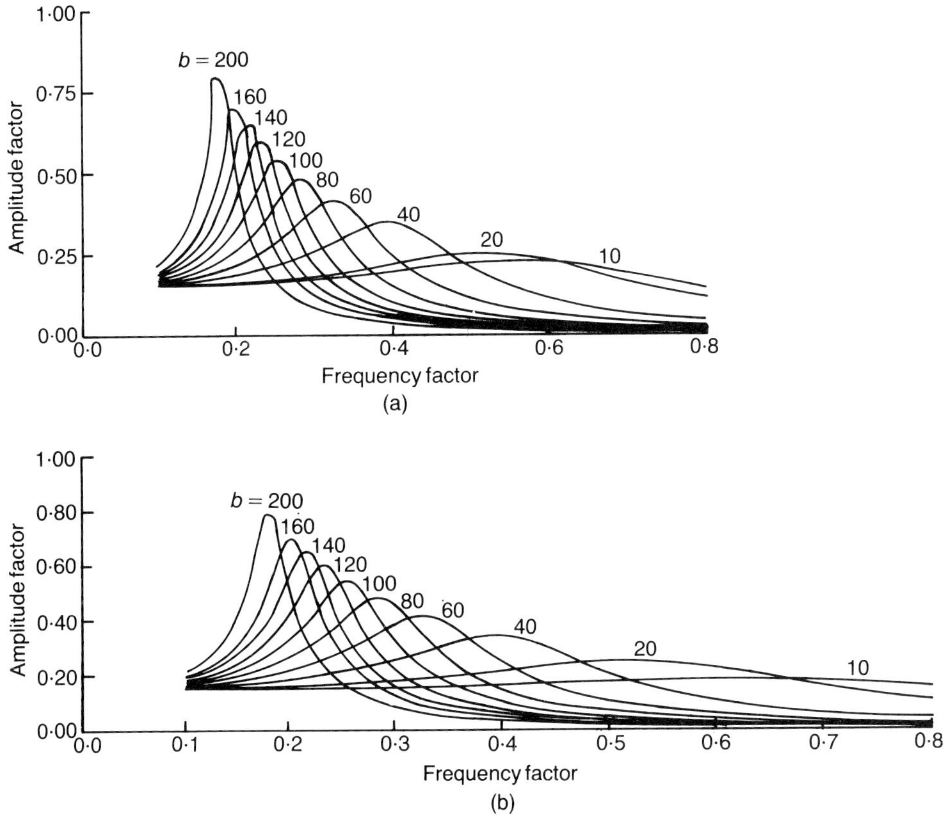

Fig. 9.5. Frequency response curves for different mass ratios for vertical motion for: (a) a square base; (b) a rectangular base, side ratio 1/2

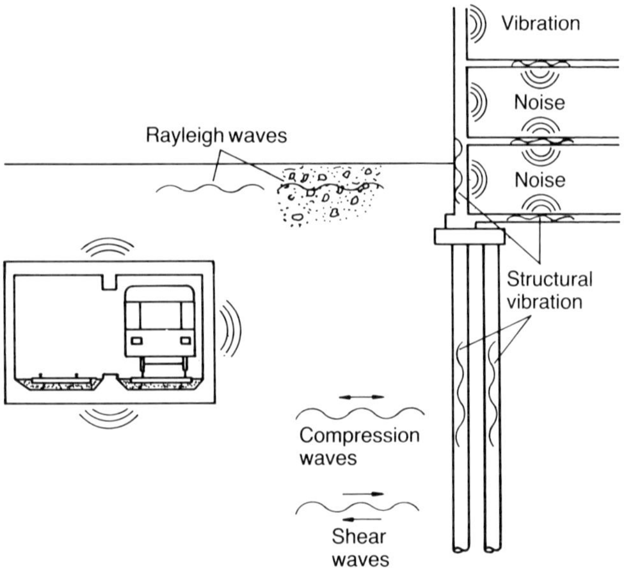

Fig. 9.6. Transmission paths for ground-borne vibration from a railway

factor $= pd/c_2$, where p = excitation frequency (rad/s), d = half width of one side and c_2 = velocity of shear waves in the soil.

A strong response at resonance can occur with a high soil loading whereas there is hardly any response for a low mass foundation with a wide base and low stresses in the soil ($b < 20$). This is because with a low mass ratio all the vibrational energy is radiated away into the half space, whereas with a concentrated mass some interchange between strain energy in the soil and kinetic energy of the mass can take place to build up a state of resonance. The curves in Fig. 9.5 show clearly the effect of radiation damping; no allowance was made in the analysis for damping by conversion into heat. These effects have been verified by experiment and are borne out in practice.

9.7. Location of resilient bearings

A diagrammatic representation of the transmission path from a railway tunnel into a building is shown in Fig. 9.6. The wheel–rail interaction produces vibrations in the track and the frequency spectrum depends upon many parameters including the method of track support. Where there are discrete rail-support points, such as sleepers, there will be a wheel-passing frequency which depends upon the speed of the train and on the spacing of the sleepers. Typical frequencies are 15–30 Hz. The track is normally laid on a concrete infill at the base of the tunnel and the track vibrations are transmitted directly into the surrounding ground. Acoustic waves are set up inside the tunnel and these strike the tunnel walls and are then transmitted as ground-borne vibrations. The three most important waves in the ground are the compression, shear and Rayleigh waves, each travelling at a different speed. The foundations of an adjacent building will receive these waves and the vibrational energy can then be transmitted throughout the building as illustrated in Fig. 9.6. The effect will be twofold; the vibration can be felt, especially when there is a floor resonance with a predominant frequency of excitation such as the wheel–sleeper passing frequency. The second effect is that of a low frequency rumble, i.e. the air-borne noise resulting from the vibration of the floor and of the walls. The level of this noise will depend upon the size of the room and on the sound absorption of the finishes. The relationship between the noise level and the velocity of the structural vibration is now well established[9.8,9.9] and will not be given here.

One of the best locations for inserting vibration isolation is inside the tunnel below the track, the so-called floating track slab.[9.10] The track can be of conventional construction, laid on a concrete slab or bridge deck which is supported on resilient bearings. A section through a cut-and-cover installation is shown in Fig. 9.7 which has been used for part of the Piccadilly Line extension to Heathrow in London where the tunnel passes through a residential area. The floating track slab has been damped in order to reduce the undesirable effects of resonance. A bituminous damping layer has been introduced within the slab, making it a composite construction. Any flexural vibration in the slab will induce a shear deformation in the damping layer which increases the vibration damping considerably.[9.11] An example of a floating track slab inside a bored tunnel is shown in Fig. 9.8; this is the type installed recently in Hong Kong and elsewhere. The slab is much smaller (than that shown in Fig. 9.7) as no increase in the diameter of the tunnel was

Fig. 9.7. Floating track slab for a section of the Piccadilly Line to Heathrow

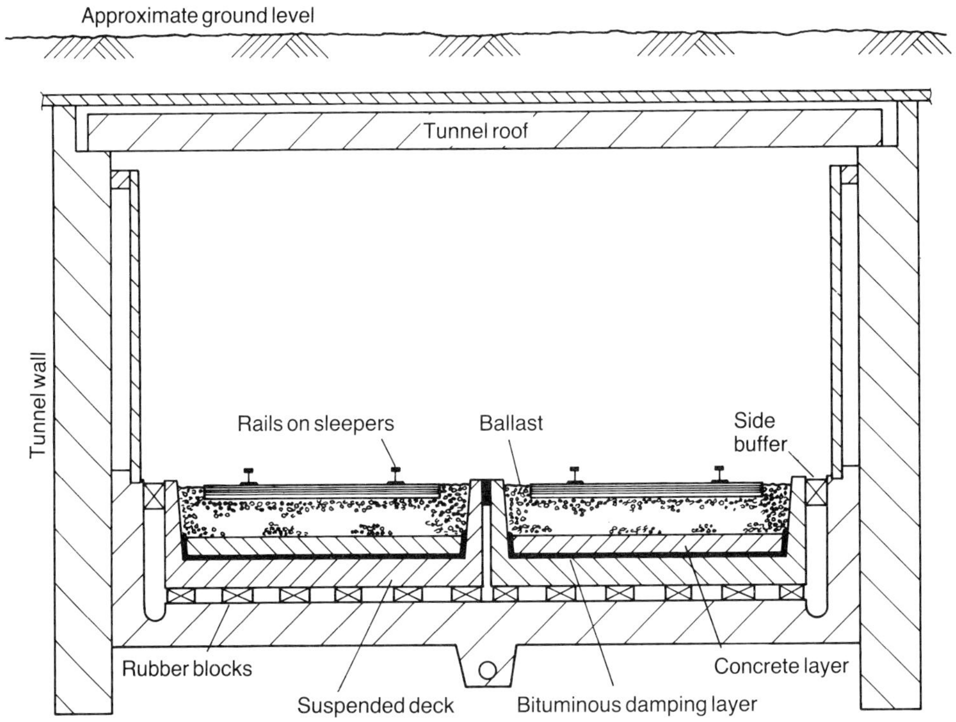

Fig. 9.8. Floating track slab in a bored tunnel; design for the Hong Kong Metro

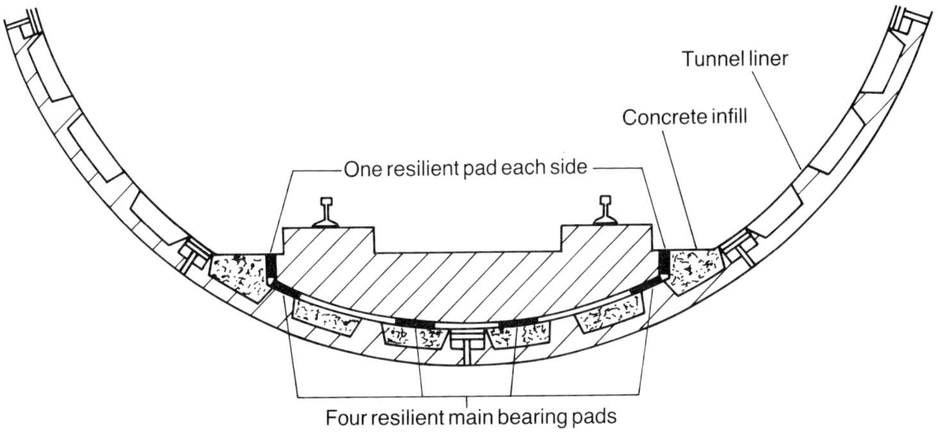

permitted for economic reasons. An alternative, albeit not so effective, method is to install a resilient mat under the ballast. This can attenuate the higher frequencies, above say 40 Hz.

Vibration isolation inside a tunnel can be installed readily only for new construction. Where the railway exists, the only approach for developing alongside or over it is by vibration isolation of the building. This can be done by separating the building from the foundations by inserting resilient bearings on top of the pile caps or on the footings. A sketch of such an installation is given in Fig. 9.9. The bearings, about 100 mm thick when unloaded, are bonded with a self-levelling epoxy grout to the pile cap or to a ground beam. A precast beam is placed on top to distribute the loads. The surface Rayleigh waves are isolated by embedding the foundation with a loose infill against a flexible skin such as Plastazote, Flexel or Aerofil. A failsafe upstand can be placed in between some of the pads to take the load in the event of a complete pad failure. Where a large load has to be supported on a single pile cap, a cluster of 10–30 rubber pads is used, with failsafe strips in between. It is important to leave sufficient clearance around the pads to permit free bulging and the failsafe upstand must be low enough to permit the full compression of the pads, including creep. The design for a column with an enlarged base supported on a cluster of pads on a pile cap is shown in Fig. 9.10. A plan view showing the pads with the steel failsafe grillage in between is included in the figure. The photograph in Fig. 9.11 shows the placing of a cluster of pads on a pile cap for Cresta House along Finchley Road in London. An advantage of a buried installation is that it is free from possible damage by fire or by hooligans. A disadvantage is that inspection is difficult and that the Rayleigh waves can perhaps be transmitted into the structure.

An alternative is to install the resilient bearings higher up in the building at the top of enlarged column heads. A simple sketch of this type of separation is given in Fig. 9.12. Protection against fire can be wrapped round the bearings as shown. Constructing an in situ floor slab on top of the bearings is not very

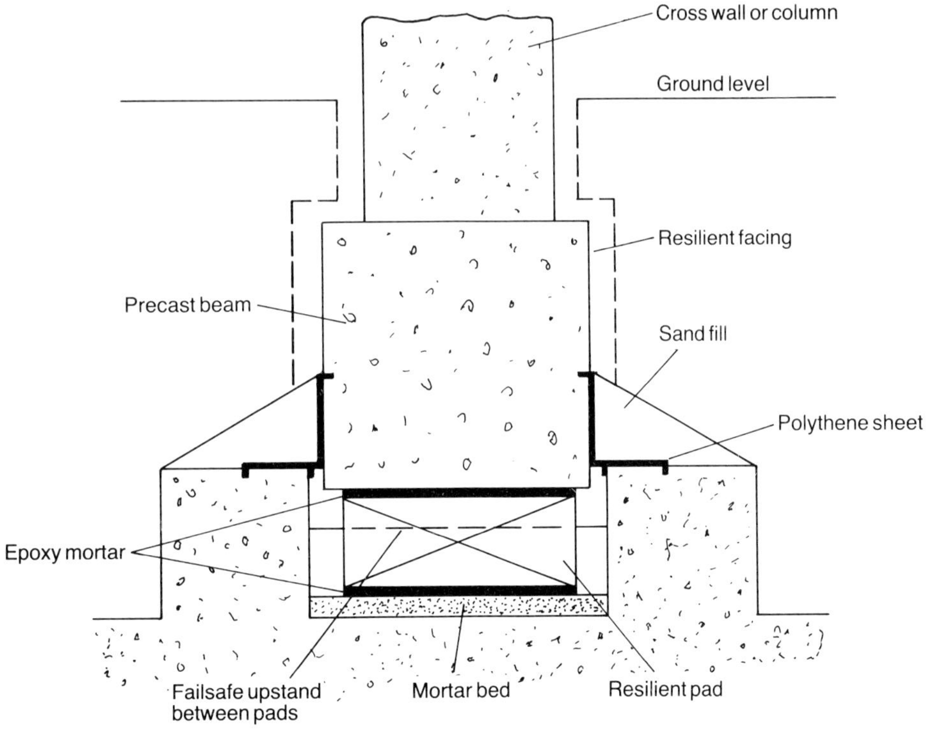

Fig. 9.9. Resilient mounting at foundation level—the Alexandra Road development

difficult. A technique is illustrated in Fig. 9.13 where a temporary sand infill between the bearings and the failsafe upstands is used to support the green slab to let it set. A thick polythene sheet is placed on top to avoid penetration by wet grout. When a large cluster of pads is installed on a single column head, it may be better to lay a steel plate over the pads as a permanent shutter. The sand can be blown out or washed out after the timber shuttering has been struck. It is advisable to use dry, sharp sand to make the removal easier.

Installing the separation higher up the building makes it possible to isolate only a part, such as the conference area or sensitive offices. Detailed attention

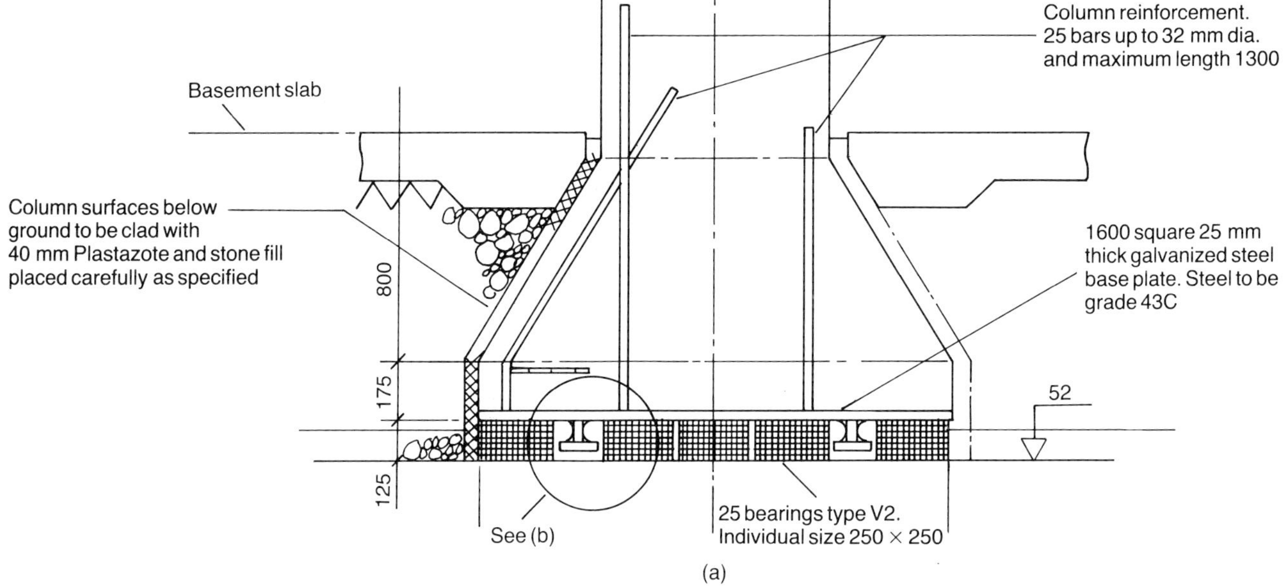

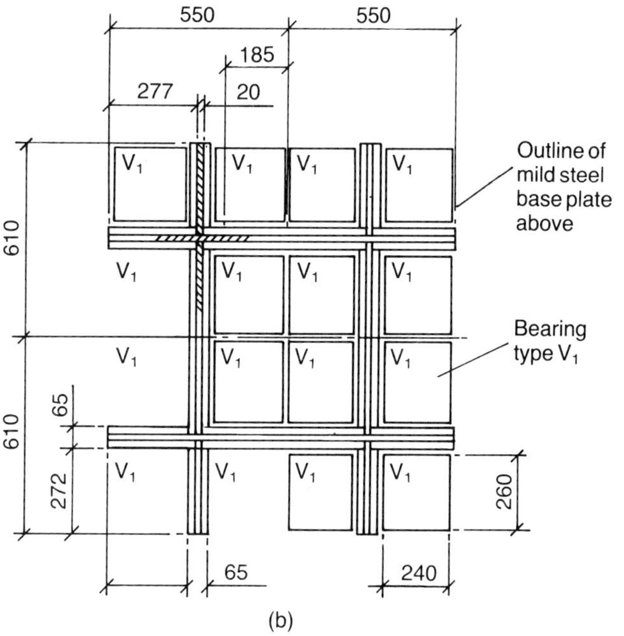

Fig. 9.10. Enlarged column base supported on a cluster of pads for Cresta House (dimensions in mm): (a) elevation; (b) plan view

Fig. 9.11. The placing of pads on top of a pile cap at Cresta House

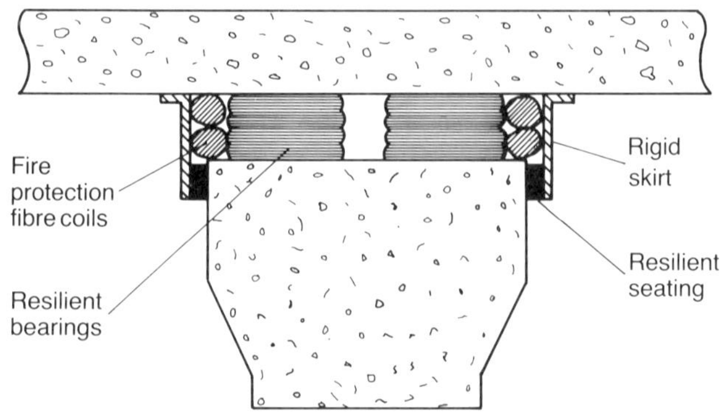

Fig. 9.12. Sketch of resilient pads supporting a floor slab on an enlarged column head

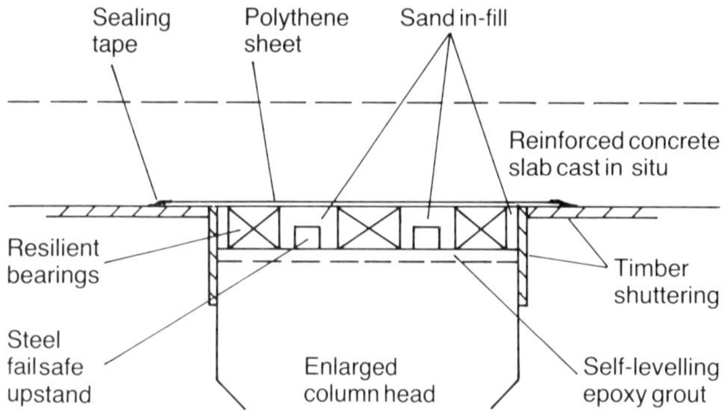

Fig. 9.13. Method of construction for an in situ cast slab on a cluster of pads

must be given to ensure that there are adequate gaps surrounding the isolated part. These gaps can, of course, transmit sound.

The location of the bearings can depend upon the type of construction of the building. For example, 2–3 storey houses of conventional construction in brick and timber or concrete floors are best mounted on a "floating raft" with the bearings on the footings or pile caps and ground beams. The additional cost of isolation is then in the somewhat more substantial raft instead of a ground slab. In a framed building, it is advisable to have as few shear walls as possible even if this requires a stiffening of the floor slabs and somewhat larger columns. It is therefore essential to consider vibration isolation at the very beginning of a design. With some care and ingenuity the additional costs need not be large and may be only a few per cent of the total cost of the building. However, to introduce separation into a design which has been frozen can lead to excessive expenditure. The siting of staircases and of lift shafts in relation to the more sensitive areas can be important. For tall buildings, some horizontal bearings may be needed to resist wind loading and the location of side bearings to act against strong points around the building can affect the rest of the design.

9.8. Selection and performance of elastomeric bearings

Large blocks of natural rubber reinforced with steel plates have been in use for bridge bearings for over 30 years. The main purpose is to accommodate thermal expansion and contraction. The bearings have to be flexible in shear and stiff in compression. For vibration isolation, somewhat different characteristics are required, i.e. a low dynamic stiffness in compression with a high static load-carrying capacity and a reasonable resistance to deformation in shear. The specification for the properties of the rubber and the methods of testing the basic compounds are similar to those for bridge bearings. There is now sufficient practical experience in the mounting of buildings on elastomeric bearings for a British Standard to have been issued in 1982.[9.12]

The load-carrying capacity and the dynamic stiffness of a bearing will depend upon many parameters, one of which is the 'shape factor' defined in BS 6177[9.12] as the ratio of the loaded area to the area free to bulge. There can be an appreciable difference between the stiffness measured from the slope of a static load deflection curve and the dynamic stiffness determined by a vibration test. This makes it essential to test dynamically samples as near as possible to full size with the actual static compressive stress superimposed. This is not at all easy and there have been tests which have ended up giving the dynamic characteristics of the test rig rather than of the bearing. Hydraulic

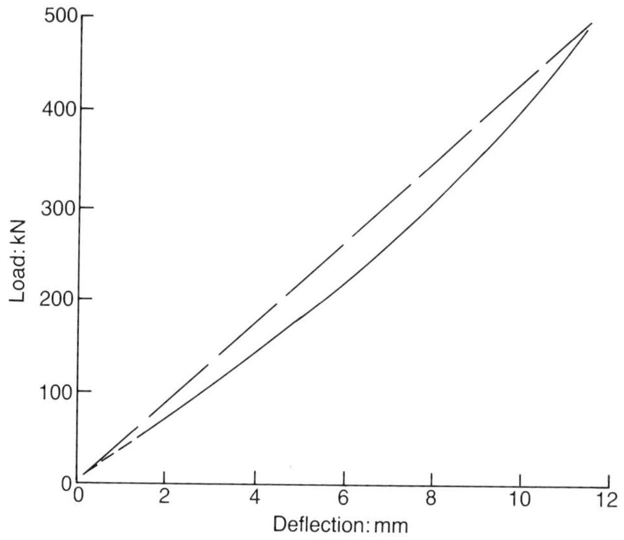

Fig. 9.14. Load deflection curve for Type A, test bearing 125 mm thick

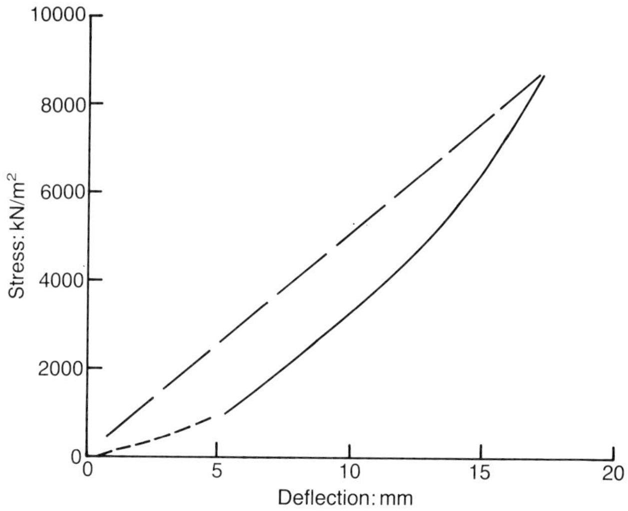

Fig. 9.15. Stress deflection curve a Type B pad, 100 mm thick

testing machines can be used as long as the dynamic stiffness of the machine is many times greater than that of the test bearing. A description of a vibrating rig, for measuring the resonant frequency of a bearing supporting a large mass, is given in reference 9.13. At least one manufacturer[9.14] has constructed a large-scale version of the rig.

A uniform rubber pad without any reinforcement is normally not suitable, except where the loads are very small. There are really only two types of pad construction in use today.

(*a*) Type A, normally a uniform block of rubber, often carbon-loaded natural rubber, reinforced and stiffened with steel plates. The flexibility in compression is controlled almost entirely by bulging at the free surfaces. This type is a variation on the designs used for bridge bearings.

(*b*) Type B, a laminated block of synthetic rubber made more flexible by loading with cork particles and reinforced with layers of woven cloth, often nylon.

A combination of the two types of design, whereby a block of natural rubber is reinforced with an aramid material, is still in development.[9.15]

The static load deflection characteristics for the two types are fairly similar: an example is given in Fig. 9.14 for a test bearing of Type A, made by Andre Ltd, 250 × 250 × 125 mm thick and reinforced with six steel plates 3 mm thick; and for Type B in Fig. 9.15 for a CV/CA pad 200 × 200 × 100 mm thick made by TICO Manufacturing Ltd. The bearing of Type A had been subjected to a preload of 50 kN and the Type B pad has been bedded in on a layer of grit paper at the bottom and top faces by precycling between 0 and 50 kN several times. Both static load deflection curves show a certain amount of non-linearity; a dashed straight line has been drawn on each graph between 0 and the maximum design stress point, which is 8000 kN/m^2 on Fig. 9.14 and 7500 kN/m^2 on Fig. 9.15. The total static deflection for Type B is somewhat greater because initially the cavities formed by the cork particles deform more easily than a bulging at the free surfaces.

A somewhat lower total deflection for the same design stress is advantageous for construction on site because the overall settlement of the building will be less as construction proceeds up to the full weight. However, a greater static flexibility has advantages when a cluster of pads is installed close together on a pile cap or on a column head. Some slight unevenness or rotation of the structure above the pads during construction causing a difference in compression between pads at one side of the cluster relative to the other side of say 2–3 mm will produce less of an overload for Type B than with Type A.

The non-linear deflection characteristic of both types is good in that the pads become stiffer when overloaded. The stiffness which determines the local resonant frequency of a bearing is the dynamic stiffness at the design stress condition, which is greater than the slope of the curve for the static deflections at the design point. The slope of the dashed straight line on each graph is of no significance whatsoever. The ratio of dynamic to static stiffness for natural rubber at these stress levels and at a frequency of say 10 Hz is between 1·7 and 2. The ratio for synthetic rubber is somewhat higher. There is a marked effect of the shape factor on the natural frequency of a pad design; pads with the same thickness and with a smaller plan area will normally have a lower natural frequency for the same stress. A better isolation can therefore be obtained by installing a large cluster of small pads than only one or two very large ones. There is, of course, a limit to this idea because the stability of the pads must be maintained.

Another consideration is the amount of overload capacity a pad will have under abnormal conditions such as whilst deformed in shear due to a local settlement of part of the foundation or due to excessive wind loading. The Type A pads must be manufactured with the steel plates absolutely in parallel, otherwise a wedge-shaped extrusion effect can occur during deformation in shear. A failure of the bond between a plate and the rubber matrix would then permit a free extrusion of the rubber with a consequential loss in carrying capacity for the pad. There are normally many more layers of woven cloth in the Type B design than steel plates in Type A. Hence a failure of one layer of cloth will hardly affect the carrying capacity. A distortion in shear has little effect because the woven reinforcement is very flexible in bending.

Although some typical maximum design stresses have been quoted above, these should be used only for industrial mountings where the total mass to be isolated is known accurately. A lower value should be used for building isolation. Many other factors need to be considered and these are referred to in the British Standard.[9.12] Protection against damage by fire must be provided when the mountings are above ground. Rubber has a very low thermal conductivity and the core of a pad will not be affected by a burning of the outer surface, except that with the Type A design the steel plates can conduct heat throughout the rubber matrix. A failure of the bond can then have a catastrophic effect. There is no such heat transfer with the Type B pads, although nylon can melt rather easily at the outside faces. A noninflammable and low-conductivity aramid material would be desirable.

A series of tests have been carried out[9.16] to determine the fire resistance of a natural rubber bearing reinforced with steel plates. There was a good resistance to flame sweep with only light charring on the exposed surfaces. A load deflection measurement of a bearing before and after exposure to heat showed a marked reduction in the static stiffness, i.e. a deflection under the same load of more than double in some tests. What matters in practice is that the load-carrying capacity is maintained when the bearings are subjected to an intensive heat-soaking period. A failure of the bonding of the steel plates would result in a serious distortion and possibly a collapse. The failsafe upstands would then come into operation and carry the structure.

A typical transmissibility frequency curve is given later (Fig. 9.22) for a sample of TICO CV/CA material. The thickness was kept small (25 mm) in

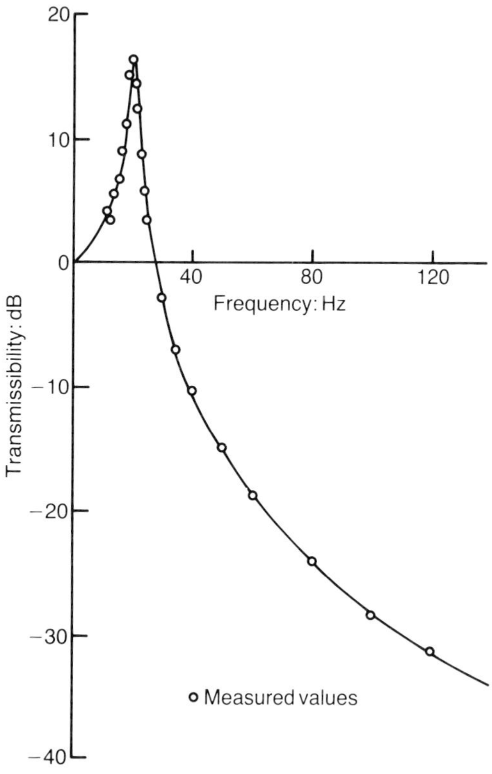

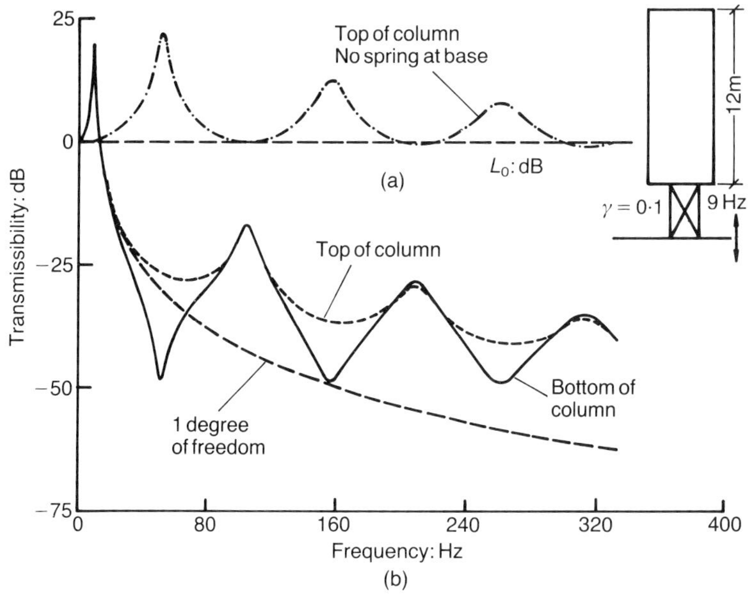

Fig. 9.17. Transmissibility frequency curves to show: (a) the effect of the dynamic response of a structure above a bearing; (b) the level which would result when no resilient bearing is installed

Fig. 9.16. Transmissibility frequency curve for a type B pad, 25 mm thick—full-line theoretical for a loss factor = 0·12

order to maintain a realistic shape factor and compressive stress on the small area of sample that could be tested. The method of testing was as described in reference 9.13 and with a load cell under each of the three pads supporting the vibrating mass. The transmissibility values plotted in Fig. 9.16 are the actual force ratios in dB. The continuous curve is the theoretical ratio from equation (9.1) with a loss factor = 0·12 to fit the resonance level at 20 Hz. This corresponds to a damping ratio of 6%. The agreement is very close, indicating that the material performance was as expected. A natural frequency below 10 Hz can be obtained for a pad with a thickness of 100 mm.

In a real installation, the resonant response of the structure above the bearings can reduce the effectiveness of the isolation at higher frequencies. As an example, consider a concrete column, 12 m tall, supported on a rubber bearing such that the natural frequency of the mass of the column on the spring is 9 Hz. The transmissibility ratio in dB at the bottom and at the top of the column relative to a base excitation of the bearing has been calculated, using the Imperial College, London, Couple 87 Program. The results are shown in Fig. 9.17. A common loss factor of 0·1 has been taken for the concrete and the rubber, which has produced rather sharp resonant peaks (for the free-free mode of the column) and deep anti-resonances (for the grounded column modes). In reality the structure connected to the column would take off some vibrational energy, thereby reducing the peaks. The attenuation through the peaks is about 12 dB/octave as expected from theory.

At first sight it looks most disappointing to have an attenuation at certain frequencies of less than half that predicted by the one degree-of-freedom theory for vibration isolation. However, the motion at the top of the structure would have been much greater if no resilient bearing had been installed. A further calculation has been made for base excitation at the bottom of the columns, i.e. no isolating spring, and the result is shown also in Fig. 9.17. There is, of course, no attenuation with the transmissibility >0 dB at all frequencies. Comparing the two curves for the motion at the top of the column makes it very clear how beneficial the installation of a resilient bearing at the base can be for a resonant structure. It is also obvious from this analysis that the more structural damping the better, but regrettably modern construction practice is in the opposite direction.

9.9. Applications of composite construction for vibration isolation and movement control

9.9.1. Floating track slab

It has been pointed out already that wherever practicable the railway should be isolated from the ground. The floating track slab construction has been described (Figs 9.7 and 9.8). That there can be a reduction in the effectiveness of the isolation because of resonant responses of the structure above the bearings has been discussed for buildings in the previous section, see Fig. 9.17. A similar transmissibility response will be obtained when the disturbing force acts on the floating slab and the ground has to be isolated, as established by the theorem of reciprocity. It is therefore important to have as much damping in the floating slab as is practicable. A composite type of construction with a high damping (viscoelastic) layer inserted in the slab can be very effective. A sketch of the damped floating track slab of the Barbican Redevelopment in London is shown in Fig. 9.18. The main deck was constructed with prestressed beams and had very little damping. Applying the constrained layer viscoelastic damping layer improved the damping by at least five times.[9.3] A similar type of constrained damping layer has been used in the Piccadilly Line extension to Heathrow, as described in Fig. 9.7. This has also proved to be very succesful.

It is intrinsically better to incorporate precautions against noise and vibration at the design and construction stages than to leave it till later on. An example where no proper precautions were taken is the Dockland Light Railway in east London, which came into operation in 1987. The noise level from the steel-framed elevated sections is well above the specified limits[9.17] and there have been many complaints.

9.9.2. Buildings on springs

There are many buildings which are supported totally on elastomeric bearings, many over underground railways with the piles straddling the tunnels. There are also large developments over main line stations, such as at Charing Cross and at Victoria Stations in London, where the columns for part or the whole of the structures are supported on bearings. The nominal design stress on a bearing pad is often taken as between 7 and 8 N/mm^2. A pad of area 250 × 250 mm can carry about 45 t. Many pads have to be installed

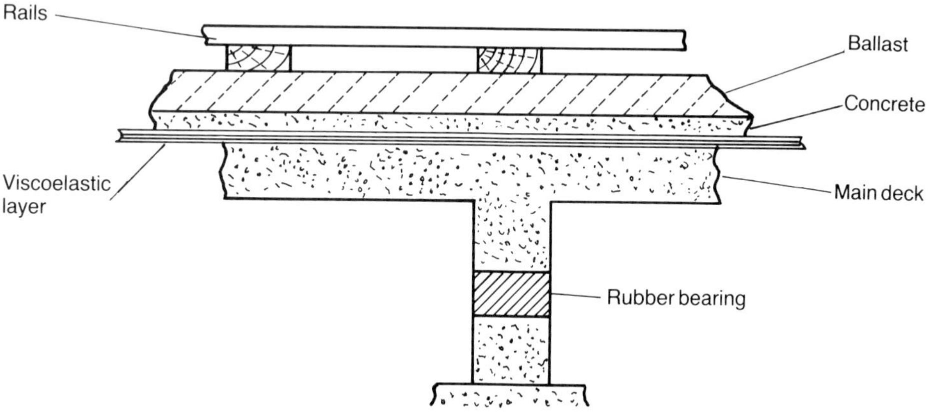

Fig. 9.18. Composite construction of a floating track slab with a damping layer

under a large building. A major consideration where judgement and experience can play an important part is in establishing the most likely live load and not taking the maximum specified in the design codes. Otherwise the bearings will be understressed and the isolation effectiveness will be reduced at the lower frequencies. The non-linear characteristics of most bearing designs provide for an increase in stiffness when the loading increases substantially and this provides a degree of failsafe (Figs 9.14 and 9.15).

There are a few early examples in the USA of the vibration isolation of buildings on lead and asbestos pads, but these were not successful because the pads exhibited a very high dynamic stiffness. The first building in London to be mounted on rubber bearings was Albany Court, a block of flats over St James' Park Station, completed in 1965.[9.18] Since then, a great variety of buildings have been mounted on bearings, including some large "noise barrier blocks" alongside the main line tracks out of Euston Station, which have a capacity of more than 1000 trains per 24 h. The barrier blocks are to shield the rest of a housing development from the air-borne noise from the trains and they are the more effective the closer they are to the railway. However, unacceptable levels of vibration and structure-borne noise will prevail inside such a barrier block if it is not fully isolated from the ground-borne vibrations.[9.19] A detail of the bearing design for the barrier block at the Alexandra Road Development in the London Borough of Camden is shown in Fig. 9.9. The main line tracks are only some 5 m away.

The elastomeric bearings can offer some resistance to lateral forces, arising from wind loading or from uneven settlement of the foundations. With low-rise buildings there are usually sufficient pads to accommodate all the wind loading. With tall buildings some additional lateral constraints will have to be provided which should not, of course, act as transmission paths for the ground-borne vibrations. Bearings which act in a horizontal direction can be installed around the perimeter of the base of the building. Alternatively, dowl pins can be installed at the locations for the main bearings. As the dowl pins will traverse the separation gap some resilient isolation must be included, usually by placing a rubber sleeve over the top half of the pin which is then cast into the concrete above the bearing. An example of an installation of rubber pads with dowl pins is shown in Fig. 9.19. There are four pads, each with

Fig. 9.19. Resilient bearing on top of a pile cap with dowl pins for lateral stability (courtesy TICO Manufacturing Ltd)

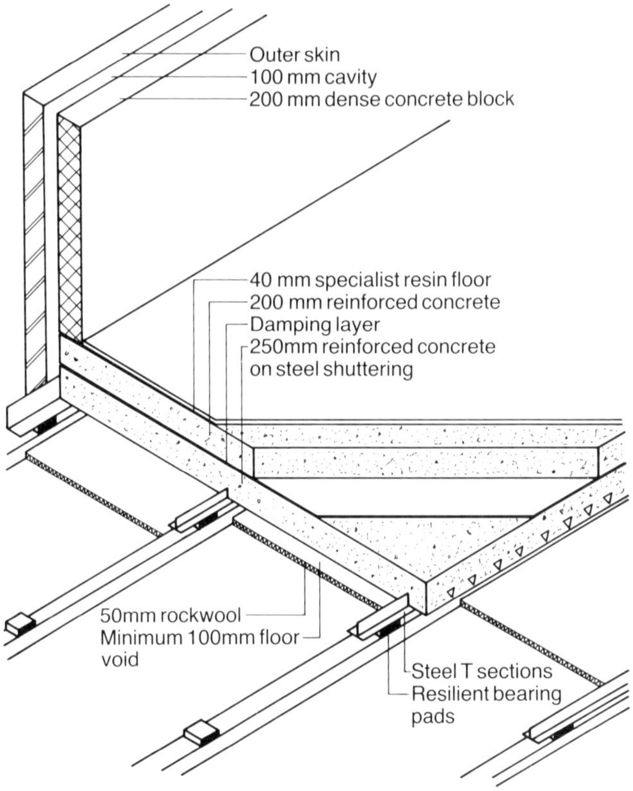

Fig. 9.20. Example of the composite construction for a TV studio on resilient bearings with a damping layer in the floor (courtesy Sandy Brown Associates)

Fig. 9.21. Offshore accommodation module mount, normal load 370 t to give a natural frequency of 12 Hz (courtesy Andre Ltd)

a dowl pin through its centre and set firmly in the pile cap. Two of the pins in the photograph have a rubber sleeve placed in position and the remaining two sleeves are lying on top of the pads. The column is then cast on top of the pads; the nest of reinforcing bars welded to a base plate can be seen standing ready in the background. It is important to allow sufficient clearance at the top of the pins to accommodate the compression and creep of the bearing once the full load is applied. The pins are usually made of stainless steel to ensure a life longer than that of the building. The installation shown in Fig. 9.19 is under the 20 storey office tower of Rank Hovis McDougal at Vauxhall Bridge, London, where some of the piles are within 1 m on either side of a tunnel of the Victoria Line.

9.9.3. Recording and TV studios

An application where any noise intrusion is usually unacceptable is in broadcasting and recording studios. The BBC in Broadcasting House, Langham Place, have suffered for many years from the structure-borne noise from the underground trains. Several of the studios in the basement have been mounted on springs. The accepted form of construction for studios is "a box within a box", with the inner box mounted on low frequency springs. It is important to introduce as much as possible structural damping in the construction of the isolated box to reduce the effects of structural resonances. The design for a TV recording studio is given in Fig. 9.20. The inner box has a mass of about 250 t, a height of 6 m and a floor area of 125 m^2. The box has been mounted on TICO CV/LF/N rubber pads with a suspension natural frequency of less than 8 Hz. Damping has been introduced into the floor by making it in two halves with a constrained bituminous damping layer. The size of the air cavity surrounding the inner box in relation to the vibration frequency spectrum in the adjacent structure has to be considered rather carefully.

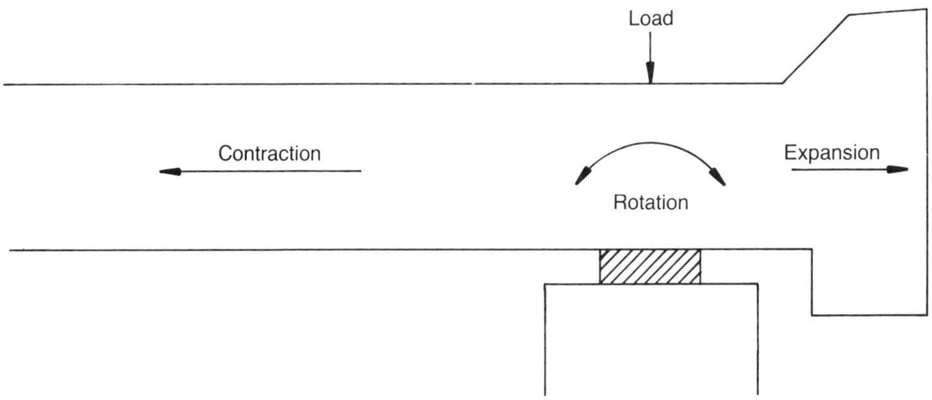

Fig. 9.22. Use of a resilient seating to take up structural motion in rotation and in translation

9.9.4. Accommodation units on offshore platforms

The need to provide a restful accommodation environment for the off-duty crew members on a drilling platform has led to the development of isolated modules which can be lifted on the platform as a unit. Some of these modules are very large with only a few attachment points. The resilient bearings tend also to be large and must provide lateral and vertical constraint. An example of such a bearing is shown in Fig. 9.21 for the Britoil Beatrice B platform. The bearing has a nominal load-carrying capacity of 370 t with an isolation natural frequency of about 12 Hz.

9.9.5. Resilient seatings

Elastomeric materials are sometimes introduced into a structure to accommodate excessive movements due to thermal contraction and expansion or from ground settlement. Local stress concentrations can be eliminated. A sketch of a single seating at the supporting point for a floor slab is given in Fig. 9.22. Movement in rotation and in translation can be accommodated readily. The lateral stability of the structure as a whole can be provided by shear walls and by the central cores for lift shafts and the stairs. The introduction of resilient seatings can eliminate stresses due to movement and can lead to a more economic design. Precautions against a collapse of the supporting structure due to an explosion or impact loading can be built in with dowl pins, similar to those shown in Fig. 9.19.

9.10 Conclusions

The use of elastomeric bearings for the vibration isolation of buildings, recording studios and of large structures is well established. With careful design and selection of materials, the bearings will have a life longer than the building. Railways, which are often the source of the ground-borne vibrations can be isolated with the floating track slab form of construction. The best results are obtained with a composite construction including a constrained damping layer.

Modern trends in building construction tend to make the transmission of structure-borne noise and vibration much easier. Trends to light-weight floor slabs of very large span and with no movement joints can make whole structures dynamically very lively. The use of welded steel framed structures in place of in situ concrete can also lead to an extensive transmission throughout a building. Excellent isolation can be obtained with resilient bearings with an isolation natural frequency of between 8 and 12 Hz, but the effectiveness is much reduced by the resonant responses of parts of the structure above the bearings. Such structural responses are stronger the less damping there is in the structure. There is thus an urgent need to introduce into modern construction some effective vibration damping. This can be done with composite construction including a viscoelastic damping layer.

The confidence in mounting large buildings solely on rubber bearings was greatly boosted in October 1987 when London and the eastern part of England were struck by a hurricane, and thousands of trees were uprooted and widespread structural damage was caused. There have been no reports that any of the many buildings mounted on resilient bearings suffered damage as a result of excessive movement during the storm. However, the decision on whether to mount a building on bearings at a low frequency and on the location of those bearings, with the subsequent effects on other structural design details, is not one to be left to the faint-hearted, but is best handled by the experts.

References

9.1. Derham C.J. *et al.* Vibration isolation and earthquake protection of buildings by natural rubber mountings, *NR Technology*, 1975, **6**, No. 2, 21–33.

9.2. Clayden K.J.G. Noise and vibration attenuation in the Barbican scheme. In *Proc. Rubber in Engineering Conf.*, paper 16. Natural Rubber Producers' Research Association, London, 1966.

9.3. Grootenhuis P. The attenuation of noise and ground vibrations from railways. *J. Env. Sci.*, 1967, **10**, April, 14.

9.4. Parsons K.C. and Griffin M.J. Whole-body vibration perception thresholds, *J. Sound and Vib.*, 1988, **121**, No. 2, March, 237–258.

9.5. Grootenhuis P. The effect of vibration on buildings and the occupants. *Inter-noise*. Institute of Acoustics, London, 1983.

9.6. Awojobi A.O. and Grootenhuis P. Vibration of rigid bodies on semi-infinite elastic media. *Proc. Roy. Soc. A*, 1965, **287**, 27–63.

9.7. Hamidzadeh-Eraghi H.R. and Grootenhuis P. The dynamics of a rigid foundation on the surface of an elastic half-space, *Earthquake Eng. and Struc. Dyn.*, 1981, **9**, 501–515.

9.8. Grootenhuis P. and Allaway P.H. Noise and vibration nuisance within buildings. In *Proc. Symp. Environmental Aspects of Pollution*, Society of Environmental Engineers, London, 1971.

9.9. Grootenhuis P. Vibration isolation of buildings—some basic principles. In *Proc. Conf. NR for Earthquake Protection of Buildings, Kuala Lumpur, 1982*, 14.

9.10. Grootenhuis P. Floating track slab isolation for railways. *J. Sound and Vib.*, 1977, **51**, No. 3, 443–448.

9.11. Grootenhuis P. Sandwich damping treatment applied to concrete structures. *Trans. Roy. Soc. A*, 1968, **263**, 455–459.

9.12. British Standards Institution. *Guide to selection and use of elastomeric bearings for vibration isolation of buildings.* BS6177. British Standards Institution, London, 1982.

9.13. Grootenhuis P. Measurement of the dynamic properties of damping materials. In: *Proc. Int. Symp. on the Damping of the Vibration of Plates by Means of a Layer*. Assoc. Belge des Acousticiens, Leuven, 1967.

9.14. Liquorish A.D. Development and application of non-metallic reinforced resilient bearings, *Trans. I. Mar. E. (C)*, **97**, conf. 2, paper 5, 35–40.

9.15. Kamarudin Ab-Malek. Development of natural rubber bearings laminated with non-metallic materials. In: *Proc. Int. Conf. Natural Rubber, Jakarta*, 1987.

9.16. Derham C.J. and Plunkett A.P. Fire resistance of steel-laminated natural rubber bearings. *NR Technology*, 1976, **7**, No. 2, 29–37.

9.17. Shield B.M. *et al.* A survey of noise levels from the Docklands Light Railway. *New sources and problems in urban transportation noise*, 17–23. Institute of Acoustics, London, 1988.

9.18. Waller R.A. *Building on springs*, Pergamon Press, Oxford, 1966.

9.19. Grootenhuis P. Resilient mountings of large structures. In: *Elastomers: criteria for engineering design*, chapter 13. Applied Science, London, 1979.

10 Geosynthetics

C. R. Lawson

10.1. Introduction: geotextiles, geomembranes, geosynthetics

One of the major advances of polymers in the last 20 years has been their use in the geotechnical engineering industry. Perhaps the most commonly known material is the geotextile. A simple definition for geotextile would be 'a textile material used in a soil (*geo*) environment'. These polymeric materials which are known as geotextiles have followed an evolutionary path, not only in terms of material development, but also with respect to the terminology associated with these materials. In the early 1970s, these materials were referred to as 'civil engineering fabrics' (they were used in civil engineering applications) or 'filter fabrics' (where their primary use was for filter applications). By the mid-1970s, they were being referred to as 'geotechnical fabrics' (their primary use was in geotechnical/soil engineering applications). It was not until the late 1970s that the term 'geotextile' (a textile used in a soil environment) was suggested as a suitable name for these polymer materials. This name was slow to take off, however, and it was not until 1982 (at the time of the Second International Conference on Geotextiles) that the name became sufficiently well known so as to be commonly identifiable with textile materials used in a soil environment.

Parallelling the development of geotextiles, impermeable polymeric membranes have also been used on an increasing scale. These materials became known as 'geomembranes', impermeable membranes used in a geotechnical environment.

By the mid-1980s, many types of polymeric-based finished materials were being used in the geotechnical engineering industry, a number of which could not be classified as being either a geotextile or a geomembrane. To encompass all of these polymeric materials a new term, namely 'geosynthetic', was derived. A simple definition for geosynthetic would be 'a synthetic (polymeric) material used in a soil environment'.

The estimated world-wide annual useage of the major geosynthetic material varieties is shown in Fig. 10.1. Geotextiles, by virtue of their relatively early emergence, comprise the most mature material type, with geomembranes and the remaining geosynthetic materials at earlier stages of use.

The reasons why the use of geosynthetics has shown such high growth are twofold. Firstly, geosynthetics have made it possible to solve geotechnical problems which could not be solved using conventional techniques. Secondly, geosynthetics have proven to be highly successful in situations where the use of conventional materials has resulted in poor performance due to either their inherent variability or their difficulty of installation.

10.2. Classification and general uses

10.2.1. Classification

Geosynthetics can be divided into five broadly based categories.

Geotextiles: these are polymeric textile materials used in geotechnical engineering applications. These materials are essentially permeable to the passage of water.

Geogrids: these are open, mesh-like polymeric structures.
Geomembranes: these are polymeric materials (in sheet form) which are essentially impermeable to the passage of water. They are normally used as impermeable liners in ponds, dams, etc.
Geo-linear elements: these are long, slender, polymeric materials normally used as reinforcing tendons in soils and rocks.
Geocomposites: this category covers all polymeric materials used in a soil environment not covered by the above four categories. Commonly, they consist of materials formed from a combination of the above four categories, or from a material from one of the above four categories combined with a completely different material. This category currently covers many different polymeric materials; some examples are prefabricated subsurface drains, linear reinforcing elements set onto a geotextile substrate, and concrete blocks set on to a geotextile substrate.

Some representative properties of geosynthetic materials are listed in Table 10.1.

10.2.2. General uses

Geosynthetics are used in a wide variety of civil engineering applications. A general description for each geosynthetic category is given below.

10.2.2.1. Geotextiles

Geotextiles are used predominantly in situations where the controlled passage of soil water is required. General applications include the following.
Temporary pavements. Here the geotextile is placed between a soft subgrade and the granular material of the pavement to prevent intermixing of the soft subgrade with the granular material.
Permanent pavements. Here again the geotextile is placed between a soft subgrade and the granular material in the pavement to prevent intermixing of the soft subgrade with the granular material.
Asphalt overlays. In this application the geotextile is placed at the underside of an asphalt overlay to prevent (reflective) cracking from migrating to the new surface of the pavement.

Table 10.1. Range of values for some representative properties of available geosynthetics

Geosynthetic	Tensile strength: kN/m	Maximum extension: %	Apparent opening size: mm	Volume water permeability: $l/m^2/s$*	Unit weight: g/m^2
Geotextiles					
Wovens	8–800	9–35	0·07–2·5	5–2000	90–1300
Nonwovens	3–90	20–80	0·01–0·35	20–200	70–2000
Knitteds	2–120	12–600	0·04–1·2	60–2000	120–500
Stitch-bondeds	30–1000	8–30	0·07–0·5	30–80	250–1200
Geogrids	10–150	3–25	25–150	Very high	150–900
Geomembranes					
Nonreinforced	10–50	100–500	0	0	300–1500
Reinforced	20–200	10–30	0	0	600–2000
Geo-linear elements	50–500†	3–15	N/A	N/A	N/A

* Normal to the plane of the geosynthetic with 10 cm constant head.
† Measured in kN—not kN/m.

Railways. In this application the geotextile is placed between the ballast layer and the foundation soil to prevent intermixing of the two.

Tailings reclamation. Tailings, which are a mixture of solid particles and water (and, in some cases, toxic chemicals) are the waste by-product of mining operations. They are normally stored in large tailings impoundment areas. Geotextiles have been used as one of the techniques employed to reclaim these tailings impoundments by facilitating the construction of soil capping layers.

Embankments. Geotextiles are used beneath embankments constructed over soft foundation soils to ensure intermixing between the embankment material and the soft foundation soil does not occur. If the geotextile is strong enough it can also be used also to reinforce the base of the embankment to prevent instability.

Reinforced slopes. Geotextiles may be used to reinforce soil slopes to prevent them from failing.

Erosion control. Geotextiles are used in two types of erosion control structure. Firstly, they are used as a filter beneath rock or pattern-placed armour units to prevent erosion of the soil. Secondly, they are used as a flexible form to retain concrete until it has hardened.

Subsurface drainage. In this application the geotextile is used as a filter to enable water to be drained from the soil.

10.2.2.2. Geogrids

Geogrids are used for a variety of soil reinforcing applications, the major ones being as follows.

Reinforced soil retaining walls. In this application the geogrid is used to reinforce the backfill of a retaining wall, thus preventing the retaining wall from collapsing.

Reinforced slopes. In this application the geogrid reinforces the soil slope to prevent it from failing.

Asphalt overlays. The geogrid is placed at the underside of the asphalt overlay to prevent cracking of the surface of the new overlay.

10.2.2.3. Geomembranes

Geomembranes are used in geotechnical applications where it is necessary to prevent the loss of water or toxic materials. Applications include the following.

Liquid containment. The geomembrane is used to prvent the loss of liquids from a containment structure.

Waste containment. The geomembrane is used to prevent the loss of toxic chemicals from waste impoundment structures.

Pavements. Geomembranes have been used to control moisture movements in pavements thus enabling them to maintain their structural integrity.

10.2.2.4. Geo-linear elements

Geo-linear elements are used for two types of applications.

Reinforced soil retaining walls. In this application they are used as reinforcing strips within the backfill to prevent the facing panels of the wall from collapsing.

Anchors. In this application they may be used to anchor sheet wall structures to the in situ soil.

10.2.2.5. Geocomposites

Geocomposites can be used for a wide variety of applications depending on the constituent components used. Examples include: prefabricated drains, flexible skins, artificial seaweed, expedient surfacings, etc.

10.3. Material types and characteristics

10.3.1. Geotextiles

The manufacture of geotextiles follows the general steps set out in Fig. 10.2. These steps are usually condensed in the geotextile literature into a two-stage process: the first stage consists of making linear elements such as filaments, fibres, tapes and yarns; the second stage consists of combining these linear elements via an appropriate textile-forming technology into the geotextile. The various environmental properties of a geotextile are primarily determined by the fibres produced at the end of the first stage. The fibres are obviously the basic load-carrying elements in a geotextile. The textile-forming technology adopted, together with any subsequent finishing treatments, determines the structure of the geotextile in terms of the packing geometry of the constituent fibres. This determines the physical characteristics of the geotextile

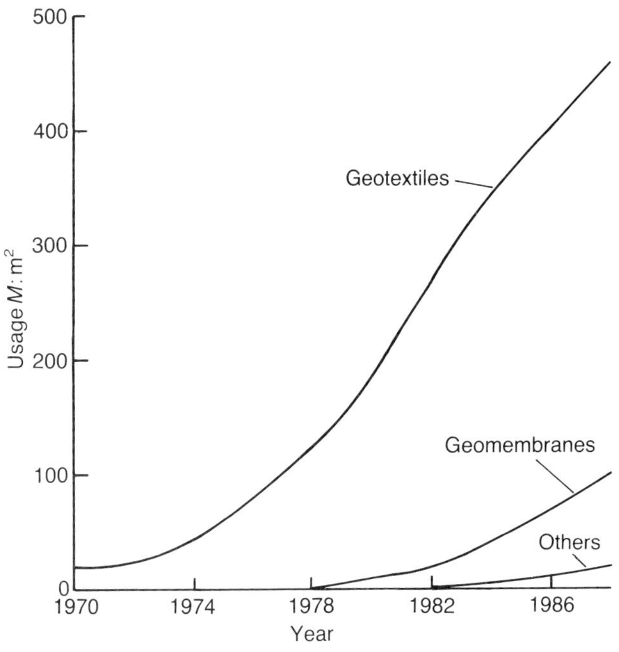

Fig. 10.1. Estimated usage of geosynthetics

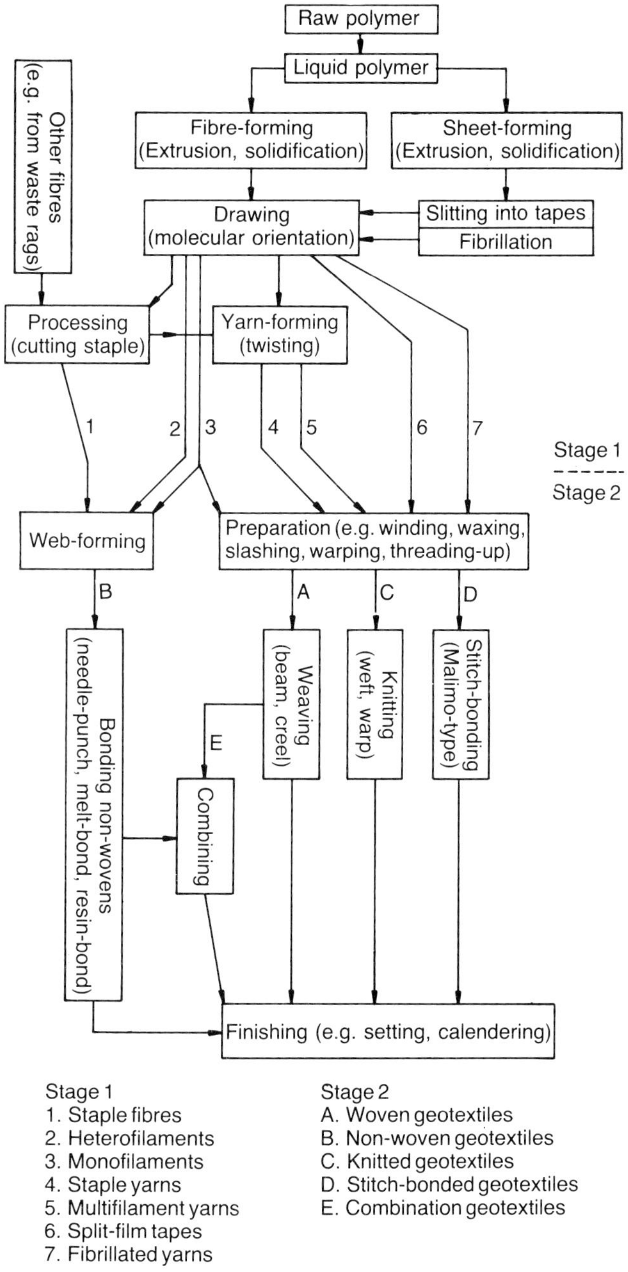

Fig. 10.2. Simplified outline of steps in the production of geotextiles[10.1]

(e.g. unit weight, thickness, and size of openings) and also the means by which loads are transmitted to, and supported by, the constituent fibres.

10.3.1.1. Fibre characteristics

Many fibres have been used in the manufacture of textile materials; however, the main ones used in geotextiles are the synthetic fibres—more importantly, polypropylene, polyethylene, polyester and polyamide. This is because synthetic fibres are the only fibres which can be engineered chemically, physically, and mechanically to suit particular geotechnical engineering applications. Natural fibres (e.g. cotton, jute) and the majority of regenerated fibres (e.g. cellulose, rayon) are seldom used to make geotextiles because they are biodegradable; however, geotextiles made from natural fibres and even paper (another fibrous product) may serve temporary functions where biodegradation is desirable (e.g. temporary erosion control). The general properties of synthetic fibres are listed in Table 10.2.

In choosing a fibre type, a manufacturer would hope to balance their properties and the relative cost of supply and manufacture with the technical requirements of the geotextile user. Further, more than one fibre type can be used in a blend in order to optimize the geotextile properties and/or to consolidate the geotextile structure.

Table 10.2. Some properties of synthetic fibres (after reference 10.2)

Fibre type	Ultimate tensile strength: MN/m^2	Extension at Break: %	Specific gravity	Melting point: °C	Maximum working temp: °C
Acrylics	300–400	20–40	1·14–1·18	Decomp.	180
Chlorofibres					
Polyvinyl chloride	340	20	1·38–1·40	160	80
Polyvinylidene chloride	300–400	20–40	1·70	160	80
Fluorofibres	105–175	40–100	2·30	Decomp.	220
Modacrylics	300–400	20–40	1·31–1·37	Decomp.	170
Nylons					
Nylon 6	450–750	20–40	1·13	215	70
Nylon 66	450–1000	12–30	1·14	260	130
Aramids	500–2900	2–12	1·38–1·45	Decomp.	450
Polyolefins					
Polypropylene	400–700	10–20	0·90	165	100
Polyethylene					
Low density	80–120	25–50	0·92	110	55
High density	350–500	20–30	0·95	135	65
Ultra high density	2400–2700	3–4	0·96	180	90
Polyesters	500–1400	8–30	1·38	260	170
Polyurethane elastomers	70–110	400–800	1·00–1·32	230	150
Vinylals	200–700	10–15	1·30	230	140
Glass	1750–3500	2–4	2·50	1300	1000
Prestressing steel	1500–2000	2	7·80		

The manufacture of synthetic fibres follows the first stage set out in Fig. 10.2. The first step involves the transformation of the raw polymer from solid to liquid form. Depending on the polymer, this is normally done in one of two ways: either by dissolving the polymer in solution or by heating the polymer until it is molten. Of the synthetic polymers listed in Table 10.2, it is common for the acrylic, modacrylic, aramid, and vinylal polymers to be dissolved into solution, with the polyamide (nylon), polyolefin (polypropylene and polyethylene), and polyester polymers being transformed into molten liquids. The chlorofibre polymers can be transformed into a liquid by either means. The liquid polymer is then extruded through a spinneret, a device consisting of many fine holes, and solidified into continuous strands (filaments).

Once the filaments are formed they are drawn (i.e. stretched along their longitudinal axis) and this increases greatly the orientation of the molecular chains within the filament structure, thus altering its stress–strain characteristics, namely higher ultimate tensile strength and lower elongation at maximum load. The amount of drawing performed is termed the 'draw ratio' and this is the amount by which the filament is extended as compared to its original length. Draw ratios between 5 : 1 and 15 : 1 are not uncommon for industrial synthetic filaments.

The production steps result in seven different types of synthetic fibres or yarns: monofilament fibres, heterofilament fibres, multifilament yarns, staple fibres, staple yarns, split-film tapes, and fibrillated yarns.

10.3.1.2. Geotextile types

Various textile-forming technologies have been used to manufacture geotextiles and these can be classified into four major groups (see Fig. 10.2 and Table 10.1) each exhibiting particular mechanical, physical and hydraulic characteristics.

(a) Woven geotextiles. Woven geotextiles are manufactured by interlacing two sets of monofilaments, tapes or yarns perpendicular to each other (see Fig. 10.3(a)). Two terms are used to describe the alignment of these principal directions of fabric construction: warp (or ends) and weft (or picks or fill). The warp direction is along the direction of manufacture of the geotextile while the weft is inserted perpendicular to the warp.

Table 10.3. Range of values for some representative properties of available geotextiles (after reference 10.2)

Geotextile construction	Tensile strength: kN/m	Maximum extension: %	Apparent opening size: mm	Volume water permeability: $l/m^2/s$	Unit weight: g/m^2
Wovens					
Monofilament	20–80	9–35	0·07–2·5	25–2000	150–300
Multifilament*	40–800	9–30	0·2–0·9	20–80	250–1300
Tape	8–90	15–20	0·05–0·1	5–15	90–250
Nonwovens					
Melt-bonded	3–25	20–60	0·01–0·35	25–150	70–350
Needle-punched	7–90	50–80	0·02–0·15	25–200	150–2000
Resin-bonded	4–30	30–50	0·01–0·35	20–100	130–800
Knitteds					
Weft	2–5	300–600	0·1–1·2	60–2000	
Warp	20–120	12–15			
Stitch-bonded	30–1000	8–30	0·07–0·5	30–80	250–1200

* Fibrillated tapes are included in this category.

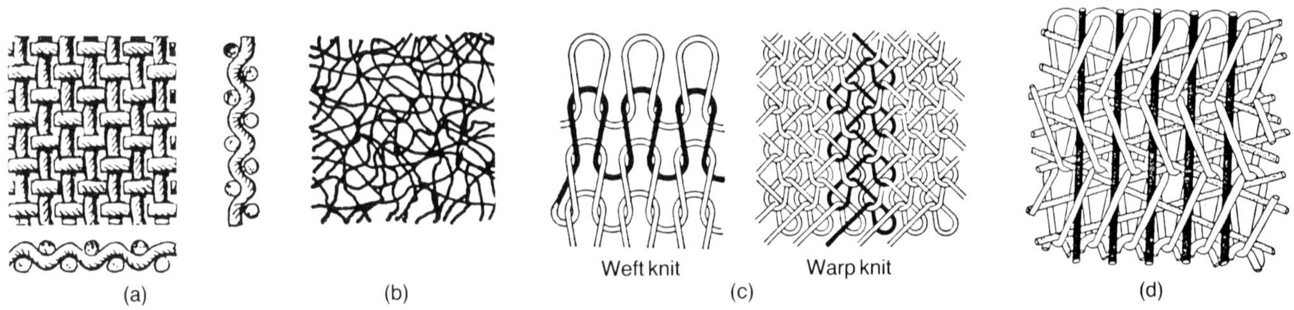

Fig. 10.3. Different geotextile constructions: (a) woven; (b) nonwoven; (c) knitted—(i) weft and (ii) warp knit; (d) stitch-bonded

Woven geotextiles can be divided into four subgroups according to the types of yarns, etc., used in their manufacture—monofilament wovens, multifilament wovens, fibrillated tape wovens, and tape wovens. Some indicative properties are given in Table 10.3.

(b) Nonwoven geotextiles. Nonwoven geotextiles are manufactured by first forming a loose web or layer(s) of staple fibres or continuous filaments, arranged in an oriented or random pattern, and then subjecting this to some form of bonding, thus achieving a cohesive planar structure. The general structure of the material is shown in Fig. 10.3(b). The method of web formation is important as it is primarily responsible for the distribution and orientation of the fibres within the final geotextile product. Following bonding, some form of finishing treatment may be applied in order to enhance some of the geotextile's properties, e.g. calendering with heated and/or textured rollers to achieve particular surface characteristics. The calendering technique has been described in section 2.6.4. The general aim is to produce a uniform and dimensionally stable geotextile which is essentially isotropic in its properties.

Nonwoven geotextiles can be subdivided into three groups, depending on the bonding method used: melt-bonded; needle-punched; resin-bonded. Melt-bonded and needle-punched nonwoven geotextiles are the most common nonwoven varieties available. Some indicative properties of nonwoven geotextiles are listed in Table 10.3.

(c) Knitted geotextiles. Knitted geotextiles are produced by mechanically interlooping yarns or monofilaments by forming loops of these fibrous elements with hooked needles and then drawing new loops through those previously formed. Although a wide variety of knitted structures can be obtained, as may be seen in apparel wear and furnishings, they have found only limited use as geotextiles. Knitted geotextiles can be subdivided into two groups depending on the method of manufacture—weft knitteds and warp knitteds (see Fig. 10.3(c)). Table 10.3 reports some indicative properties.

(d) Stich-bonded geotextiles. In the textile industry, there are several different types of stitch-bonding techniques. Essentially the method consists of stitching yarns or monofilaments through a fibrous web, thus obtaining a cohesive fabric. Alternatively, cross-laid yarns may be stitched together; this is the Malimo fabric construction illustrated in Fig. 10.3(d). The Malimo fabric construction exhibits load-extension properties similar to that of a woven geotextile because the loads are carried by the warp and weft fibres (the same as for wovens); the stitching thread only binding the structure into a stable mass. Some indicative properties of stitch-bonded geotextiles are listed in Table 10.3.

10.3.2. Geogrids

Geogrids are open, grid-like structures. As such, they develop a very efficient bond with the surrounding soil. Geogrids may be divided into two

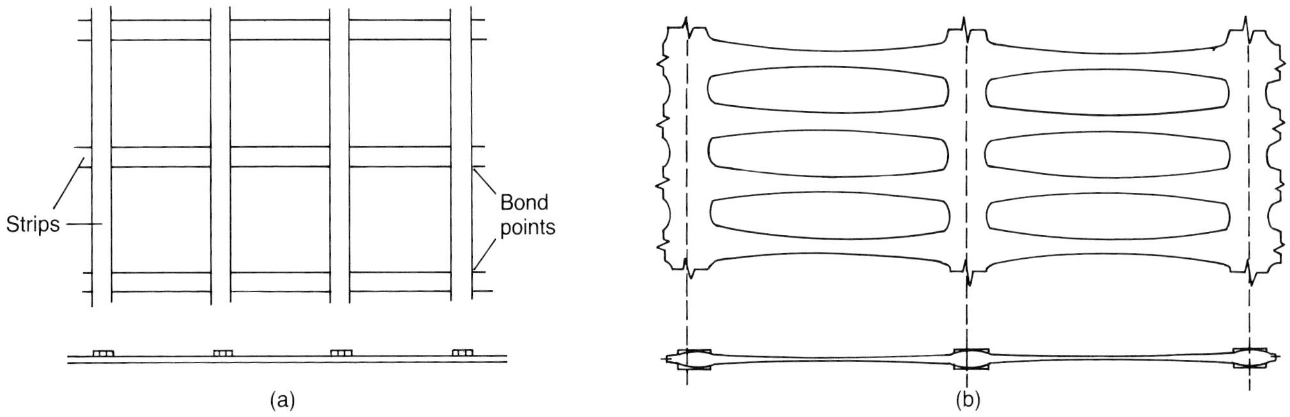

Fig. 10.4. Different geogrid structures: (a) cross-laid strip geogrids; (b) punched sheet geogrids

Table 10.4. Range of properties of available geogrids and geo-linear elements

Type	Tensile strength: kN/m	Maximum extension: %	Apparent opening size: mm	Unit weight: g/m^2
Geogrids				
Cross-laid strips	25–200	3–20	50–300	300–1200
Punched sheets	10–110	11–30	40–150	200–1100
Geo-linear elements				
Geostrips	20–150*	3–20	N/A	N/A
Geobars	20–500*	3–20	N/A	N/A

* Measured in kN (not kN/m).

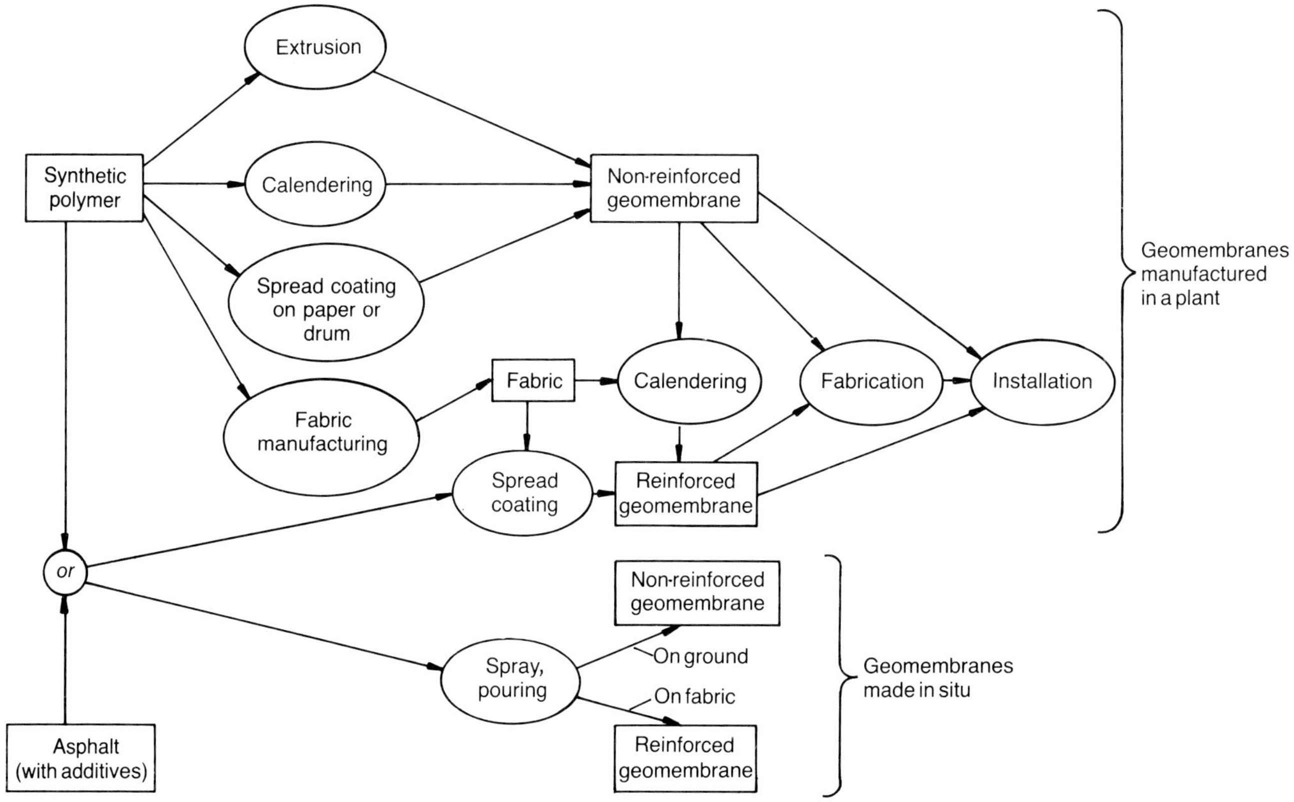

Fig. 10.5. Production of geomembranes[10.4]

categories—those formed from cross-laid strips and those formed from punched polymer sheets.

10.3.2.1. Cross-laid strips

Grids falling under this category are formed by the cross laying and bonding of geostrips (geo-linear elements). The geostrips are bonded together at their cross-over points (thus forming the geogrid). An example of a geogrid formed by this procedure is shown in Fig. 10.4(a). The resultant strength of the grid is a function of the strength of the individual strips and the spacing between the strips. Some typical properties of geogrids formed from cross-laid strips are listed in Table 10.4.

10.3.2.2. Punched sheets

The basic polymer materials used in the manufacture of these geogrids are either polypropylene or high density polyethylene. The geogrid is formed by the punching of holes on a regular pattern into a preformed polymer sheet. The punched sheet is then drawn either uniaxially or biaxially. An example of a uniaxially drawn structure is shown in Fig. 10.4(b). Some indicative properties of these types of materials are listed in Table 10.4. Section 2.6.1 has discussed the manufacturing techniques of extruded highly orientated grids.

10.3.3. Geomembranes

In general, four different categories of geomembrane exist which may be classified according to their manufacturing process. Of the four different categories there are two broad classes: geomembranes manufactured in situ (i.e. at the construction site) and geomembranes manufactured in a production facility. In each class, two categories of geomembranes can be identified: nonreinforced (sometimes called homogeneous) and those reinforced with a textile.

The different manufacturing variables used are listed in Fig. 10.5. Two different types of materials are used to impart impermeability: either synthetic polymers or bituminous materials. Bituminous materials are used either for geomembranes constructed in situ (both reinforced and nonreinforced) or for

Table 10.5. Major types of geomembrane polymers in use (after reference 10.3)

Plastics (thermoplastic polymers)
Polyvinyl chloride (PVC)
Polyethylene (LDPE, LLDPE, HDPE)
Chlorinated polyethylene (CPE)
Elasticized polyolefin (3110)
Polyamide (PA)
Rubbers (thermoset polymers)
Isoprene–isobutylene (Butyl)
Epichlorohydrin rubber
Ethylene propylene diene monomer (EPDM)
Polychloroprene (Neoprene)
Ethylene propylene terpolymer (EPT)
Ethylene vinyl acetate (EVA)
Combinations
PVC-nitrile rubber
PE-EPDM
PVC-ethyl vinyl acetate
Cross-linked CPE
Chlorosulphonated polyethylene (CSPE)

reinforced geomembranes manufactured in a production facility. Synthetic polymers are used for nonreinforced and reinforced geomembranes manufactured in a production facility. The types of synthetic polymers used for geomembranes are listed in Table 10.5.

Textile reinforcement may be used in geomembranes for a variety of reasons: to impart stability to the polymer compound during manufacture, to provide dimensional stability to compounds that would excessively shrink or expand as a result of temperature changes, to increase the strength of the geomembrane, and to reduce the elongation of the geomembrane when it is subjected to stresses. There is normally one of two types of textile used with geomembranes. For geomembranes manufactured by a saturating process the textile used is a nonwoven fabric which provides a substrate for the application of the saturant. For geomembranes manufactured by a coating process the textile used is a woven scrim fabric.

10.3.4. Geo-linear elements

Geo-linear elements are long, slender strips (geostrips) and bars (geobars) generally consisting of a continuous filament fibre core with a plastic or resin sheath. The fibres provide the strength-extension characteristics of the material while the sheath provides the shape of the material and its bond characteristics with the surrounding soil. The fibres normally consist of relatively high modulus materials such as polyester, aramid, or glass. The sheath normally consists of a very low permeability material such as low density polyethylene, or a resin. Because these materials are long and slender, they are quite flexible, and being made from synthetic materials they are quite light; hence they are easy to handle. Examples of the two types of geo-linear elements are shown in Fig. 10.6. Some general properties of these materials are listed in Table 10.4.

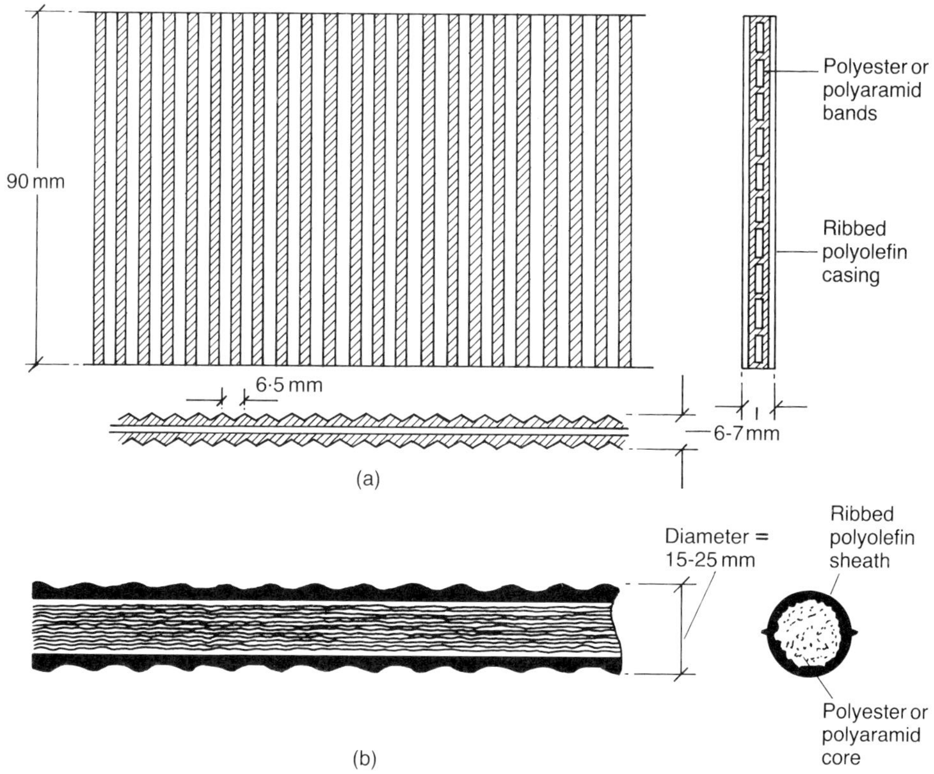

Fig. 10.6. Two geo-linear types: (a) geostrips; (b) geobars

10.3.5. Geocomposites

Geocomposites consist of two or more different polymer structures which are combined together to produce a hybrid material structure. As such, there are a multitude of different geocomposite material structures possible. To cover all of the geocomposite structures already in existence would be outside the scope of this book, thus it is proposed to concentrate on one major material type only—the prefabricated drain.

Prefabricated drains are geocomposite materials comprising a three dimensional polymer core surrounded by a geotextile filter. The polymer core acts as the drainage conduit along which water flows after it has been filtered by the geotextile skin. Two categories of prefabricated drain are in use: vertical (or band) drains and fin drains.

Vertical drains are strip-like in appearance. Their role is to act as drains to increase the rate of consolidation, when inserted into soft foundation soils. Table 10.6 lists the various vertical drains available.

Fin drains are sheet like in appearance. Their role is to replace conventional subsurface drainage structures, primarily behind concrete retaining structures. Table 10.7 lists the variety of fin drains currently available.

10.4. Applications

Geosynthetics are used in a wide range of geotechnical engineering applications. To cover every application would be outside the scope of this book. Therefore, only four major application areas will be discussed. These are filtration, separation, soil reinforcement, and containment.

10.4.1. Filtration

Filters enable the controlled passage of water out of soil masses. Filters must be permeable enough to allow easy passage of the soil water, while at the same time maintain soil particle retention to ensure the soil is not eroded.

Table 10.6. Details of variously available vertical drains (after reference 10.3; reproduced courtesy of Prentice-Hall)

Type	Manufacturer or agent	Core material	Core shape	Number of channels	Filter material (cover)	Size: mm
Kjellman	SGI	Paper		10	Paper	100 × 3·5
Geodrain	SGI	Plastic	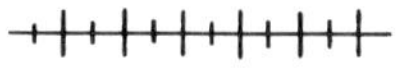	27	Paper	92 × 4·0
Alidrain	SGI	Plastic		Studs	Geotextile	100 × 6·0
Castle	Japanese	Plastic		36	Unitized	94 × 2·6
Bando	Japanese	Vinyl chloride		48	Unitized	96 × 2·9
Mebra drain	Dutch	Polyethylene		38	Geotextile	–
Ester					Polyester	95 × 3·3
Pap					Paper	95 × 3·2
Prop					Polypropylene	95 × 3·4
Sol					Polypropylene	95 × 2·0
Colbond [18]		Geotextile	xxxxxxxxxxxxx	∞	Geotextile	100 × 6·0
PVC		PVC		11	Unitized	100 × 1·6
Amerdrain	ICE	Polyethylene		8	Geotextile	92 × 10
Hitek	Burcan	Polyethylene		Studs	Geotextile	100 × 6·0
Desol	Recosol	Polyolefine	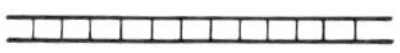	24	Unitized	92 × 2·0

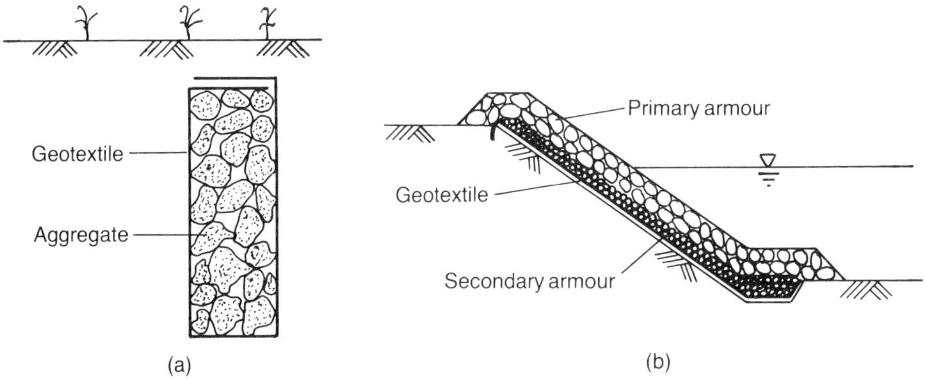

Fig. 10.7. Two applications where geotextiles are used as filters for: (a) subsurface drainage; (b) erosion control

These two conflicting flow requirements place considerable restrictions on the specification and use of filter materials.

Geosynthetics, in the form of geotextiles, are used extensively as filters. The reason for this is that they are produced in a factory under controlled conditions, thus ensuring consistent properties. Geotextiles can be engineered to be highly permeable, while at the same time exhibiting good soil-retention characteristics.

Geotextiles are used as filters for two major applications: for subsurface drainage, and for erosion control. For subsurface drainage, geotextiles are wrapped around the exterior of permeable drainage media (aggregate and/or porous pipes, or polymer spacers). The role of the geotextile is to allow seepage water to pass through into the drain but to prevent the passage of soil. This application is shown in Fig. 10.7(a). For erosion control, geotextiles are placed beneath the rock armour layers of revetments. The role of the geotextile is to allow water to pass through while preventing the erosion of the soil through the rock layers in the revetment. This application is shown in Fig. 10.7(b).

Table 10.7. Summary of available fin drains (after reference 10.3; reproduced courtesy of Prentice-Hall)

Trade name	Manufacturer	Core characteristics			Geotextile filter
		Thickness: mm	Shape	Material	
Amerdrain	International Construction Equipment, Inc.	8	6 mm corrugated ribs	Polyethylene	Nonwoven, heat bonded
Cordrain	Burcan Mfgr Co.	16·5	Waffle	Polystyrene	Nonwoven, heat bonded
Enkadrain	American Enka Corp.	10 and 20	Monofilament web	Nylon	Nonwoven, needle-punched
Filtram	ICI Fibres	5	8 mm diamond grid	Polyethylene	Nonwoven, heat bonded
Geotech drainage panel	Geotech Systems Corp.	51	Beads with bitumen binder	10 mm expanded polystyrene	Nonwoven, needle-punched
Hydraway	Monsanto	25·4	6 mm cylinders	Polyethylene	Nonwoven, needle-punched
Miradrain	Mirafi Corp.	16	Waffle	Polystyrene	Nonwoven, needle-punched and heat set

10.4.1.1. Subsurface drainage

10.4.1.1.1. Geotextile filters. Geotextiles have several advantages over granular filters in that they offer improved economy, have consistent and well defined properties, and are easy to install. The use of geotextiles has also enabled a greater flexibility in the types of subsurface drainage systems available. This is evidenced by the trend over the last 5 years toward completely prefabricated subsurface drainage systems; primarily because of ease of installation, and hence reduced labour costs.

In order for it to perform as a filter, the geotextile must meet specific criteria governing its ability to act as a filter and its ability to survive throughout its required design life. The hydraulic properties of the geotextile determine its ability to act as a filter, while its mechanical integrity properties govern its ability to survive throughout its required design life.

In subsurface drainage applications the movement of ground-water through the geotextile filter occurs in one direction only—from the soil, through the geotextile filter, and into the drainage medium. This mode of water movement can result in the soil immediately adjacent to the geotextile having a major influence on the filtration performance of the drainage system.

For a geotextile to act as a filter it is essential that a condition of equilibrium is established at the soil/geotextile interface as soon as possible after installation to prevent soil particles from being piped indefinitely through the geotextile. The time taken to establish equilibrium conditions with geotextile filters varies, but is normally between 1 and 6 months. Because of the mode of water movement through the geotextile, the structure or stratification of the soil immediately adjacent to the geotextile at the onset of equilibrium conditions dictates the filtering efficiency of the system. This stratification is dependent on the type of soil being filtered, the size and frequency of the pores in the geotextile, and the magnitude of the seepage forces present.

Figure 10.8 depicts the typical stratification which can occur with three different soil types—single-sized granular soils; well-graded, predominantly granular, soils; and cohesive soils. For granular soils where the particles are virtually all the same size (single-sized soils), very little rearrangement of the soil occurs at the interface (Fig. 10.8(a)). The function of the geotextile, therefore, is simply to maintain the existing soil state. When the soil is well graded, considerable rearrangement of the soil takes place (Fig. 10.8(b)). At equilibrium, three zones may be identified: the undisturbed soil, the "soil filter", and a bridging layer. Once the stratification processes is complete, it is actually the "soil filter" layer which actively filters the soil. For soils exhibiting considerable cohesion the soil particles can actually bridge the pores of the geotextile, thus forming a stable structure (Fig. 10.8(c)). In such cases, the pores of the geotextile can be many times the size of the individual soil particles without affecting the equilibrium.

It should be noted that the size and frequency of pores in the geotextile have a marked effect on the geotextile's performance as a filter. The size and

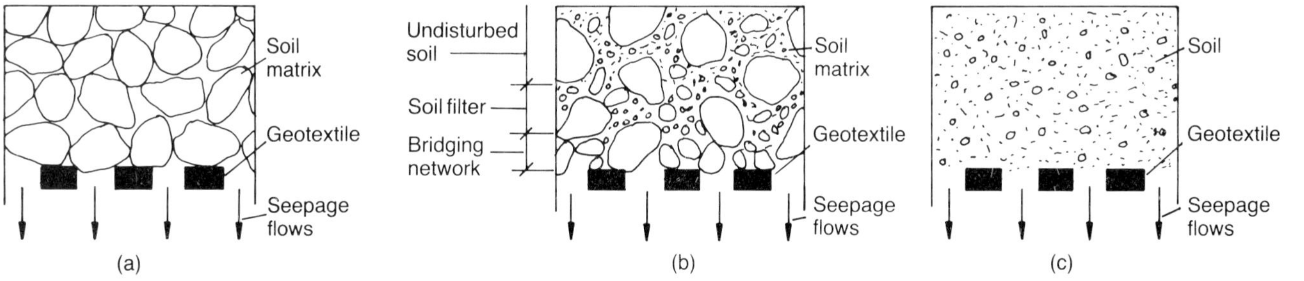

Fig. 10.8. Idealized interface conditions at equilibrium between three different soil types and geotextile filter (reference 10.12): (a) single-sized granular soils; (b) well-graded, predominantly granular soils; (c) cohesive soils

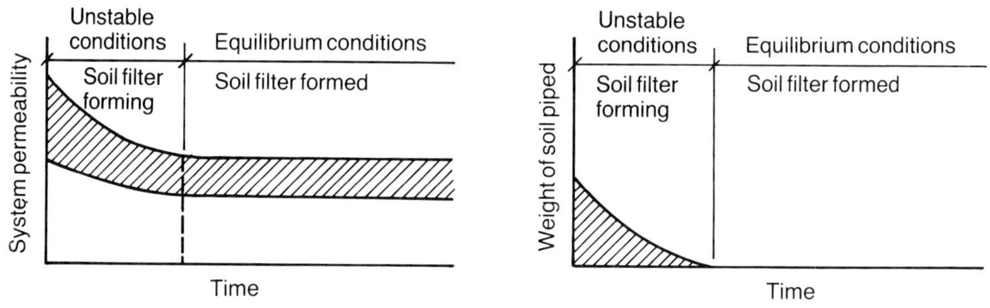

Fig. 10.9. Overall requirements for optimal filter performance[10.11]

frequency of pores govern the amount of soil that can be piped through the geotextile, and hence the ability to attain equilibrium. The frequency of pores (the number of pores per unit surface area) also ensures that adequate pore channels exist in the geotextile to remove all seepage flows without pressure build-up.

While the actual mechanics of soil/geotextile filtration are quite complex, the overall performance requirements governing geotextile filters are quite simple. Lawson[10.11] lists two overall criteria which must be satisfied in order to achieve optimal filter performance.

(*a*) Following an initial period of instability (which occurs before equilibrium conditions are established), the overall permeability of the base soil/geotextile system should remain relatively constant with time.
(*b*) Following an initial period of soil piping (which may occur before equilibrium conditions are established), no further in situ soil should be piped out of the filter system.

These two performance criteria are shown in Fig. 10.9. The time taken to reach constant "system permeability" conditions should be greater than or equal to the time taken for zero soil piping to be effected. If the permeability performance criterion is not adhered to, then the movement of water through the filter system may be critically impeded. If the piping performance criterion is not adhered to, then the filter system may allow continual piping of soil.

To ensure the overall geotextile filter performance requirements are satisfied, filter criteria for geotextiles have been proposed by a number of authors. A selection of the most commonly used geotextile filter criteria is listed in Table 10.8. In all cases the filter criteria were developed using soils with

Table 10.8. Variety of geotextile filter criteria used for unidirectional water flows

Reference	Lower limit (permeability)*	Upper limit (piping)*
Calhoun[10.5]	4% $\leqslant$ POA POA $\leqslant$ 40%	$O_{95} \leqslant D_{85}$
Ogink[10.6]	$k_g > k_s$	$O_{90} \leqslant D_{90}$
McKeand[10.7]	$k_g > k_s$	$O_{50} \leqslant D_{85}$
Schober & Teindl[10.8]	$k_g > k_s$	$O_{90} \leqslant BD_{50}$
Giroud[10.9]	$k_g > 0{\cdot}1\,k_s$	$O_{95} \leqslant ED_{50}$
Tan *et al.*[10.10]	–	$O_{95} \leqslant 2D_{85}$
Lawson[10.11]	$O_{90} \geqslant C_1 D_n$	$O_{90} \leqslant C_2 D_n$

* POA is the percent open area of the geotextile; k_g is the coefficient of permeability of the geotextile; k_s is the coefficient of permeability of the soil; O_{50}, O_{90}, and O_{95} are the apparent opening sizes of the geotextiles; D_{50}, D_{85}, D_{90}, and D_n are particular size fractions of the soil; B, E, C_1 and C_2 are coefficients.

appreciable granular fractions (and little to no cohesive fractions). Thus, some care should be exercised when rigidly applying these relationships to different soil types (including, even, different granular soils).

While the filter criteria presented in Table 10.8 refer specifically to predominantly granular soils, comparison of the calculated geotextile properties using each of the listed criteria does show a significant variation. Consequently, it is recommended that the following criteria be used as a more universal approach to geotextile filters.

For predominantly granular soils

Piping criterion

$$O_{90} \leqslant D_{85} \text{ soil} \tag{10.1}$$

Permeability criterion

$$k_g \geqslant k_s, \text{ and water permeability} \geqslant 40\,\text{l/m}^2\text{/sec at } 10\,\text{cm constant applied water head} \tag{10.2}$$

For predominantly cohesive soils

Piping criterion

$$O_{90} \leqslant 0{\cdot}12\,\text{mm} \tag{10.3}$$

Permeability criterion

$$O_{90} \geqslant 0{\cdot}05\,\text{mm, and water permeability} \geqslant 30\,\text{l/m}^2\text{/sec at } 10\,\text{cm constant applied water head} \tag{10.4}$$

where O_{90} is the apparent opening size of the geotextile measured according to ASTM D4751 or BS6906: Part 2, D_{85} is the particle size fraction below which lies 85% of the soil, k_g is the coefficient of permeability of the geotextile, k_s is the coefficient of permeability of the soil.

Equations (10.1) and (10.2) agree closely with the criteria of Calhoun[10.5], Ogink,[10.6] McKeand[10.7] and Lawson[10.11] (listed in Table 10.8). Equations (10.3) and (10.4) assume that extensive particle bridging occurs with the cohesive soil (see Fig. 10.8(c)). For the vast majority of compacted cohesive soils this will be the case. However, for cohesive soils which have dispersive characteristics, equations (10.3) and (10.4) should be used with some caution.

While the primary function of the geotextile is to filter the soil (and hence its hydraulic properties are of major importance), the geotextile also has to survive the mechanical stresses imparted during installation and throughout the design life of the material. For subsurface drainage applications, the

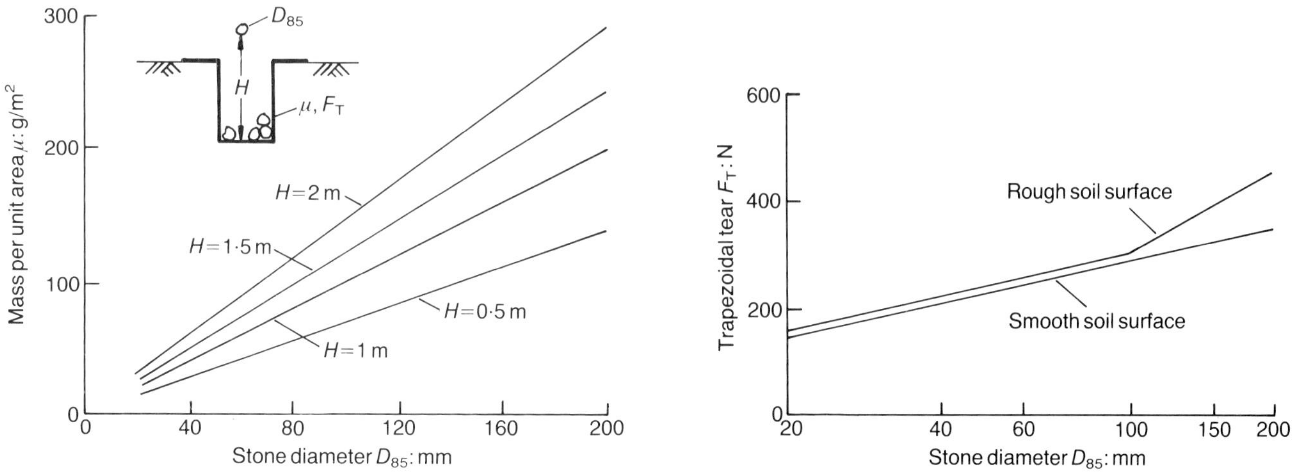

Fig. 10.10. Mechanical integrity requirements for geotextiles used in trench drains (after Lawson)[10.12]

greatest mechanical stresses are imparted during the installation of the geotextile filter. Lawson[10.12] has presented mechanical integrity requirements in terms of mass per unit area and "trapezoidal" tear resistance for geotextile filters used in trench drains. These are shown in Fig. 10.10. The mass per unit area of the geotextile is measured according to ASTM D1910 and the trapezoidal tear resistance is measured according to ASTM D4533. Mechanical integrity requirements for other subsurface drainage configurations have also been developed.

10.4.1.1.2. Fin drains. As discussed earlier, prefabricated drainage structures are increasingly used in geotechnical engineering. The reasons for this are primarily threefold. Firstly, the prefabricated drains are manufactured in factories where good quality control can be exerted. Secondly, the use of prefabricated drains reduces markedly the labour costs associated with drain installation. Thirdly, prefabricated drains have enabled drainage techniques to be used successfully in situations where the use of conventional drainage materials (e.g. granular materials) has proved unfeasible practically.

Tables 10.6 and 10.7 list the variety of prefabricated drains available. However, in this section it is proposed to limit discussion to prefabricated fin drains only.

Fin drains, which are manufactured in sheet or roll form, consist of a polymer spacer core wrapped, or laminated, in a geotextile filter. The function of the geotextile is to filter the soil water, while the function of the polymer spacer is to act as a drainage medium enabling water to be freely transported within its structure (Fig. 10.11). Conceptually, there are two types of fin drain system available—those that require the use of a pipe, and those that do not. For those requiring the use of a pipe to transport the water longitudinally, the role of the polymer spacer is to permit the transport of water (vertically) over short distances to the pipe. For those systems which do not require pipes, the polymer spacer has to be permeable enough to additionally transport the water longitudinally over considerable distances to drainage outlet points.

Polymer spacers must also be structurally stable because in many of the applications for which they are used they can be subjected to appreciable lateral soil pressures. As a consequence of having to maintain good in-plane flow characteristics under a variety of lateral earth pressures, the types of polymer spacer structures which readily fulfil these requirements are limited. Three of the most commonly used polymer spacer structures are shown in Fig. 10.12—mesh, waffle and alveolite structures.

Because of their ease of installation, fin drains are used for a wide variety of subsurface drainage applications. Examples include drainage behind retaining walls (Fig. 10.13(a)), drainage against basement walls (Fig.

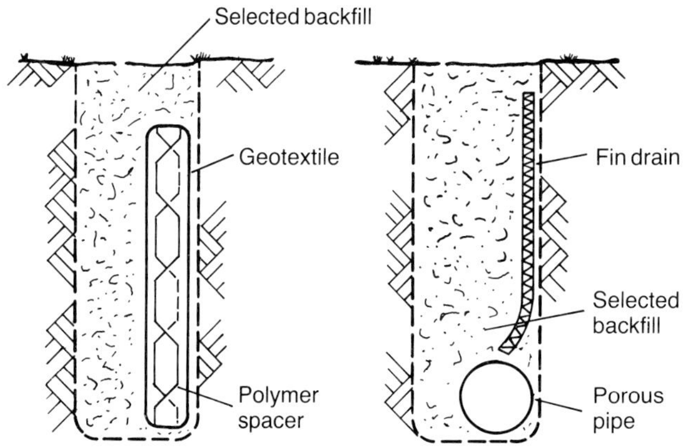

Fig. 10.11. Two varieties of fin drain

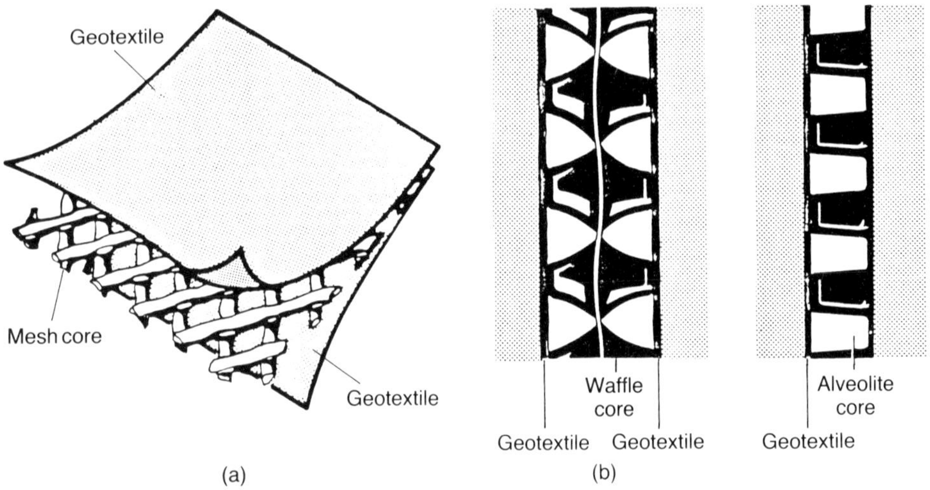

Fig. 10.12. Three different polymer spacer structures: (a) mesh core structure; (b) waffle core structure; (c) alveolite core structure

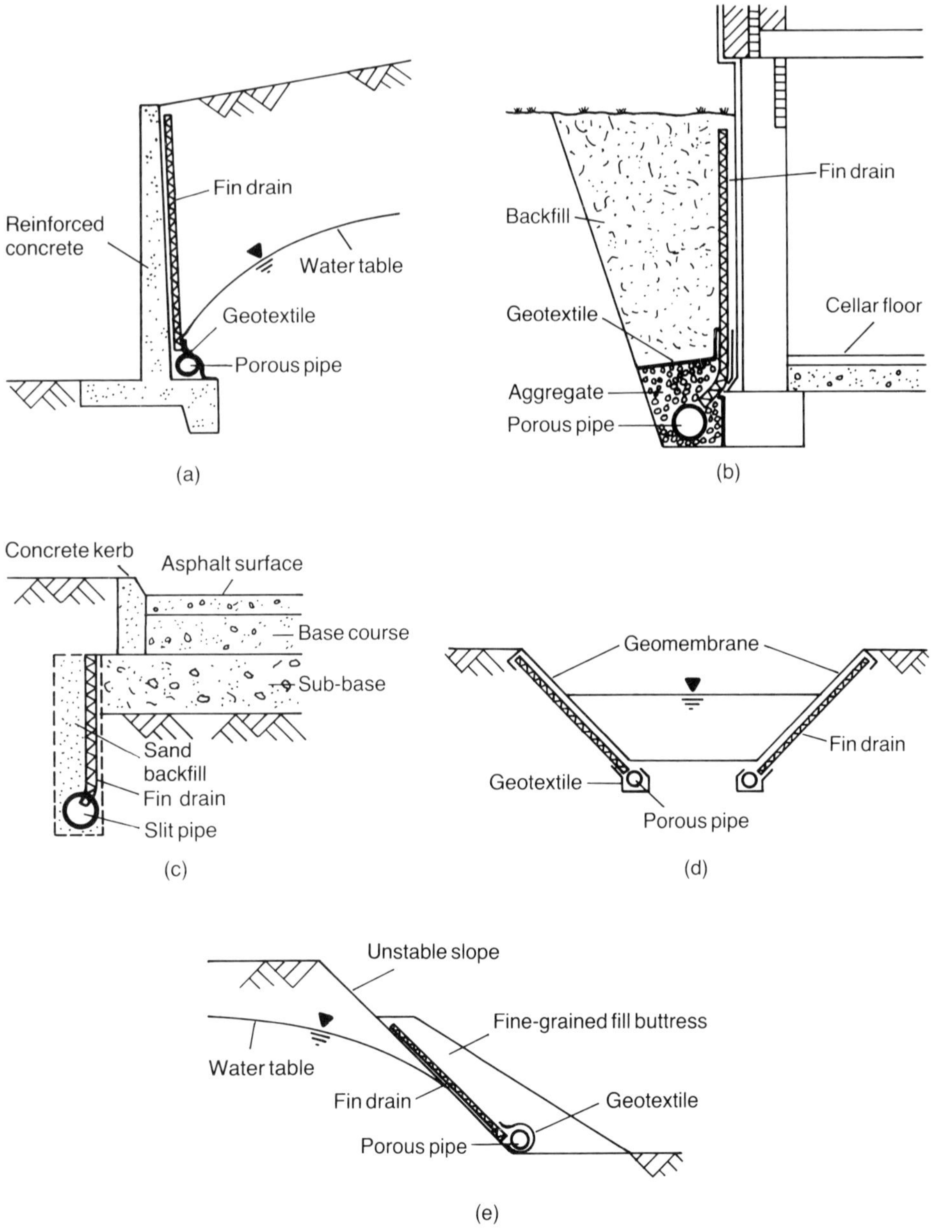

Fig. 10.13. Typical applications for fin drains: (a) drainage behind retaining walls; (b) drainage against basement walls; (c) pavement edge drains; (d) drainage beneath impermeable linings; (e) drainage for slope stability

Table 10.9. Types of erosion control systems

Classification	Type	Examples
Geometrical	Alter geometry	Flatten exposed slopes
Stabilized	*En masse*	Soil replacement
	External—rigid	Soil-cement stabilization Concrete slabs Rigid geotextile forms Cement grouted armour
	External—semi-rigid	Bitumen grouted armour
	External—flexible	Dumped rock Concrete units Gabions and mattresses Integrated armour Flexible fabric forms Synthetic skin

10.13(b)), pavement edge drains (Fig. 10.13(c)), drainage beneath impermeable linings (Fig. 10.13(d)), and drainage for slope stability (Fig. 10.13(e)).

10.4.1.2. Erosion control

There are a number of protection techniques which can be applied to soil surfaces to prevent their erosion. Table 10.9 classifies these techniques. Of particular interest as far as geotextiles are concerned are the flexible externally stabilized systems, for it is these which use geotextiles as an integral element of the overall protection system.

While six types of flexible externally stabilized system are listed in Table 10.9, these are normally condensed into four main groups for the purpose of classifying different geotextile application requirements. These groups are

(*a*) the use of geotextiles in combination with inert armour units such as dumped rock, concrete units, gabions and mattresses*
(*b*) the use of geotextiles as part of integrated armour systems
(*c*) the use of geotextiles as forms to contain concrete and soil
(*d*) the use of geotextiles as flexible synthetic skin.

10.4.1.2.1. Use of geotextiles in conjunction with inert armour units. For this application the primary function of the geotextile, once installed, is to act as a filter to prevent the foundation soil from eroding through the interstices of the armour layers which have been placed above the geotextile. The theory and the efficacy of using geotextiles for this application may be described in terms of the following.

When highly permeable discrete armour units are used to dissipate the energy of a hydraulic force, turbulence occurs within the interstices of the revetment structure. This turbulence can cause erosion of the foundation soil through the pores in the highly permeable armour facing. To prevent this erosion, it has been the convention to place layers of granular materials (filters) beneath the highly permeable armour facing. The design requirements for granular filters are such that the filter particles have to be small enough to

* Strictly speaking, gabions and mattresses are pattern-placed units, and as such, it is possible to classify them as integrated armour systems. However, since they are constructed by filling (with rock) in situ, they have been grouped in the inert armour units category. The same rules governing geotextile selection for inert armour unit systems apply to gabions and mattresses.

prevent erosion of the layer beneath but also large enough to prevent the filter particles themselves from eroding through the granular layer above. Because of these physical constraints placed on the size of granular filter materials in revetment structures, in many cases, multiple granular filter layers are required to satisfy the overall filter criteria. For example, if the base soil is highly erodible (e.g. a silty sand) and if the erosive forces are quite large, to achieve complete filter compliance using wholly granular materials could require three granular filter layers beneath the primary armour layer. The resultant structure would look similar to that shown in Fig. 10.14(a). This type of structure is obviously expensive, and is difficult to construct to the standards required for good in-service performance. Unfortunately, in many cases, one or more of the filter layers have to be deleted because of cost considerations. This invariably results in less than optimal in-service performance and, in many instances, premature failure.

Over the last 30 years, geotextiles have been used as an economic alternative to granular filter layers in revetment structures. Their relatively small pore sizes combined with robust mechanical properties have enabled their use as substitutes for one or more of the granular filter layers. For example, the equivalent solutions utilizing geotextiles, which are alternative to the wholly granular case depicted in Fig. 10.14(a), are shown in Figs 10.14(b), (c) and (d). The preferred solution in Fig. 10.14 will depend on a number of factors, namely, the characteristics of the base soil, the mechanical, hydraulic and durability properties of the geotextiles available, the type(s) and magnitude(s) of the hydraulic forces acting on the structure, and the cost of each alternative. It is evident that, for each of the three alternative solutions shown in Figs 10.14(b), (c) and (d), the hydraulic requirements for the appropriate geotextile(s) are the same (hydraulically the geotextile has only to prevent the migration of the same base soil), but the mechanical requirements will differ (mechanically the geotextile has to resist the stresses imposed by the granular layer(s) on top) because of the different-sized granular materials placed on top of the geotextile in each case.

Fig. 10.14. Revetment structure utilizing wholly granular materials, and equivalent structures utilizing geotextiles:[10.14] (a) revetment utilizing wholly granular materials; (b) geotextile substituting for the fine granular filter layer; (c) geotextile substituting for the fine and coarse filter layers; (d) geotextile substituting for fine, coarse, and secondary armour layers

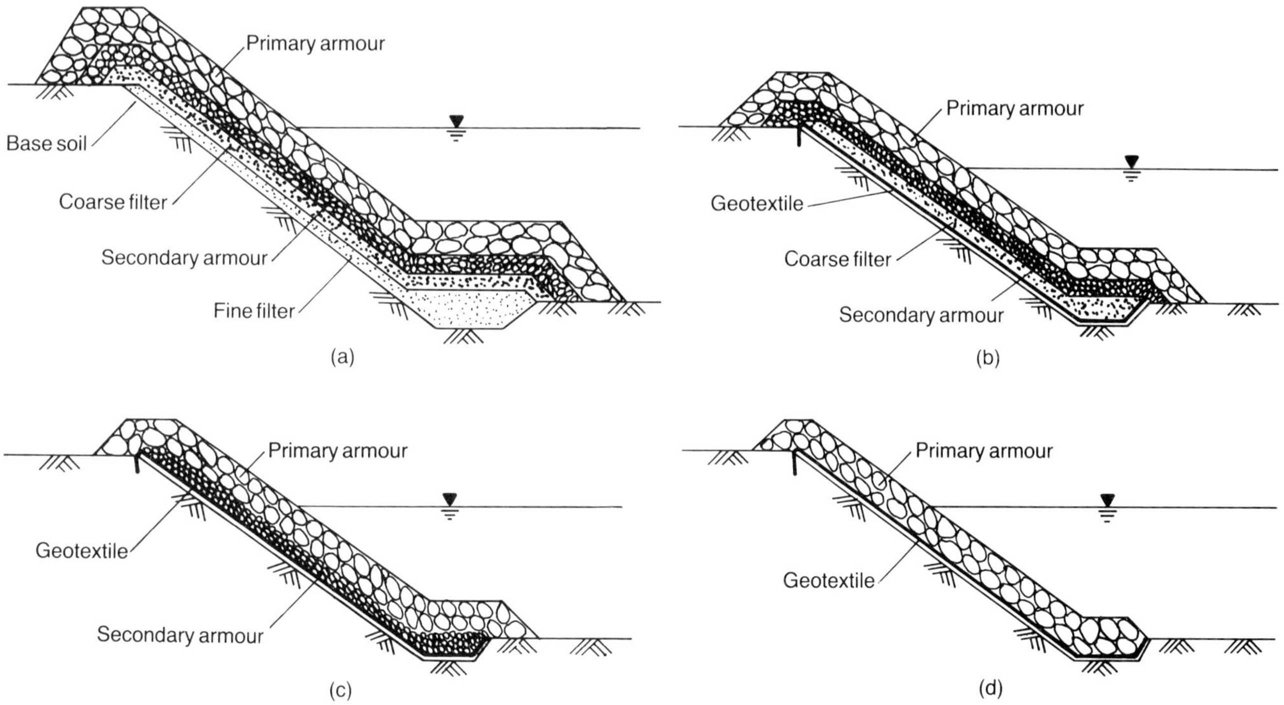

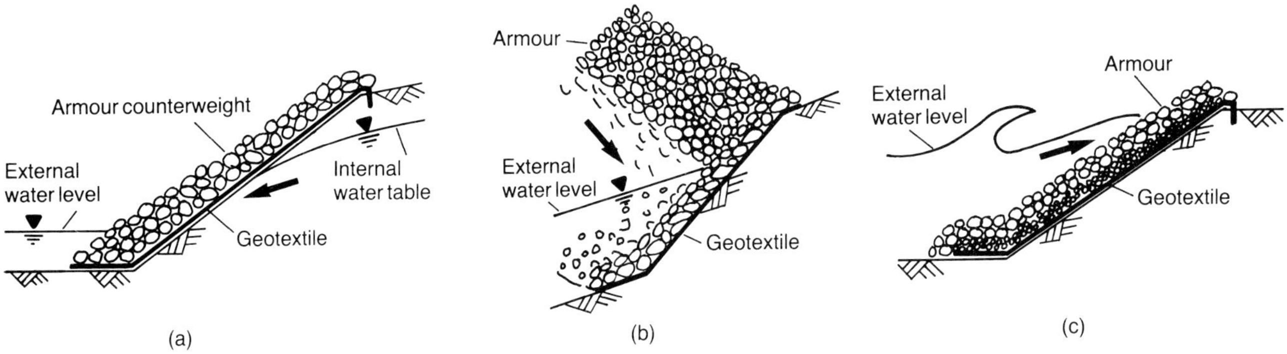

Fig. 10.15. Modes of water movement causing erosion: (a) unidirectional water flow internal to the structure; (b) parallel water flow external to the structure; (c) inclined water flow external to the structure

To establish a rational design procedure for the use of geotextile filters in revetment structures it is necessary to quantify relevant performance parameters. These performance parameters must relate to the relevant hydraulic, mechanical, and durability properties of the geotextile filter.

As far as the filter function is concerned, the geotextile may be subjected to quite complex water-flow regimes (certainly more complex than the relatively simple unidirectional flows occurring in subsurface drainage). This leads to a conservative approach to the design of the geotextile filter.

Figure 10.15 displays three possible water-flow regimes which can cause erosion—unidirectional seepage flows, water velocities parallel to the protected slope, and wave forces. In selecting the appropriate geotextile filter properties for a particular site, the most severe hydraulic conditions occurring need to be considered.

For unidirectional flow cases (Fig. 10.15(a)), the geotextile filter criteria listed in Table 10.8 or equations (10.1), (10.2), (10.3) and (10.4) (for subsurface drainage) are applicable in determining the appropriate geotextile filter properties. For cases where external water velocities and/or wave action dominate, then one set of the filter criteria presented in Table 10.10 should be used to determine the appropriate hydraulic properties of the geotextile filter. The various filter criteria presented in Table 10.10 assume that no soil particle bridging occurs across the pores of the geotextile filter under the influence of

Table 10.10. Variety of geotextile filter criteria for complex water flows

Reference	Lower limit (permeability)	Upper limit (piping)
Ragutzki[10.13]		For wovens: $O \leqslant (0{\cdot}5 - 1{\cdot}3)D_{50}$ For nonwovens: $O \leqslant (0{\cdot}5 - 1{\cdot}5)D_{50}$
Ogink[10.6]		$O_{98} \leqslant D_{15}$
Schober and Teindl[10.8]	$k_g \geqslant k_s$	$O_{90} \leqslant AD_{50}$
Lawson[10.14]	$O_{90} \geqslant D_{15}$	$O_{90} \leqslant D_{50}$
Heerten[10.15]	$\eta k_g \geqslant k_s$	Non-cohesive soils: $O_{90} \leqslant D_{50}$ Cohesive soils: $O_{90} \leqslant 10D_{50}$ $O_{90} \leqslant D_{90}$ $O_{90} \leqslant 0{\cdot}1$ mm

external or reversing water flows (unlike the assumptions made for the unidirectional flow case). Consequently, the criteria presented in Table 10.10 result in more conservative hydraulic property requirements than those presented in Table 10.8.

The required mechanical integrity properties of the geotextile filter also have to be determined. This is necessary to ensure that the mechanical integrity of the geotextile is maintained both during the installation procedure and throughout the design life of the structure. Lawson[10.14] has presented relationships for determining the required mechanical properties of geotextile filters for revetment structures. These relationships are presented in graphical form in Fig. 10.16. The two geotextile properties of interest are mass per unit area (measured according to ASTM D1910) and trapezoidal tear resistance (measured according to ASTM D4533).

All synthetic fibres are susceptible to some extent to degradation by ultra-violet light. Thus, it is essential when designing and constructing revetment structures to ensure the geotextile is not exposed for long periods of time. Needless to say, once the geotextile is covered adequately by soil or rock no further degradation due to ultra-violet (UV) light occurs. It is common practice (and should be looked upon as essential) for the more highly susceptible fibres (polypropylene, polyethylene, and polyamide) to have a UV stabilizer incorporated in the fibre at the point of manufacture. It should be noted that while some synthetic fibres exhibit good natural UV resistance, and UV stabilizers may be incorporated in other fibre types, conventional geotextiles should not be left exposed for prolonged periods as the UV stability (either natural or synthetic) only delays the degradation, it does not prevent it.

Once the geotextile is buried in the soil no further UV degradation occurs. Thus, for good long-term performance it is essential that the geotextile be utilized at such a depth below the top armour layer so as to prevent long-term exposure of the geotextile to UV radiation. This is quite critical, especially if the geotextile is to be used as a filter directly beneath the primary armour layer. In this instance, the thickness of the primary armour layer should be a minimum of twice its characteristic rock size (i.e. $>2 \times D_{85}$ armour) to protect the geotextile from UV degradation. If the geotextile is to be used

Fig. 10.16. Mechanical integrity requirements for geotextile filters used in revetments (after Lawson)[10.14]

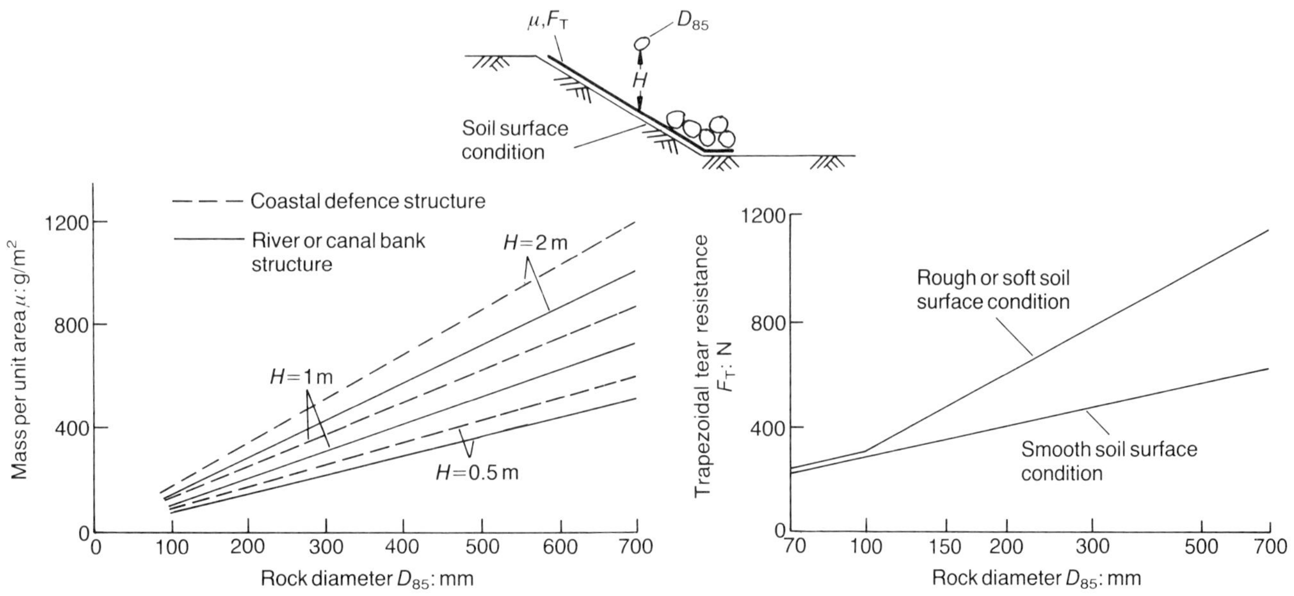

deeper in the revetment structure then no special precautions need be taken with regard to layer thicknesses over and above those requirements stipulated by conventional design procedures.

In specific instances it is possible for the geotextile to be subjected to abrasion where high hydraulic impact forces occur, and where there is extensive movement of soil sediment in the water. Unfortunately, there are no appropriate abrasion test requirements appropriate for geotextiles. Therefore, it is recommended that, until such time as appropriate abrasion test requirements are developed, a conservative approach be taken to ensure abrasion does not occur, by limiting the maximum size of the rock in the layer immediately adjacent to the geotextile to 500 mm diameter.

10.4.1.2.2. Use of geotextiles with integrated armour units. Although the most common form of revetment structure using geotextiles utilizes rock armour units, in recent years there has been considerable development in semi-prefabricated or fully prefabricated revetment structures which consist of a geotextile filter and pattern-placed armour units (so-called integrated armour systems). The armour units are manufactured in a factory to such a geometry that, when placed in a predetermined pattern, they exhibit very high damage coefficients (high stability). To enable cost-effective pattern placement, the individual armour units are normally fixed together by cables or attached to a geotextile substrate in the factory before transport to site. A number of proprietry integrated armour systems are currently in use.

10.4.1.2.3. Use of geotextiles as forms. One erosion control technique which shows considerable promise is the use of geotextiles to provide a formwork for the in situ pouring of concrete which, once hardened, provides protection against erosion. While the first patent for using fabric as a forming system was issued in 1906, the system was not developed properly until the late 1960s following advances in fabric and concrete technology.

To satisfy different hydraulic, soil, and location requirements several formwork shapes are available. These include the following.

(a) Concrete revetments. In this instance the geotextile form is composed of two parallel geotextile layers connected together in a specific pattern, such that when filled with concrete they attain a predesigned shape. Three specific shapes predominate—Filterpoint, Unimat, or Articulating Block.

(b) Concrete bags and sausages. In this instance the geotextile is sewn into either a bag or a sausage shape prior to filling with concrete. These structures have been used as retaining walls, pipe saddles, void/cavity forms, and in spillways.

(c) Concrete pile jackets. In this instance the geotextile is made into a sleeve which can be fitted around a deteriorated pile before filling with concrete.

The concrete to be placed in the geotextile form has to be fine grained and relatively fluid because the spaces between the geotextile layers may be quite small, and because concrete pumping is required to fill the geotextile form. Some interesting examples of the use of concrete forms are given by Koerner and Welsh,[10.16] Welsh and Dominske,[10.17] and Lupton.[10.18]

Woven geotextiles are preferred for forms because they have a greater resistance to localized stresses (which occur at the joining points of the geotextiles during the concrete injection process), and maintain their shape better during the curing of the concrete.

It must be remembered that the outer geotextile skin is left exposed to the atmosphere after installation, hence it will degrade due to ultra-violet light. In most cases this degradation is not detrimental to the structure because once the concrete has hardened the geotextile does not play a structural role in the system. In some instances, UV degradation may cause problems however, especially with filterpoint and articulating block mattresses which rely on the connections of the two parallel geotextile layers to provide a filter for the dissipation of pore water from behind the mattress. In locations which are prone to high UV radiation it is prudent to use an additional geotextile layer beneath the concrete mattress as the filter.

Until now, all discussion on the use of geotextiles as forms has concentrated on the use of concrete as the filling material. While this is the most common filling material, other materials, such as sand, can also be used. The geotextile forms which are used to retain sand fill are either sausage shaped (e.g. Koerner and Welsh[10.16] or tubular mattress shaped (e.g. Tutuarima and Van Wijk.)[10.19] Unlike concrete, sand filling results in a fairly flexible structure which is able to deform readily with changing foundation levels. However, also unlike concrete, the sand fill cannot solidify, and thus the geotextile skin has constantly to retain the sand fill throughout the design life of the structure. This places considerable property constraints on the geotextile, for not only does it have to resist mechanical stresses during installation and throughout its design life (woven geotextiles are generally best suited for this application), but it also has to be highly durable, especially with regard to UV resistance. High UV resistance is essential for the geotextile skin. For this application a maximum design life of no greater than 15 years should be assumed (in areas subject to high UV radiation levels the design life may be considerably less than 15 years).

10.4.1.2.4. Use of geosynthetics as flexible synthetic skin elements. For this technique the geosynthetic itself acts to prevent erosion, both by absorbing the water forces and by preventing soil loss. Normally it is only specially manufactured geocomposite materials which can fulfil these specific requirements. Two types of geosynthetic systems are in use—flexible skin revetments and submerged scour protection systems (commonly known as artificial seaweed).

Flexible skin revetments comprise heavy geocomposite mats which are placed directly on the unstable soil slope to prevent erosion. The flexible skin may be either permeable or impermeable depending on the requirements. In practice, to meet the requirements of both water force dissipation and erosion prevention the resultant skin element normally consists of a webbing face (to dissipate the water force) with a geotextile (if a permeable skin is required), or geomembrane (if an impermeable skin is required) backing.

Submerged scour protection systems utilize a geotextile mat base onto which is attached synthetic fronds. The basic principle of operation is as follows. The scour protection mat is placed on the seabed, or streambed, which has been subject to scour. The fronds, which have a specific gravity less than water, float above the geotextile mat slowing down water velocities to such an extent that water-borne sediment is deposited, thus enabling a build-up of sediment on top of the mats.

The performance of these artificial seaweed devices is primarily dependent on the correct placement of the geotextile mat and the length, specific gravity and connection density of the fronds. The mat should be firmly anchored to the seabed in such a way to ensure that no part of the mat can lift off the bottom. The synthetic fronds normally consist of polypropylene (specific gravity 0·90) or polypropylene/polyester fibre blends (to achieve a more even water velocity distribution within the fronds).

10.4.2. Separations

The placement of geotextiles between soft saturated soils and granular fill materials to prevent their intermixing was one of the earliest applications for geotextiles. In these applications the geotextile acts as a permeable separation layer allowing easy cross-flow of water, but preventing intermixing of the two soil types.

Geotextiles are used as separation layers for a variety of applications, e.g. pavements constructed over soft subgrades, working platforms for structures over soft soils, railway tracks over soft soils, and soil platforms over disused waste sites. For the purposes of this book, however, it is proposed to cover only three of the major applications—the use of geotextiles as separation layers for temporary pavements, for permanent pavements, and for the reclamation of very soft soil deposits.

10.4.2.1. Temporary pavements

Temporary pavements (e.g. access roads) are, in essence, an expedient solution to the problem of transporting goods, or materials, overland to a particular site over a limited time period. As such, they are normally basic in structure because once the particular task is completed they do not provide any further economic benefit.

Because the quality of the pavement is commensurate with minimum required riding qualities over its maximum required design life, a greater degree of flexibity concerning the structural design of these pavements is possible than is the case for permanent pavements. In many instances, it is possible to tolerate significant plastic deformations at the surface of temporary pavements (as well as the normal elastic deformations).

Geotextiles are used in temporary pavements where they are placed on a soft subgrade, and then granular materials are placed on top (see Fig. 10.17(a)). In addition to preventing the intermixing of the granular material with the soft subgrade (the geotextile acting as a separator), it is possible to utilize the structural characteristics of the geotextile by reducing the thickness of the granular material on top of the geotextile to such an extent that the geotextile undergoes deformation at each vehicle pass. By deforming

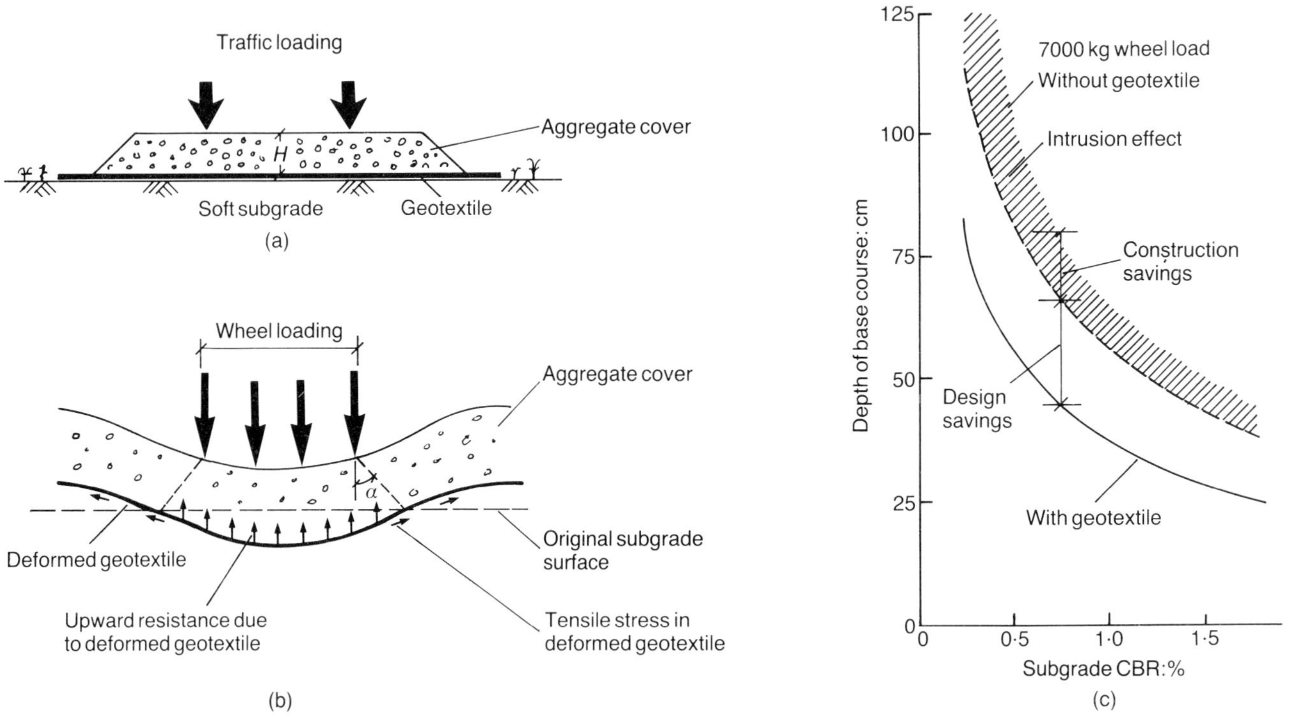

Fig. 10.17. Geotextile mechanism and realized economies for temporary pavements: (a) use of geotextiles in temporary pavements; (b) geotextile response mechanism in a deformable pavement; (c) design curve for unbound deformable pavements (after Barenberg et al.)[10.16]

at each vehicle pass the geotextile acts to confine the granular fill, thus contributing an increase in support to the pavement structure. This membrane-support mechanism is shown in Fig. 10.17(b). It should be emphasized that the added benefit of this membrane support is only realized if the geotextile undergoes deformation at each vehicle pass. In practical terms, ensuring a realistic geotextile deformation (and thus the benefit of the membrane support) requires a rut depth of 25 mm minimum at the surface of the pavement at each vehicle pass.

Barenberg *et al.*[10.20] attempted to quantify the possible benefits of the use of geotextiles in deformable temporary pavements constructed over very soft subgrades. He was able to divide these benefits, which he quantified in terms of a saving in the required thickness of granular materials, into two components—those due to the role of the geotextile as a separator, and those due to the effects of geotextile deformation. He then presented his findings in the form of a series of design curves for deformable temporary pavements. One of the design curves, produced in Fig. 10.17(c), shows the economies due to the separation role of the geotextile (in terms of a construction saving), and that due to the membrane support (in terms of a design saving). When the two geotextile functions act in concert, total realized savings in the thickness of granular materials may approach 40% of the total thickness normally required when no geotextile is used at all.

Since Barenberg's early work, other researchers have studied the geotextile tensioned-membrane effect in more detail. Kinney[10.21] developed a fabric tension model in the laboratory to study the tensioned-membrane effect of geotextiles in unbound deformable pavements. Kinney was also the first person to attempt to identify the geotextile properties which contribute to improved performance. He concluded that geotextile stress–strain characteristics, creep, geotextile/soil friction, and method of geotextile placement (geotextile should be placed flat with few creases to obtain maximum benefit) were all important considerations in how the pavement performed. Subsequently, Giroud and Noiray,[10.22] Raumann[10.23] and Sellmeijer *et al.*[10.24] have all investigated, in one form or another, the tensioned-membrane effect of geotextiles in deformable pavements.

10.4.2.2. Permanent pavements

Permanent pavements are characterized by negligible plastic deformations (i.e. rutting) and low to medium elastic deformations (i.e. deflections). Because of the constraint on the level of tolerable plastic deformation it is impossible to realize any benefit from the membrane-support effect of geotextiles (unlike in the case of temporary pavements). Thus, the only benefit imparted by geotextiles for permanent pavements is one of separation (i.e.

Fig. 10.18. Use of geotextile separation layer in permanent pavement construction

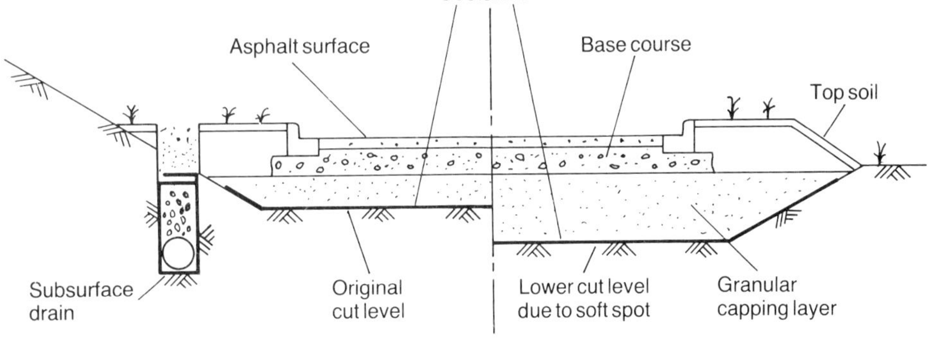

preventing the intermixing of the granular pavement layers with the soft subgrade). A typical permanent pavement cross-section depicting the use of a geotextile is shown in Fig. 10.18.

The separation benefits of geotextiles in permanent pavements are realized both during the construction phase of the pavement and throughout its design life. The construction benefit of the separator lies in the ability of the geotextile to facilitate the construction of the pavement over a soft subgrade. The long-term benefit of the separation layer is much more difficult to quantify. However, it may be safe to assume that once the geotextile separator is installed it will continue to prevent the intermixing of the soft subgrade with the granular pavement materials throughout the full design life of the pavement.

10.4.2.3. Reclamation of very soft soil deposits

In certain situations it may be required to construct soil capping layers over very soft soil deposits. These deposits may consist of dredged spoil from mining or dredging operations, or sediments at a current stage of deposition. These deposits are normally very soft, with undrained shear strengths ranging between 1 to 5 kN/m^2, and may cover relatively large areas. The problems associated with physically constructing a soil capping layer over such soft soil deposits have been alleviated by the use of geotextiles as a separation layer between the soft deposits and the soil capping layer.

The technique normally used is shown in Fig. 10.19. The geotextile sheet, normally prefabricated into large areas before installation, is laid across the very soft soil deposit. Once the geotextile has been installed, earth berms are placed over the extremities of the geotextile sheet to anchor it securely at the edges of the deposit. Next, a series of parallel earth bunds are constructed on top of the geotextile in order to stress it evenly across the surface of the soft soil deposit. Once the stressing of the geotextile has been performed the areas between the bunds can be filled with earth, thereby reclaiming the very soft soil deposit.

While the geotextile has to have sufficient tensile strength for installation and stressing to proceed smoothly, if vertical drains are also to be used to consolidate the soft deposits, the geotextile cannot be too robust otherwise it will be very difficult to install the vertical drains through the geotextile.

Techniques for reclaimining very soft soil deposits, similar to that described above, were first proposed by Yano *et al.*[10.25] and Nishibayashi,[10.26] and later by Broms.[10.27]

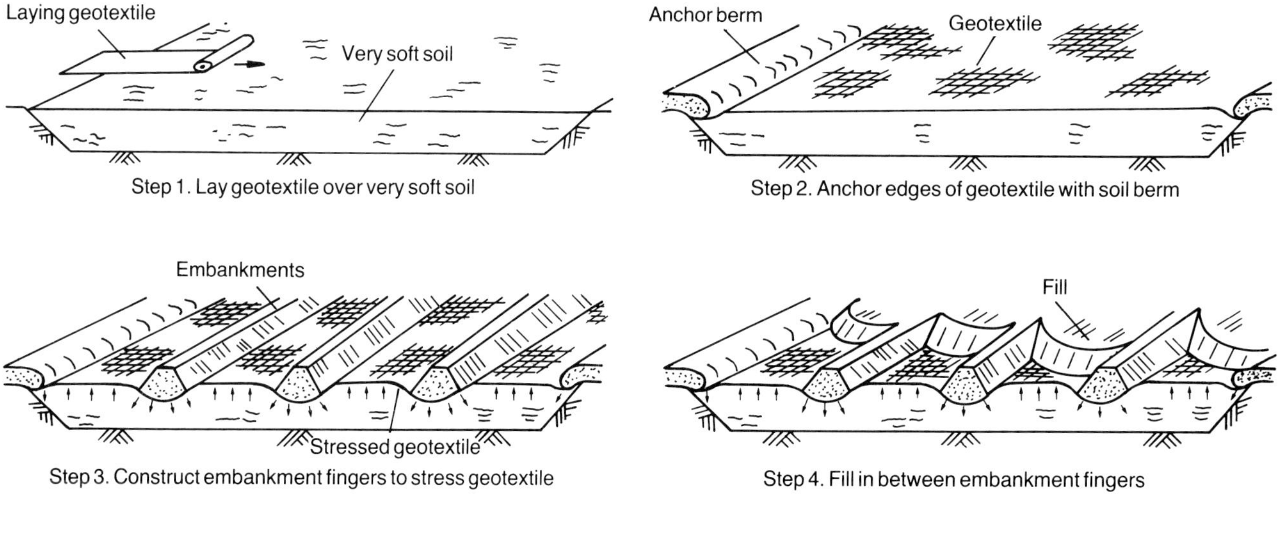

Fig. 10.19. Technique used to reclaim very soft soil deposits: step 1—lay geotextile over very soft soil; step 2—anchor edges of geotextile with soil berm; step 3—construct embankment fingers to stress geotextile; and step 4—fill in between embankment fingers

10.4.3. Soil reinforcement

The concept of reinforced soil is not a new one. As early as 1400 BC ziggurats (temples) were constructed by the Babylonians with the aid of vines as reinforcement for the soil bricks.[10.28] In the modern context, however, reinforced soil really came into being when Vidal[10.29] obtained a patent for the design and construction of retaining walls using a thin metal facing connected to horizontal steel reinforcing strips set into the backfill (now known as reinforced earth).

Since the mid-1970s geosynthetics have been increasingly used to reinforce soils. The reasons for their popularity are numerous, but are mainly attributable to their good durability in a wide range of soil types and environments, their ability to be easily engineered for different applications, and their mechanical compatibility with soils.

The use of geosynthetics for soil reinforcement distinguishes itself from other geosynthetic applications in one major detail—the geosynthetic reinforcement is required to carry tensile loads over extended time periods (up to 120 years) and undergo relatively small deformations. This fundamental requirement restricts greatly the choice of suitable polymer types and geosynthetic constructions. The types of geosynthetics used to reinforce soils are woven multifilament geotextiles, warp knit geotextiles, geogrids, and geo-linear elements.

The most critical reinforced soil design aspect, as far as the geosynthetic reinforcement is concerned, is the necessity to determine the long-term load-carrying capability of the geosynthetic (its design load), and its short-term and long-term extensions. To do this, special attention must be given to the initial and the time-dependent characteristics of the geosynthetic reinforcement.

Tables 10.3 and 10.4 list typical tensile properties (ultimate tensile strength and extension) of the range of geosynthetics. However, these properties only provide a starting point for the determination of the appropriate long-term load-carrying characteristics of geosynthetics. Since geosynthetic reinforcements are required to carry loads over long periods of time in a soil environment, the effects of installation damage, creep, and temperature differentials have to be taken into account.

For geosynthetic reinforcements consisting of fibres as the structural elements (the most common variety of geosynthetic reinforcements), the effect of site installation damage is to reduce the ultimate tensile strength of the material without affecting its maximum extension (Fig. 10.20(a)). Thus, it is possible to represent the effects of site installation damage by the use of a partial factor f_d applied to the ultimate tensile strength of the "as manufactured" (i.e. undamaged) material. For reinforcements which have fibres as their structural elements, the effect of installation damage is only immediate

Table 10.11. Partial factors for installation damage f_d for various geosynthetic reinforcement types

Geosynthetic type	Soil type		
	Sand	Gravel	Cobbles
Woven geotextiles	1·00–1·17*	1·05–1·50	1·40*–1·80
Geogrids			
Cross-laid strips	1·00–1·15	1·00–1·50	1·30–1·80
Punched sheets	1·10–1·25*	1·30–1·60*	–
Geo-linear elements			
Strips	1·00–1·15	1·00–1·30	1·00–1·50

* After Jewell and Greenwood.[10.31]

(i.e. only during installation), and it has no influence on the long-term properties of the geosynthetic. For geosynthetics which consist of punched polymer sheets, installation damage reduces the maximum extension of the material as well as the ultimate tensile strength. Installation damage of these types of materials may also affect their long-term load-carrying capabilities, due to the effect of notch and crack growth (caused by the installation damage) under load. Typical partial factors f_d for a range of geosynthetic reinforcements, and a range of fill types are listed in Table 10.11.

All polymeric materials are prone to creep at ambient temperatures, although the degree of creep varies significantly for different polymers. Chapter 7, part B has discussed the viscoelastic nature of polymers. To allow for the full effect of creep it is necessary to examine both the effect of creep on the load-carrying capability (stress-rupture curves are normally used for this), and the effect of creep on the extension (isochronous creep curves are normally used for this). In general, the effect of creep on the extension of the geosynthetic reinforcement is not a structural constraint on the choice of the reinforcement, but can be a serviceability constraint.

The use of stress-rupture curves to determine the long-term load-carrying capacity of geosynthetic reinforcements is shown in Fig. 10.20(b). The tensile strength of the geosynthetic reinforcement installed in the soil $P_R(install)$ is

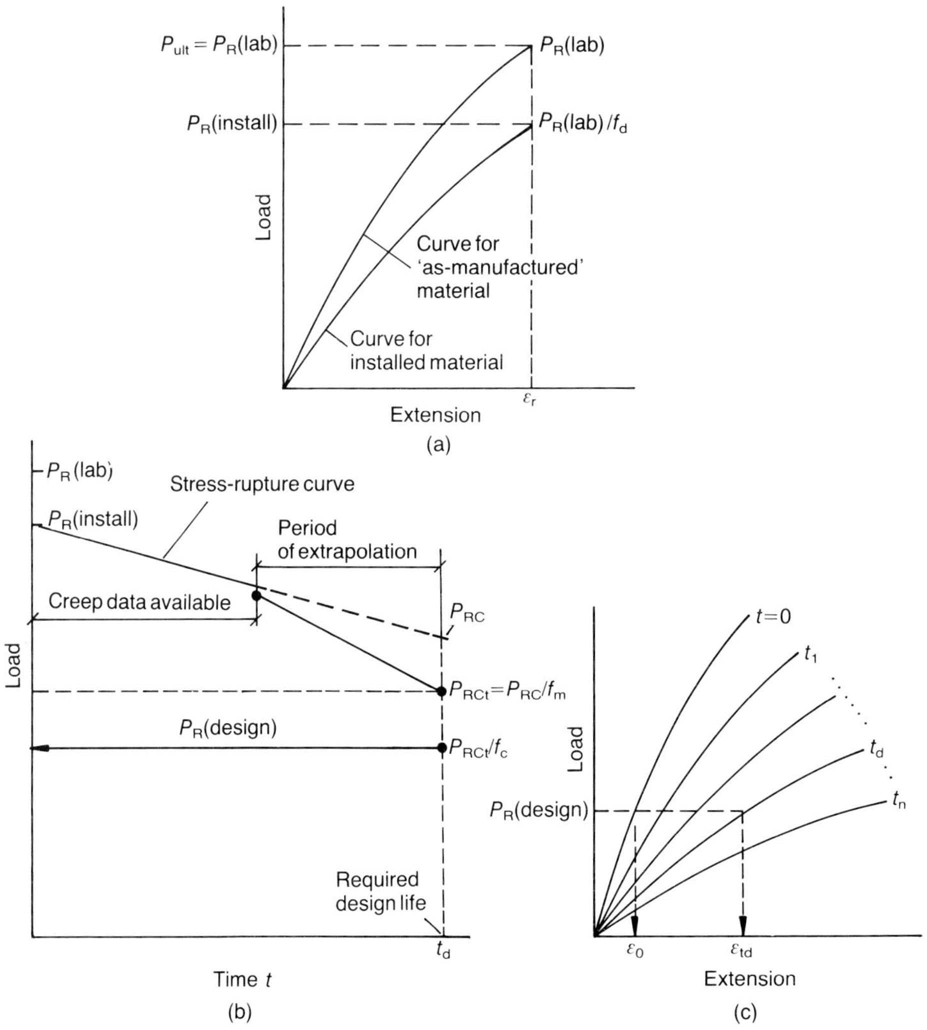

Fig. 10.20. Method of determining the design loads and extensions of geosynthetic reinforcements:[10.30] (a) effect of installation damage on the load-extension curve of geosynthetic reinforcement; (b) method of determining the geosynthetic design load from stress-rupture curves; (c) method of determining the initial and total extensions over design life of geosynthetic reinforcement by use of isochronous creep curves

Table 10.12. Partial factors of safety for extrapolation of creep data f_m over different time periods (after reference 10.31)

	No extrapolation	1 $\log_{10}$ cycle extrapolation (factor of 10)	1·5 $\log_{10}$ cycles extrapolation (factor of 30)	2 $\log_{10}$ cycles extrapolation (factor of 100)
Supported by creep test data	1·3	1·5	1·8	2·2
Not supported by creep test data	–	2·0	Not recommended	Not recommended

taken as the tensile strength at time $t = 0$. Presuming the geosynthetic reinforcement is required to carry a specific design load for a design life $t = t_d$, the following procedure, presented by Jewell and Greewood,[10.31] for determining the long-term design strength of the geotextile reinforcement is to be recommended.

Using the creep data available, the stress-rupture curve for the particular geosynthetic reinforcement is constructed using the value $P_R(install)$ as the starting point (Fig. 10.20(b)). If the available creep data does not cover the full design life of the reinforcement then extrapolation of the data is allowed, subject to the inclusion of a material partial factor of safety f_m. This partial factor of safety takes into account uncertainties regarding the extrapolation of time-dependent stress-rupture results over different time periods. Suggested values of f_m for different extrapolation periods are given in Table 10.12. This partial factor of safety is applied to the extrapolated rupture load P_{RC} (at design life t_d), such that

$$P_{RCt} = P_{RC}/f_m \tag{10.5}$$

where P_{RCt} is the "safe" load-carrying capacity of the geosynthetic reinforcement over the design life, taking into account the effects of creep.

For engineering design, very rarely is the design strength of the material equated to its rupture strength at the nominated design life (i.e. materials are not designed to fail on reaching the design life of the structure). Thus, it is prudent to use an additional partial factor of safety f_c, applied to the safe long-term geosynthetic strength P_{RCt}, to arrive at the appropriate long-term design strength of the geosynthetic reinforcement $P_R(design)$ (Fig. 10.20(b)). This additional partial factor of safety is related to the consequences of reaching the critical limit state of the structure, and is dependent upon potential danger to human life, economic consequences, etc. The value of f_c should range from 1·0, for very low risk structures, to 1·2 for structures where the result of failure at the limit state has severe consequences.[10.32] Thus, the design strength of the geosynthetic reinforcement $P_R(design)$ becomes,

$$P_R(design) = P_{RCt}/f_c \tag{10.6}$$

Once the design long-term strength of the geosynthetic reinforcement $P_R(design)$ has been calculated, isochronous creep curves can be used to determine both the initial extension of the material ε_0 (at time $t = 0$), and its total extension, ε_{td}, over time period t_d (Fig. 10.20(c)). The numerical difference between these two extension values is the creep extension of the geosynthetic reinforcement over the design life t_d. For certain applications it may be the total extension of the geosynthetic which governs the appropriate design strength, rather than its long-term load-carrying capability.

Typical stress-rupture characteristics of the various drawn, load-carrying elements used in geosynthetic reinforcements are shown in Fig. 10.21. While the stress-rupture characteristics of polyester and polyaramid fibres conform

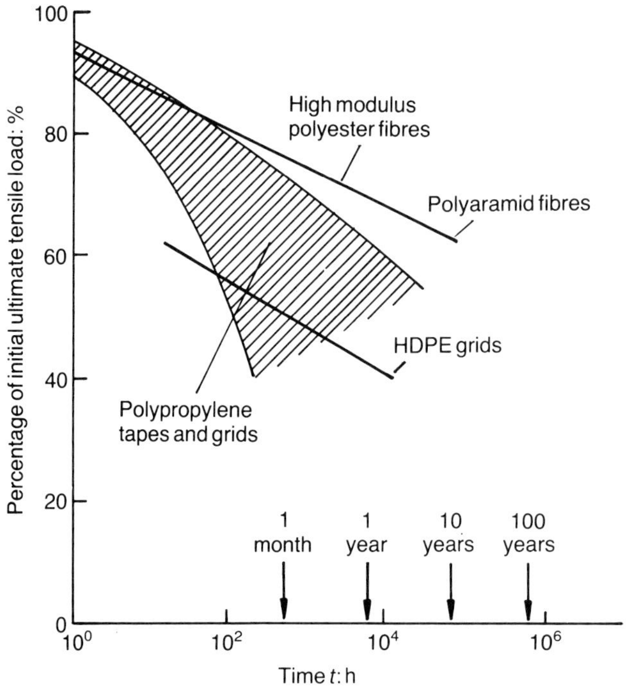

Fig. 10.21. Stress-rupture characteristics of high modulus polyester fibres, polypropylene tapes and grids, and HDPE grids, at 23°C

to a well-defined pattern, those of high density polyethylene and polypropylene can exhibit highly varying stress-rupture characteristics depending on the amount of drawing carried out during manufacture.

Comparison of the typical rate of creep of the different load-carrying elements used in geosynthetic reinforcements can be observed conveniently by the use of creep coefficients. The method of determining the creep coefficient of a material is shown in Fig. 10.22(a). The creep coefficient is a measurement of the rate of creep of a material. Fig. 10.22(b) shows the relationship between creep coefficient and applied load for a range of geosynthetic reinforcement structural elements. In general, polyaramid fibres show the lowest rate of creep while polypropylene fibres show the highest.

Since temperature affects the rate of creep and the stress-rupture characteristics of geosynthetic reinforcements, this must be taken into account if the creep data to be used in design have been obtained at a different operating temperature to that occurring when the geosynthetic reinforcements are in service. Where the creep data have been derived at higher ambient temperatures than those expected in service, conservative predictions will result if these data are used directly in the calculation. If the creep data have been derived at lower ambient temperatures than those expected in service then *unsafe* predictions would result if these data are used directly for the determination of the long-term design properties of the geotextile reinforcement.

Just as different polymer materials exhibit different creep rates at the same temperature (related to their visco-elastic state), changes in ambient temperature affect the creep of the different polymers in varying degrees. For polyaramid and polyester fibres, where normal ambient temperatures (i.e. below their visco-elastic range), the effect of increasing operating temperature will not increase their rate of creep to the same degree as it does HDPE grids and polypropylene grids and fibres (where normal ambient temperatures lie within their visco-elastic range).

Geosynthetics are used for a range of soil reinforcement applications. Fig. 10.23 lists the five major categories—reinforced soil retaining walls,

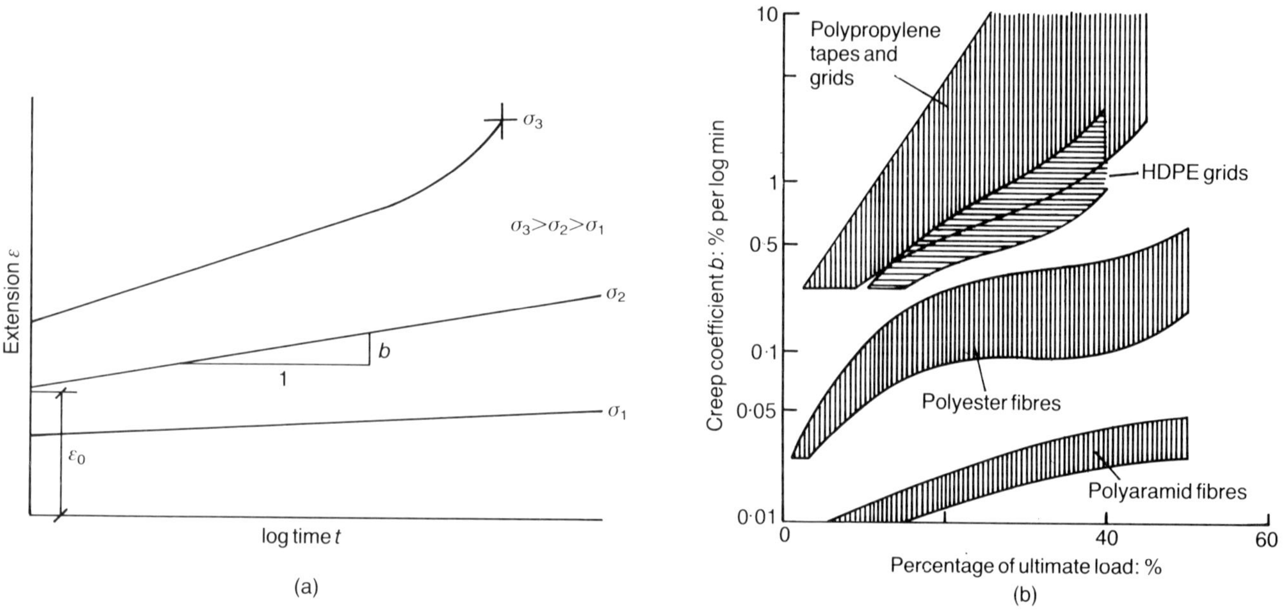

Relationship $\varepsilon_t = \varepsilon_0 + b \log t$ where ε_t is total extension at time t, ε_0 is initial extension at time $t = 0$, b is creep coefficient and σ_1, σ_2 and σ_3 are stresses applied to the geotextile element

Fig. 10.22. Rate of creep of various geosynthetic reinforcement elements: (a) determination of creep coefficient; (b) creep coefficient versus applied load for polyaramid, polyester, and polypropylene fibres, and HDPE grids, at 23°C[10.33]

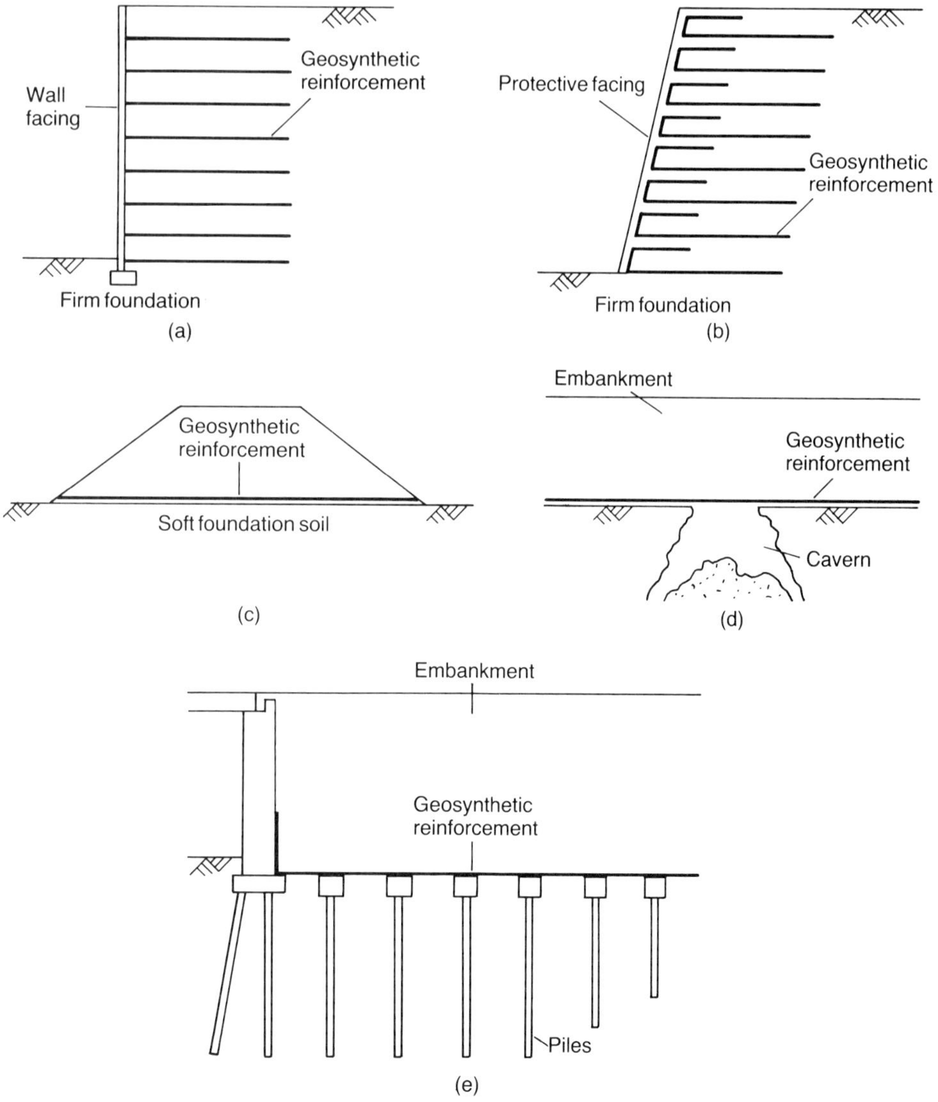

Fig. 10.23. Applications for reinforced soil: (a) reinforced soil retaining walls; (b) reinforced steep slopes; (c) reinforced embankments over soft foundation soils; (d) reinforced embankments over areas prone to subsidence; (e) reinforced piled embankments

reinforced steep slopes, reinforced embankments on soft soils, reinforced piled embankments, and reinforced embankments over areas prone to subsidence.

10.4.3.1. Reinforced soil retaining walls

In this application, shown in Fig. 10.23(a), the geosynthetic reinforcement is laid horizontally into the backfill behind the retaining wall facing. The geosynthetic reinforcing elements are connected to the wall facing, thus providing lateral restraint. The types of geosynthetic materials used for this application are, most commonly, geostrips (see Fig. 10.6(a)), but can also be geogrids.

Reinforced soil retaining walls are designed by either a tie-back wedge analysis or a coherent gravity analysis. Both methods divide into two parts: one part determines the required strength and the spacings of the reinforcement, while the other part determines the reinforcement length and friction requirements. In the UK a technical memorandum exists[10.34] which is used as the formal design method for all UK Department of Transport reinforced soil retaining wall structures.

10.4.3.2. Reinforced steep soil slopes

In this application, shown in Fig. 10.23(b), the geosynthetic reinforcement is placed in horizontal layers in the soil slope as it is being constructed. The

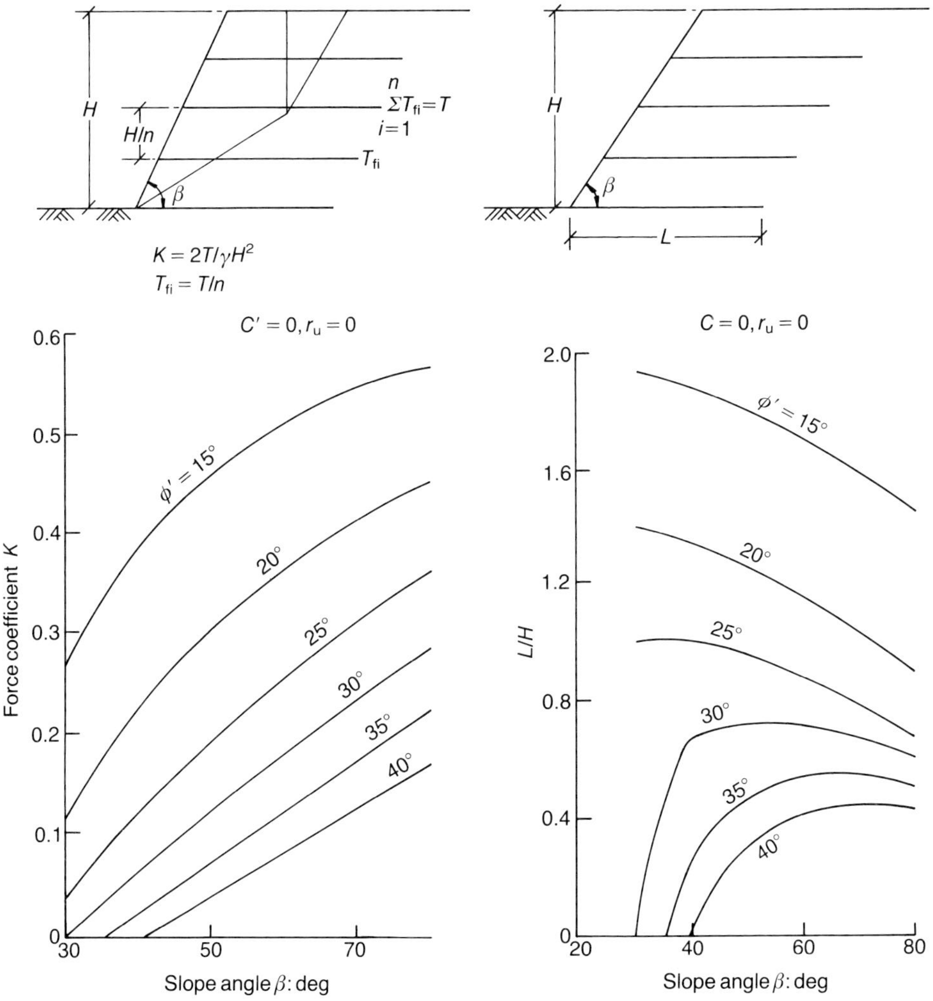

Fig. 10.24. Design curves for reinforced soil slopes (after Jewell et al.)[10.35]

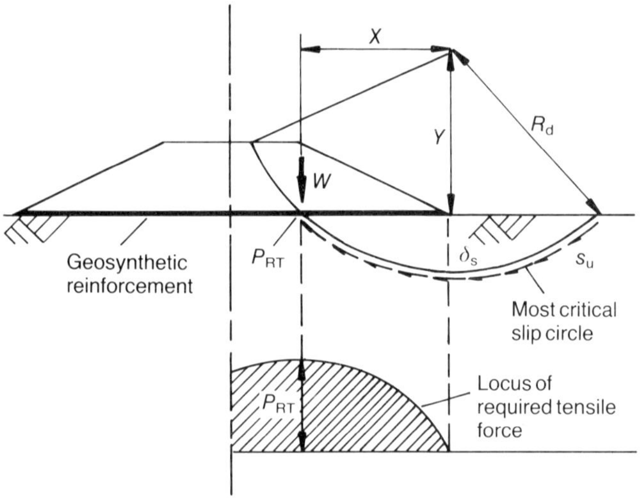

Fig. 10.25. Use of slip circle analysis to account for effect of geosynthetic reinforcement

role of the geosynthetic material is to provide lateral restraint for the slope. The types of geosynthetic materials commonly used for this application are geogrids, multifilament woven geotextiles, fibrillated tape woven geotextiles (for temporary structures), and warp knit geotextiles.

While no universally accepted standard design methods for reinforced soil slopes currently exist, the method proposed by Jewell *et al.*[10.35] based on a two-part wedge failure mechanism has become popular. A series of design curves based on the two-part wedge failure mechanism for simple reinforced soil slopes has also been published; one set of curves is shown in Fig. 10.24.

10.4.3.3. Reinforced embankments over soft foundation soils

In this application, shown in Fig. 10.23(c), the geosynthetic reinforcement is placed between the embankment fill and the soft foundation soil. The role of the geosynthetic is to provide adequate stability to the embankment until such time as the foundation soil has consolidated and gained adequate shear strength to support the weight of the embankment. Therefore, unlike reinforced soil retaining walls and slopes, the design life of the reinforcement in the embankment is less than the actual design life of the embankment. The types of geosynthetic materials commonly used for this application are multifilament woven geotextiles, stitch-bonded geotextiles, and some geocomposite materials.

Reinforced embankments are normally designed by assuming that the geosynthetic reinforcement provides an additional restoring moment to the stability of the embankment. The technique of equating this additional restoring moment is shown in Fig. 10.25. The additional restoring moment due to the geosynthetic reinforcement ΔM_R is equal to the working tensile strength of the reinforcement P_{RT} multiplied by the appropriate moment arm Y.

10.4.3.4. Reinforced piled embankments

For this application, shown in Fig. 10.23(e), the geosynthetic reinforcement is used to span between adjacent pile caps, thus enabling the total embankment load to be completely transferred onto the pile caps.

The piling of embankment foundations is normally practised where soft foundation soils occur, and where it is necessary to prevent settlement of the embankment (e.g. for embankments adjacent to piled bridge foundations). Conventionally, in order to transfer all of the embankment loading onto the

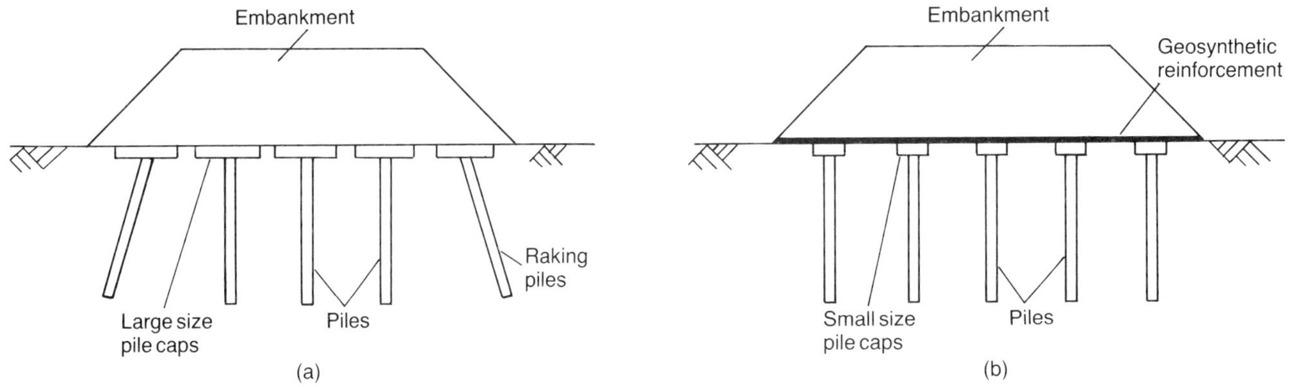

Fig. 10.26. Piled embankment configurations: (a) conventional piled embankment; (b) piled embankment using geosynthetic reinforcement

piles, large sized pile caps were required (Fig. 10.26(a)). In addition, raking piles were often required along the extremities of the piled area to counteract the lateral thrust of the embankment.

When using a geosynthetic reinforcement across the tops of the pile caps, economies can be gained by the use of smaller sized pile caps as well as by the removal of the necessity to install raking piles (Fig. 10.26). The geosynthetic reinforcement acts to transfer the vertical embankment loading onto the pile caps as well as to counteract the horizontal thrust of the embankment. Methods of analysis of geosynthetic reinforced piled embankments are given by ECGL,[10.30] and examples where this technique has been used successfully are given by Broms.[10.36]

10.4.3.5. Reinforced embankments over areas prone to subsidence

For this application, shown in Fig. 10.23(d), the geosynthetic reinforcement is used to span across a void, which may arise at some time following construction of the embankment. The role of the geosynthetic reinforcement is as shown in Fig. 10.27. If a void, or cavern, forms beneath the embankment, then the geosynthetic reinforcement has to have adequate tensile strength, and *controlled* extension, to ensure that it can support the embankment loading as well as control the amount of subsidence at the surface of the embankment to ensure the area can still be trafficked.

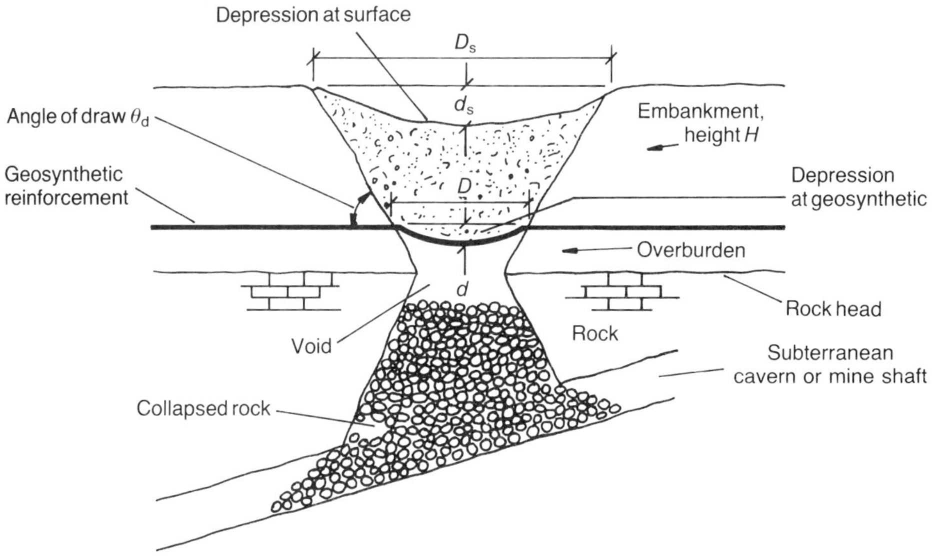

Fig. 10.27. Model used to determine the required reinforcing characteristics of the geosynthetic to limit the surface effects of subsidence[10.30]

Geosynthetic reinforcement is used to limit the amount of surface subsidence (d_s/D_s in Fig. 10.27). Parry[10.37] observes that this maximum allowable differential surface deformation should be limited to 0·01 for motorways, and 0·02 for lower class pavements. The maximum allowable surface deformation translates into a maximum allowable extension in the geosynthetic reinforcement. While the actual deflected shape of the geosynthetic forms a catenary (a hyperbola), the analysis of this is quite complex. Consequently, simplified analysis procedures have been developed based on a circular deflected shape (Giroud *et al.*[10.38]) or a parabolic shape (ECGL[10.30]). Unfortunately, the assumed deflected shape of the reinforcement greatly affects the calculated tensile force in the geosynthetic. For small geosynthetic extensions it is considered that the assumed parabolic deflected shape is more realistic.

10.4.4. Containment

Geosynthetics are used for three types of containment application

- for moisture barriers in pavements
- for the containment of water
- for the containment of waste.

In each application, the geosynthetic acts to prevent the passage of liquids, whether it is water (in liquid or vapour form), or solutions involving water.

The types of geosynthetic used to prevent the passage of liquids are the geomembranes. Geomembranes (unlike geotextiles) are impervious to the passage of liquid water, but may allow small amounts of water vapour to pass. As shown in Fig. 10.5, geomembranes can be either manufactured in situ (by saturating a geotextile with a saturant) or in a factory (by the extrusion or coating of polymers).

10.4.4.1. Moisture barriers in pavements

For the majority of cases where geosynthetics are used in pavement construction, free liquid water movement through the pavement structure is allowed, and encouraged. There are, however, cases where this is undesirable, and hence the prevention of moisture movement becomes very important to the performance of the pavement.

The applications where geomembranes are used as moisture barriers in pavements are shown in Fig. 10.28. Fig. 10.28(a) shows the use of the geomembrane as an expedient surfacing over a compacted, dry subgrade. This technique was first used during the Second World War to rapidly construct airfield runways in war zones.[10.39,10.40] Since then considerable experimental

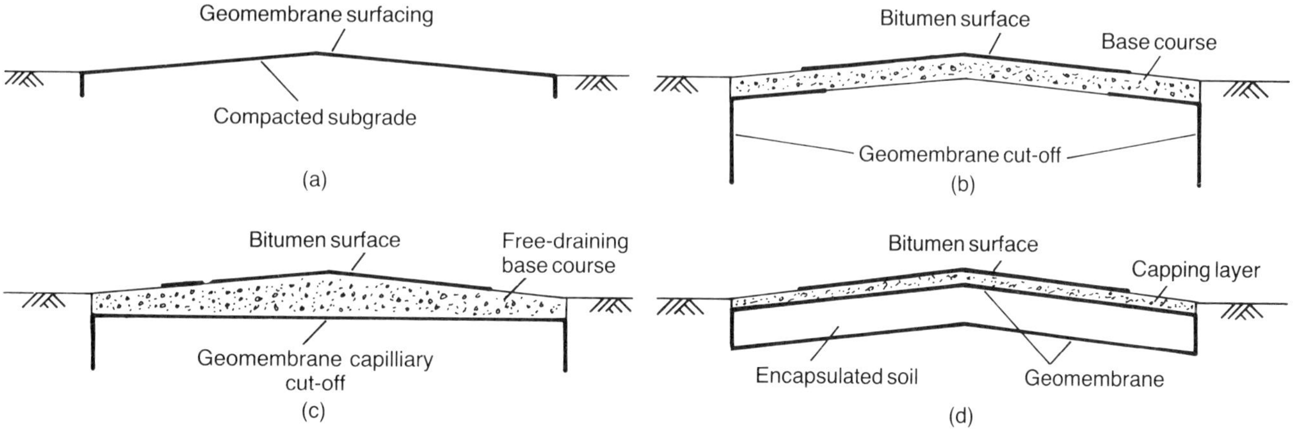

Fig. 10.28. Applications for geomembranes in pavements (after Curiskis and Lawson):[10.1] (a) geomembrane used as an expedient surfacing material; (b) geomembrane used as a moisture barrier; (c) geomembrane used as a capillary cut-off; (d) geomembrane used as a membrane encapsulated soil layer (MESL)

work has been carried out by the US Corps of Engineers to perfect this technique.[10.41] The appropriate geomembrane for this application has to be robust (because vehicles traffic directly on the geomembrane). It is considered that this application will remain in the domain of the military forces because there is little need for these expedient measures in civil applications.

Figure 10.28(b) shows a geomembrane used as a moisture cut-off to maintain a relatively constant moisture environment in the subgrade below the pavement. The geomembrane is placed to a depth below the level of seasonal moisture variation, thus ensuring that the presence of the geomembrane alone controls moisture movements in the subgrade. This technique has proved very effective in controlling volume changes in expansive clay subgrades.[10.42]

Figure 10.28(c) shows the use of a geomembrane as a capillary cut-off. In this case the geomembrane extends all the way beneath the pavement, preventing the rise of moisture into the base course layer. This technique has been applied to the treatment of frost-susceptible soils, the purpose of the geomembrane being to prevent the rise of moisture into the pavement where it can freeze and swell during the winter.

Figure 10.28(d) shows the use of a geomembrane to encapsulate a soil layer completely, to protect it from excessive moisture variations. The resultant structure is known as a "membrane encapsulated soil layer" (MESL). The MESL technique has been used successfully to stabilize frost-susceptible soils,[10.43] to control expansive clay subgrades,[10.44] to stabilize normally "sub-standard" materials to enable their use as structural layers in pavements[10.45] and as a protection against random events (e.g. flooding).

10.4.4.2. Containment of water

The loss of water through the foundation soil from water impoundment and canal structures can be a major problem (from an economical viewpoint). For foundation soils without an appreciable clay content it is normal practice to construct a water-repellent lining around the extremity of the containment structure to prevent excessive water losses due to infiltration. Historically, clay, asphalt or concrete liners have been utilized. However, problems with the cost and performance of these conventional liners have led to the extensive use of geosynthetic liners in the form of geomembranes (by themselves or in combination with other geosynthetic materials). A number of different geosynthetic lining systems have been proposed for water-containment structures. Two of the most common are shown in Fig. 10.29. The decision regarding which system to use depends on a number of factors, namely, type of foundation soil, types of suitable geomembranes, and how the water containment structure is to function.

There are two dominant components which determine how effective the geomembrane is in preventing water losses. These are the integrity of the geomembrane itself, and the integrity of the geomembrane seams (joins). Specification of the correct geomembrane polymer composition and mechanical properties ensures the integrity of the geomembrane itself. However, very close attention to detail regarding the quality of the seaming technique on site is critical to the performance of the geomembrane liner.

The most basic form of geosynthetic system is to use a geomembrane liner by itself. The geomembrane can be either left exposed, or covered over (for protection) with an earth covering. For temporary impoundment structures the geomembrane can be left exposed, while for permanent structures the geomembrane should be covered with earth. When the geomembrane is used as the only component in the lining system, considerable care has to be taken to ensure that the foundation soil below the geomembrane is carefully prepared to ensure the geomembrane is not punctured during installation. In

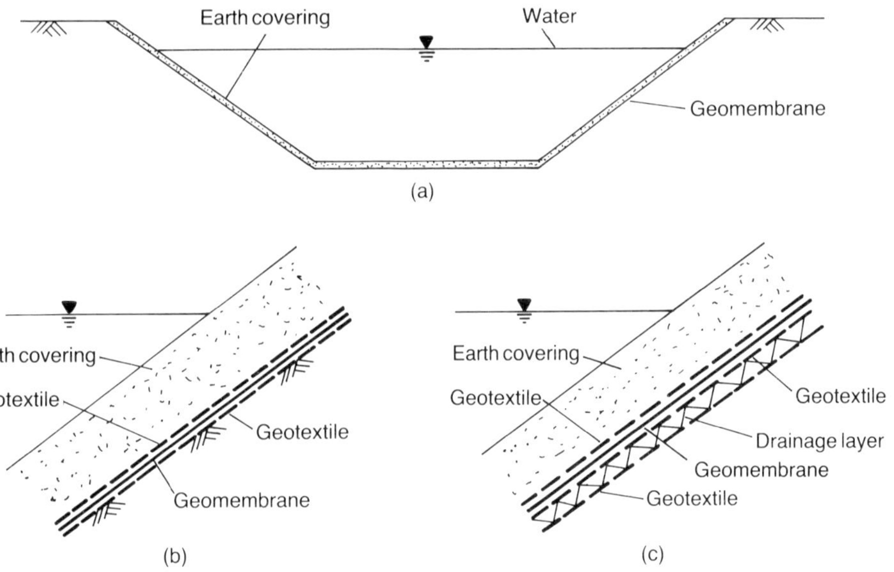

Fig. 10.29. Use of geomembranes for water containment: (a) water containment structure using a geomembrane liner; (b) geomembrane detail on free-draining foundation soil; (c) geomembrane detail on fine foundation soil

addition, this type of system requires a robust geomembrane (which normally means the use of a thick or reinforced geomembrane).

For the vast majority of applications it is impossible to ensure a smooth foundation surface before geomembrane installation. Consequently, a cushioning layer, in the form of a thick geotextile, is normally placed on the prepared foundation prior to installation of the geomembrane. This technique is shown in Fig. 10.29(b). To protect the geomembrane on the upper side a geotextile cushion layer is placed on top of the geomembrane prior to placement of the earth covering. When geotextiles are used to protect the geomembrane liner less robust geomembranes may be used than is the case if the geomembrane is used by itself.

The use of the geomembrane lining system shown in Fig. 10.29(b) is very effective if the foundation soil is free draining, and if the level of the water in the containment structure is to be kept at fairly constant depth (with only slow rises and falls). However, if the foundation soil has a low permeability (but not low enough to dispense with the need for a geomembrane) and/or the level of the water in the containment structure undergoes severe drawdown, then a more sophisticated liner system should be employed (Fig. 10.29(c)). The liner system shown in Fig. 10.29(c) contains a drainage layer beneath the geomembrane. The purpose of the drainage layer is to prevent the build-up of hydrostatic pressure behind the geomembrane during rapid drawdown of the impounded water.

The drainage layer shown in Fig. 10.29(c) can consist of either a geotextile filter with a granular drainage layer or, more commonly, a geocomposite fin drain.

10.4.4.3. Containment of toxic waste

The containment of toxic and land-fill waste presents a number of key problems. The waste is normally placed in the containment area in a relatively untreated form. Consequently, many of the chemicals in the waste, whether in isolation or in the presence of water, remain toxic for long periods of time. These toxic chemical compounds have to be prevented from leaching out of the waste and polluting the natural groundwater. The presence of free water in the waste provides the means of transport for the chemicals into the natural groundwater. Thus, the movement of water into and out of the containment structure must be prevented.

Historically, thick clay linings were used to contain waste. However, doubts concerning the quality of clay linings on their own (it has been shown that the presence of some toxic metal ions actually increases the permeability of clay) have led to the adoption of clay/geomembrane or geomembrane lining systems.

Due to the long-term toxicity of waste impoundments, and the consequent level of pollution which would occur if the lining of the waste impoundment failed, geomembrane *systems* are used to seal the impoundment, rather than just geomembranes on their own. This is because a geomembrane system provides much greater insurance against a potential failure of the lining than would be the case if only an individual geomembrane liner were utilized.

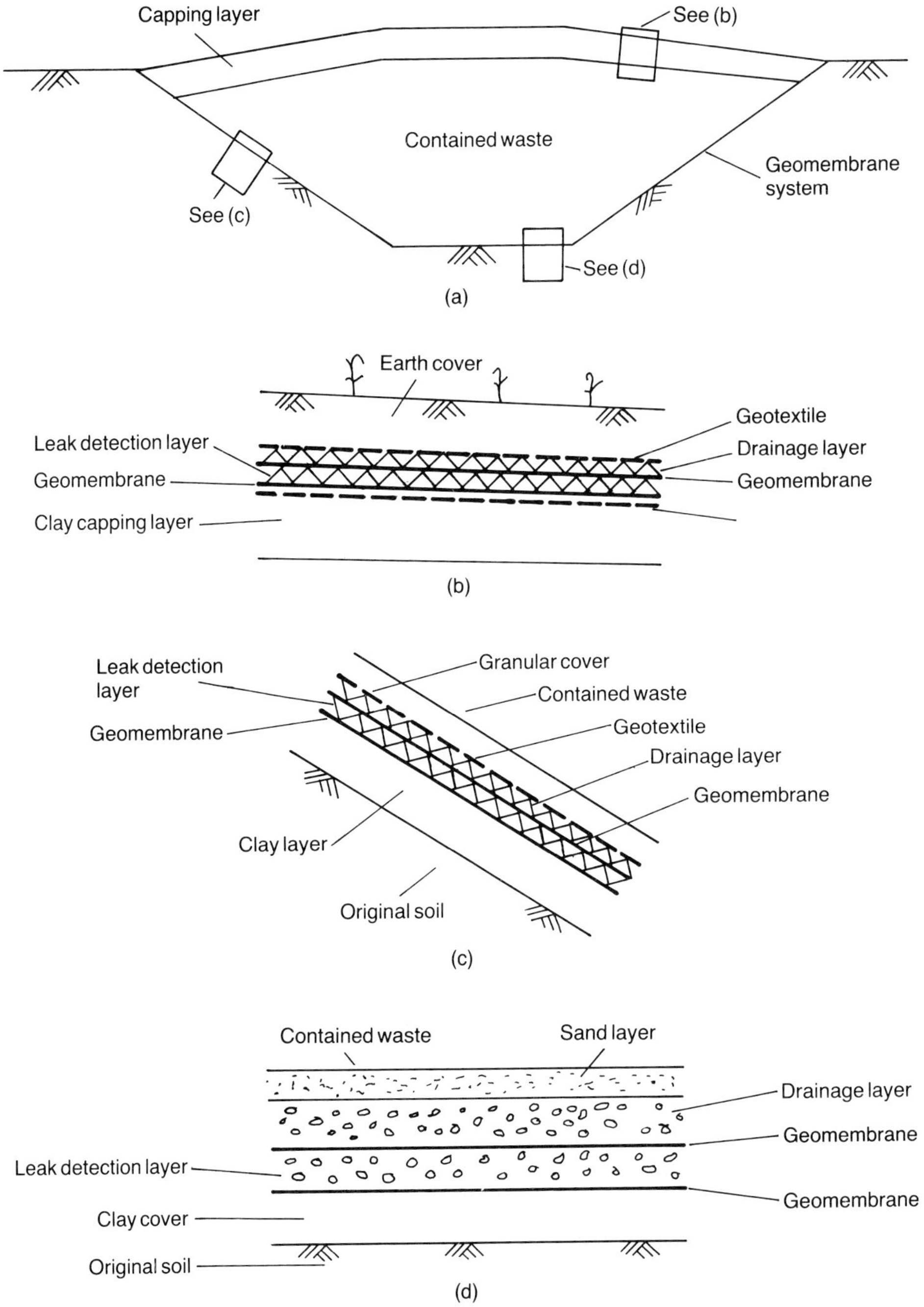

Fig. 10.30. Use of geomembrane systems for the containment of toxic waste (after Koerner and Richardson):[10.46]
(a) application of geomembrane system to contain toxic waste;
(b) details of the capping protection layer; (c) details of the side protection layer;
(d) details of the base protection layer

Fig. 10.30 shows the types of geomembrane systems that are used to line various locations in a waste containment structure.

Because it is impossible to ensure a perfect long-term seal of the geomembrane liner, the fundamental concept for design is to assume that at some stage during the design life of the lining there will be some leakage. (Outwardly, this concept for design would appear a contradiction since the geomembrane is used to stop leakages; however, adoption of this approach leads to safe and conservative lining systems.) Once having established this fundamental principle, additional components can be added to the lining (at the design stage) to ensure a safe and conservative lining system. A safe lining system is normally composed of combinations of geomembranes, geotextiles, drainage layers, and (maybe) clay.

Figure 10.30(a) shows the general layout of a completed waste containment structure, with the capping layer having been installed to seal the structure. The types of lining systems used at each location in the containment structure are dictated by the long-term changes in the environment of the contained waste and the functional characteristics of the protection materials used. For the containment structure shown in Fig. 10.30(a) three different lining systems are utilized.

Details of the capping layer (the protective layer installed after filling of the waste structure) are shown in Fig. 10.30(b). The protective capping layer is designed to prevent surface water from entering the containment structure, as well as to allow the generated gas (in the contained waste) to be removed freely. The upper geotextile filter and drainage layer enable the surface water that passes through the earth cover to be drained away above the upper geomembrane layer. Should the upper geomembrane layer develop leakage, then this can be detected and prevented from entering the containment structure by the leak detection drainage layer and the lower geomembrane. Below the lower geomembrane, a narrow drainage layer with a geotextile filter is used to remove any gas build-ups from the contained waste. Below this, a clay capping layer is normally the first material placed on top of the contained waste.

Details of the side protection layer are shown in Fig. 10.30(c). It is normally constructed by first placing a compacted clay layer directly against the original foundation soil. To detect any leakage through the upper geomembrane layer, a lower geomembrane and a leak detection drainage layer are installed. To drain off leachate from the contained waste a drainage layer with a geotextile filter is installed on top of the upper geomembrane layer. On top of this upper geotextile filter, a granular soil layer is placed to provide protection to the lining system during placement of the contained waste.

Details of the base protection layer are shown in Fig. 10.30(d). A compacted clay layer is first placed on top of the original foundation soil (the same as it is for the side protection). The lower geomembrane layer is then installed along with a leak detection drainage layer. The upper (primary) geomembrane layer is then installed. On top of the primary geomembrane layer a drainage layer is constructed (to remove leachate with may build up at the base of the impoundment during filling with waste). Finally, a granular filter layer is placed on top of the drainage layer to prevent contamination of the drainage layer and to protect the lining system.

References

10.1. Curiskis J.I. and Lawson C.R. An introduction to geotextiles with specific reference to applications in pavement design and construction. *Proc. Symp. on Recent Developments in Pavement Des. and Constr., Canberra*, 1985, ADFA, 3.1–3.134.

10.2. Lawson C.R. Geotextiles—a short course. *Proc. Short Course on Soil Improvement Tech., Bangkok*, 1982, AIT, 1–95.

10.3. Koerner R.M. *Designing with geosynthetics*. Prentice-Hall, New York, 1986.

10.4. Giroud J.P. Geotextiles and geomembranes. *Geotextiles and Geomembranes*, 1984, **1**, No. 1, 5–40.

10.5. Calhoun C. *Development of design criteria and acceptance specifications for plastic filter cloths*. US Army Waterway Experiment Station, Vicksburg, 1972.

10.6. Ogink H.J.M. *Investigations on the hydraulic characteristics of synthetic fabrics*. Publication No. 146. Delft Hydraulics Laboratory. Delft, 1975.

10.7. McKeand E. The behaviour of non woven fabric filters in subdrainage applications. *Proc. Int. Conf. on the Use of Fabrics in Geotechnics, Paris*, 1977, **2**, 171–176.

10.8. Schober W. and Teindl H. Filter—criteria for geotextiles. *Proc. 7th Eur. Conf. on Soil Mech. and Found. Eng., Brighton*, 1979, **2**, 121–129.

10.9. Giroud J.P. Filter criteria for geotextiles. *Proc. 2nd Int. Conf. on Geotextiles, Las Vegas*, 1982, **1**, 103–108.

10.10. Tan H.H. *et al.* Hydraulic function and performance of various geotextiles in drainage and related applications. *Proc. 2nd Int. Conf. on Geotextiles, Las Vegas*, 1982, 1, 155–160.

10.11. Lawson C.R. Geotextile filter criteria for tropical residual soils. *Proc. 3rd Int. Conf. on Geotextiles, Vienna*, 1986, **2**, 557–562.

10.12. Lawson C.R. *Designing for subsurface drainage*. ICI Fibres, UK, 1988.

10.13. Ragutzki G. Beitrag zur ermittlung der Filter-Wirksamkeit durchlassiger Kunststoffe. *Forschungsstelle fur Insel und Kustenschutz, Sonderdruck aus Jahresbericht*, 1973.

10.14. Lawson C.R. Geotextile requirements for erosion control structures. *Proc. Conf. on Rec. Dev. in Ground Improvement Tech., Bangkok*, 1982, 177–192.

10.15. Heerten G. Dimensioning the filtration properties of geotextiles considering long-term conditions. *Proc. 2nd Int. Conf. on Geotextiles, Las Vegas*, 1982 **1**, 115–120.

10.16. Koerner R.M. and Welsh J.P. *Construction and geotechnical engineering using synthetic fabrics*. Wiley, New York, 1980.

10.17. Welsh J.P. and Dominske D. Synthetic fabrics as a concrete forming device. *Proc. 2nd Int. Conf. on Geotextiles, Las Vegas*, 1982, **2**, 457–461.

10.18. Lupton T.J. The advancing technique in flexible forms. *Proc. 2nd Int. Conf. on Geotextiles, Las Vegas*, 1982, **2**, 463–466.

10.19. Tutuarima W.H. and Van Wijk W. ProFix Mattresses—An Alternative Erosion Control System. *Proc. Int. Conf. on Flexible Armoured Revetments Incorporating Geotextiles*, 259–272. Institution of Civil Engineers, London, 1984.

10.20. Barenberg E.J. *et al. Evaluation of soil aggregate systems with Mirafi fabric*. Report UILU-Eng-75-2020. University of Illinois, 1975.

10.21. Kinney T.C. *Fabric induced changes in high deformation soil-fabric-aggregate systems*. PhD thesis. University of Illinois, 1979.

10.22. Giroud J.P. and Noiray L. Geotextile reinforced unpaved road design. *J. Geotech. Eng. Div. Am. Soc. Civ. Engrs*, 1981, **107**, GT9, September, 1233–1254.

10.23. Raumann G. Geotextiles in unpaved roads: design considerations. *Proc. 2nd Int. Conf. on Geotextiles, Las Vegas*, 1982, **2**, 417–422.

10.24. Sellmeijer J.B. *et al.* Calculation method for a fabric reinforced road. *Proc. 2nd Int. Conf. on Geotextiles, Las Vegas*, 1982, **2**, 393–398.

10.25. Yano K. *et al.* Earthwork on soft clay grounds using rope-netted fabrics. *Proc. Conf. on Rec. Dev. in Ground Improvement Techniques, Bangkok*, 1982, 225–238.

10.26. Nishibayashi K. Surface layer stabilisation of soft ground using synthetic chemical fibre sheet. *Proc. Conf. on Rec. Dev. in Ground Improvement Techniques, Bangkok*, 1982, 239–254.

10.27. Broms B.B. Stabilisation of very soft clay using geofabric. *Geotextiles and Geomembranes*, 1987, **5**, No. 1, 17–28.

10.28. Kérisel J. The history of geotechnical engineering up until 1700. *Proc. 11th Int. Conf. on Soil Mech. and Fdn Eng., San Francisco*, 1985, 3–95.

10.29. Vidal H. La terre armée. *Annales*, 1966, **223–4**, 888–938.

10.30. ECGL. *Designing for soil reinforcement*. Exxon Chemicals, UK, 1989.

10.31. Jewell R.A. and Greenwood J.H. Long term strength and safety in steep soil slopes reinforced by polymer materials. *Geotextiles and Geomembranes*, 1988, **7**, Nos 1&2, 81–118.

10.32. CIRIA. *Rationalisation of safety and serviceability factors in structural codes*. Report No. 63. Construction Industry Research and Information Association, London, 1977.

10.33. Lawson C.R. Geosynthetics for soil reinforcement. *Proc. Seminar on Geotextiles in Civ. Eng.* Institution of Engineers Australia, Newcastle, 1986.

10.34. Department of Transport. *Reinforced earth retaining walls and bridge abutments for embankments*. Technical memorandum BE3/78. Department of Transport, London, 1978.

10.35. Jewell R.A. *et al.* Design methods for steep reinforced embankments. In: *Polymer grid reinforcement*, 70–81. Thomas Telford, London, 1985.

10.36. Broms B.B. Embankment Piles. *Proc. 3rd Int. Geotech. Seminar on Soil Improvement Methods, Singapore*, 1985, 167–178.

10.37. Parry H.J. Coping with Fife's mining industrial heritage. *Municipal Eng.*, 1983, **110**, No. 778, 231–240.

10.38. Giroud J.P. *et al.* Load-carrying capacity of a soil layer supported by a geosynthetic overlying a void. *Proc. Conf. on Theory and Practice of Earth Reinforcement, Kyushu*, 185–190. Balkema, 1988.

10.39. Colquhoun R.S. and Brock G. The development and application of prefabricated bituminous surfacing in India and South East Asia Commands. *The civil engineer in war—a symposium of papers on war-time eingineering problems*, **1**, 274–291. The Institution of Civil Engineers, London, 1948.

10.40. Grigson R.R.W. and Lee A.R. Rapidly constructed airfields, part 2: prefabricated bituminous surfacing. *The civil engineer in war—a symposium of papers on war-time engineering problems*, 1, 251–273. The Institution of Civil Engineers, London, 1948.

10.41. Burns C.D. and Barber V.C. *Expedient surfacing and drainage of roads, streets, and parking and storage areas in theater of operations*. Miscellaneous Paper 5-69-11, Report No. 1, US Army Engineer Waterways Experiment Station, Vicksburg, 1969.

10.42. Steinberg M.L. Geomembranes used to control expansive soil. *Proc. Int. Conf. on Geomembranes, Denver*, 1984, **2**, 507–511.

10.43. Bell J.R. and Yoder E.J. Plastic moisture barriers for highway subgrade protection. *Proc. Hwy Res. Board*, 1957, 36, 713–735.

10.44. Benson J.R. The use of asphalt membranes in highway and airfield pavements. *Proc. Assn Asphalt Pavement Technologists*, 1953, **22**, 56–91.

10.45. Lawson C.R. and Ingles O.G. Long term performance of MESL road sections in Australia. *Proc. 2nd Int. Conf. on Geotextiles, Las Vegas*, 1982, **2**, 535–540.

10.46. Koerner R.M. and Richardson G.N. Design of geosynthetic systems for waste disposal. *Proc. Conf. on Geotechnical Practice for Waste Disposal*, 65–86. Geotechnical Special Publication No. 13. American Society of Civil Engineers, New York, 1987.

11 Further applications of polymers and polymer composites

L. Hollaway and C. J. Burgoyne

11.1. Introduction

Previous chapters have discussed uses of reinforced and unreinforced polymers in specific forms in the construction industry. This chapter will discuss other uses of polymers which do not fall neatly into the categories described previously.

11.2. Moisture barriers

Polymers are used extensively in the construction and building industries as ancillary materials. Plastic films serve as permanent moisture barriers, preventing the deleterious migration of water through various building and construction elements. Plastics can be used to protect or modify concrete. Silicones can effectively prevent the entry of water through porous bricks, stone or concrete. Thermoplastic lattices can be added to concrete and mortar to improve adhesion and tensile strength. Concrete formwork can be made of moulded or thermoformed plastics (see section 2.6.3) which allow design freedom, simple erection, good surface finish and compound curves.

Some of the common applications for plastics films in the construction and allied industries are for

(*a*) vapour barriers between walls and ceilings, and between rough and finished floors
(*b*) curing blankets for concrete, to retard evaporation
(*c*) underlays for foundation wall waterproofing, highways, airport runways and apron subgrades (see section 10.4.4)
(*d*) temporary protection for personnel and materials.

Polyethylene film is often used in the above applications. This has a very low water permeability and below ground its life expectancy is superior to that of its competitive materials. Polyethylene is unaffected by water salts or bacteria; its superior toughness is a result of the high elongation of the material when under load.

The Glaswall system, which is a glass fibre reinforced polyester, can be applied to a vertical surface of brickwork and concrete to give a water-resistant and hard-wearing protective surface with a high gloss finish.

11.3. Forms for concrete

Unreinforced and reinforced polymers are used in formwork, treamies, chutes and various tools; their attraction lies in their ease of handling. Glass-reinforced polyester composites manufactured by the hand lay-up, spray-up or by matched metal moulding techniques (section 2.8) are often used for these purposes and have the advantage of economy due to the repeated use of the forms. The initial cost of the mouldings may be higher than that for the timber formwork but the ability to reuse the composite forms can result in an overall lower cost. The glass fibre composite formwork is tough and cannot be easily dented, it does not rust as steel does and it does not react with alkaline constituents of concrete. In addition, if the formwork is damaged it can be repaired easily with no visible evidence on the finished concrete.

Helically wound GRP composite tubes have been used in column construction for marine structures as a replacement for steel tubes; both types of tubes would be filled with concrete. Steel used in this context has certain limitations; for instance, due to its high modulus of elasticity, the axial load taken by the column will be mainly borne by the steel. The latter's Poisson's ratio is greater than that of the contained concrete, thus there is little confinement of the concrete as the steel tube separates from the concrete core. On the other hand, a helically wound reinforced GRP composite tube is an ideal material to use because the axial load will be taken entirely by the concrete, the expansion of the composite in the circumferential direction is smaller than that of the concrete and the tensile strength of the composite is high. Consequently, the GRP casing counteracts lateral expansion of the concrete under load and when used in a short column, the axial strength of the concrete core increases over its uniaxial value and can reach a triaxial failure strength of up to four times the uniaxial strength. It has been suggested that fibre-reinforced polymer casings could be used to encase reinforced concrete beams thus improving their strength and ductibility in the compression zone and, in addition, to provide permanent formwork and corrosion protection.

11.4. Resin bonded systems

It is clear that modern forms of plywood and particle board could not exist without the polymer adhesive. The polymer used in the manufacture of plywood is only about 4% of the panel by weight and therefore wood is the dominant material.

Sandwich construction is another form of a composite system using resin bond. Honeycomb is a fabricated cellular structure made of sheet material which is assembled to form nested hexagonal voids. In structural applications it is commonly used as a low cost, low density core in sandwich construction.[11.1] Honeycomb can be fabricated from aluminium, stainless steel, reinforced plastics and, in the building industry, from kraft paper honeycomb which is impregnated with phenolic resin. Paper honeycomb is utilized in many building applications, such as partition, doors and other lightweight structures. To form the sandwich system, the core material is bonded to the face material by resin adhesive to form a stiff lightweight beam system.

11.5. Foamed polymer systems

Foamed polymers have been discussed in section 2.2.3. These types of polymers can be hard or soft, rigid or flexible or intermediate between the two extremes. Some foams are only available in the form of board (slab stock) whilst others come as liquids or particles that can be foamed or sprayed in place. The foam density and the proportion of interconnected cells generally control the physical strength properties. Thermal conductivity is a function of density, cell size and gas within the cell. Water absorption and water vapour transmission vary with the water—susceptibility characteristics of the basic polymer and the proportion of open cells.

11.5.1. Thermal insulation foams

Foamed polymers can be used as a thermal insulation material. Polymers generally have a very low thermal conductivity and are therefore ideal as thermal insulations. They have been used as space insulation, pipe, wall and roof insulator mediums.

11.5.2. Foamed polymers in road embankments

The load exerted by embankments on soft fine-grained subsoils often leads to problems of bearing capacity and settlement. Slides may occur as a result of shear failure in underlying clay and peat soils. Such subsoils often yield unacceptable settlements. One method of protecting the road against instability and troublesome settlement is to use lightweight fill material. Expanded polystyrene (EPS) has been used in this context where extremely difficult subsoil conditions exist and where other lightweight materials (which are relatively much heavier than EPS) would not provide sufficient stability. The material is required to form a stable volume with low weight and it should be sufficiently strong to take dead and live loads from the pavement and the traffic, thus reducing the loads transmitted to the weak subsoil.

The properties of EPS which are important for the construction of superlight-weight embankments are

- mechanical properties (in particular the compressive strength)
- geometrical size and uniformity
- resistance to moisture absorption
- chemical resistance (solvents, fire)
- long-term properties (ageing, disintegration, resistance against micro-organisms, etc.)

The compressive strength is important if large deflections of EPS material are to be prevented. A required amount of base material is necessary to obtain a particular compressive strength, but beyond this value the density of the material is not too important. As yet there is no British code of practice relating to the use of expanded polystyrene for sub-base to roads, but Scandinavian countries and the Netherlands do have specifications on this topic: the Norwegian Directorate of Roads has had experience in the use of expanded polystyrene for road embankments and had laid down specifications for EPS used for this purpose. These state that the compressive strength of the material may be determined by an unconfined compression test and the specimens should be $50 \times 50 \times 50$ mm. The average minimum strength should be 100 kN/m^2 at 5% compression (2.5 mm) and no single test result should be less than 80 kN/m^2.

The geometric size and uniformity of the material is of importance in order to obtain a stable construction and to facilitate placing and, in this case, the Norwegian Directorate of Roads has stated that the sides of the EPS blocks shall be cut at right angles to each other and that the maximum allowable deviation from the given dimensions (width, length and height) is $\pm 1\%$. The tolerance on the degree of flatness of horizontal surfaces shall be within 5 mm and the smallest dimension of the block shall be 0·5 mm.

The moisture-absorptive characteristics are important for long-term behaviour of the fill. Stability problems can occur if the blocks absorb too much water. However, at present there appears to be no Norwegian specification on the moisture absorption of EPS for road construction, but it seems likely that, provided the design density of 100 kg/m^3 is maintained, the absorption of water will be at a minimum.

Solvents will dissolve EPS but protection against chemical spillages can be provided by placing high density polyethylene sheet material under the road pavement. Precautions against fire are possibly by using fire-resistant EPS.

Expanded polystyrene has a resistance against chemical attack from soils and against attack from micro-organisms and rodents.

11.5.3. Concrete shuttering with EPS insulant

When concrete buildings are constructed, concrete is poured into a costly system of shuttering which is subsequently removed. When EPS board is used for external insulation (a method of conserving heat often used in West Germany, neighbouring countries and sometimes in the UK) it has to be attached to the wall either mechanically or with an adhesive in a separate operation.

A system exists in which the process is reversed. The shuttering is itself moulded EPS so that after the concrete has set, the insulation is already in place. The system was originally conceived in Austria in the early 1960s but a number of difficulties had to be overcome before it was acceptable and international status could be granted for it. It is now possible to make complete EPS shuttering elements and when assembled the modular elements form a continuous system of thermal insulation both inside and outside the building.

11.6. Sealants

Sealants are elastomeric materials which can be used for the sealing of joints against wind and water in the construction industry. In thin curtain wall construction, which employs highly effective materials to provide the heat insulation, there is no cavity for the dispersion of water that may leak through the joints on the outside; also, there is no effective baffling system to prevent air blowing directly to the inside in the event of open joints. Consequently, adhesive and elastic sealants are needed to enable this type of construction to be used efficiently.

The largest variety of sealants fall into the classification of solvent release and are composed of three component parts; these are

(*a*) the basic non-volatile vehicle (the liquid portion of the compound)
(*b*) the pigment component
(*c*) a solvent or thinner to make application easier.

The non-volatile vehicle can vary from a vegetable oil (e.g. linseed) to a synthetic elastomer; each has its particular property and specific application. The pigment component introduces opacity or colour to the material and also assists the rheology and flow control. The solvent is present merely to provide a reduced viscosity to the material to enable the sealant to be applied easily and to ensure that the correct thickness is achieved. After application the solvent evaporates thus "curing" the sealant and causing it to reach its required viscosity and to shrink by an amount equivalent to the volume of the solvent. The Butyl rubber solution and the acrylic copolymer solution fall into this category.

Another group of sealants are those which are chemically cured. The main sealants under this heading are the polysulphide compounds and the silicone base compounds. The latter is a two-part sealant and is highly dependent upon the environmental conditions for the rate of cure. If the humidity and/or temperature is low, the curing period can be very long. The chemically cured compounds do not generally contain much solvent, but do require adhesion additives in order to bond reasonably well to the surfaces to which they are applied.

The desired properties of a sealant are

- a good adhesion with the joint
- permanent elasticity
- low rate of hardening
- low rate of shrinkage.

The choice of sealants is a compromise as no one product has all the above mentioned attributes.

Most sealants will fail by adhesion rupture. If high elasticity or elastomeric properties are required, the cross-linking within the polymer will be high and this results in high cohesive forces; however, this is not consistent with a good adhesive characteristic. In addition, sealants are generally applied on site and adhesive problems are aggrevated by the construction dust and foreign matter which are deposited on the joints.

Elasticity is a difficult property to measure as the load on joints tends to be of dynamic form with low-magnitude repeated strains being applied. Because of the difficulties of setting up this type of dynamic test procedure, a simple single tensile test is usually undertaken to provide some indication of the amount of elongation before rupture. It must be mentioned that elasticity should not be improved at the expense of adhesion.

Ideally sealants should not harden, although, in practice, a small amount of hardening is not critical. If the sealant becomes too hard, it will be cohesive and will fail in cohesion. A small amount of shrinkage in sealants may be acceptable as this property will cause a corner joint to become concave. However, it is not desirable to have excessive shrinkage as the sealant will shrink into the joint, with an aesthetically unpleasing effect.

11.7. Adhesions

Chapter 6 has discussed the methods of bonding and the analyses of joints to transfer loads; this section will briefly cover the adhesives for bonding polymer building materials.

Resilient floor tile adhesives fall into one of the following types

(*a*) rubber-based mastic with water vehicle
(*b*) rubber-based mastic with alcohol solvent
(*c*) linoleum paste
(*d*) epoxy resin mastic.

There are several types of rubber-based mastics in use, including ones which use low cost reclaimed rubber with a high water resistance and flexibility. Others used for concrete applications are made from synthetic rubber-type mastic with a water vehicle and have good water and alkali resistance.

11.8. Epoxy–bitumen combinations

About 25 years ago the Transport and Road Research Laboratory discovered that calcined bauxite has exceptionally high durability and resistance to polishing of bituminous roads; this problem is caused by heavy traffic stresses and vehicle breaking on the surface of the material. As calcined bauxite was and remains in short supply it can only economically be considered as a top dressing to a conventional asphalt surface. Furthermore, standard bitumen surface dressing binders are not strong enough to hold bauxite chippings in place under conditions of high stress. Consequently, a bitumen-extended epoxy resin binder was developed by Shell Research Ltd (the material system has the trade name Shellgrip, in which calcined bauxite aggregate is embedded in a flexible bitumen Epikote which is also a trade name). The Greater London Council arranged for an experimental patch to be laid in south London in 1966. The system continues to be used in London and is widely used elsewhere in the UK and in many other countries throughout the world. In towns such treatment shows the best return on sites with a high wet-skidding rate such as road intersections, tunnels and pedestrian crossings.

11.9. Structural composites

In this section, composite structural components and composite structures that are in the development stage will be discussed. It was shown in chapter 2 that composite materials manufactured from glass fibre and polymer resin have a low modulus of elasticity compared with that of steel. It was also stated

Fig. 11.1. Space frame by Fibreforce Composites Ltd, joint design by Composite Structures Unit of the University of Surrey

Fig. 11.2. British Railway electrification gantries (courtesy BP Advanced Materials Ltd)

that the modulus and the strength are highly dependent upon the fibre volume fraction and the orientation of the fibres. Consequently, the orthotropy of the material must be considered in the analysis and design of the composite.

11.9.1. Skeletal systems for earth applications

Composite skeletal systems used in the construction industry would, at the current level of technology, be manufactured by the pultrusion technique (section 2.8.2.3), using glass fibre (or a hybrid of glass and carbon fibre) in a polyester matrix or by the geoform (GRP) technique.*

Pultrusion sections which have exclusively unidirectional reinforcement are anisotropic with extremely high axial and flexural strength but relatively low transverse strength. However, by incorporating hooped strands along a reinforcement core, the hoop strength can be improved. Mats, particularly continuous fibre mats, can also be used to improve the transverse strength. To improve the appearance, corrosion resistance and handling of the product, glass fibre and polymeric veils can be added to the laminate construction to depress the reinforcement from the surface, thus providing a polymer rich surface to the composite. The two most commonly used products are surface tissues of high alkali content glass (A glass) or chemical resistant glass (C glass), and polymeric veils of polyester if this is one of the parent materials.

The main properties of the pultruded composite which give it a special significance are

- high strength-to-weight ratio
- dimensional stability
- corrosion resistance.

As the problem of jointing skeletal composite systems has only recently been investigated and a solution found, there are not many examples available to illustrate their use. However, one example is shown in Fig. 11.1 which utilizes the composite end cap discussed in section 6.20. The members of this structure

*Geoform is a light GRP lattice system manufactured by a process in which unidirectional glass fibres are impregnated with a high performance resin system and are progressively layered into the lattice: fixings are incorporated into the structure during manufacture.

were manufactured by the pultrusion technique by Fibre Force Ltd and are 1 m long, 25 mm external diameter and have a wall thickness of 2 mm. Some research work undertaken during the development of this system has been described in reference 11.2; also given in this reference is a detailed discussion of the pultrusion developments in recent years. Fig. 11.2 illustrates the use of geoform as a lightweight lattice system.

11.9.2. Skeletal systems for space applications

It is likely that some structural engineers will, in the future, become involved in the analysis and design of space structures for communication satellites in which large diameter parabolic reflectors in excess of 50 m diameter and space platforms in excess of 300 m length will be deployed in space. It is almost certain that these structural systems will be of skeletal form and will be manufactured from composite materials because of the latter's unique combination of high specific strength and stiffness, low weight, good dimensional stability and high specific damping capacity. High modulus graphite-reinforced epoxies have been the principal composite material system used for space applications. Initially composites were selected primarily to reduce weight. However, the development of high precision structures, such as the space telescope, required high modulus composites with a coefficient of thermal expansion close to zero to enable the design requirements to be met.

Large parabolic reflectors and space stations will probably be the first large structures where composites are used in the primary structure. One of the composites being investigated for use as the material of construction for a large parabolic reflector contains graphite fibres in a matrix of polyethersulphone. Tube members have been made from this material using the film stacking and hot press moulding technique (section 2.7.2) and units of the structural system have been fabricated.[11.3] Special node joints have been designed to enable the structure to be collapsed and stored in the cargo bay of either Space Shuttle or Ariane and then to be deployed at low earth orbit; it would then be transported to geostationary orbit by either a rocket or space tug. Fig. 11.3 shows a half size model of part of the University of Surrey's tetrahedral reflector manufactured from graphite fibre and polyethersulphone matrix under test in a 3·5 m solar simulation chamber.

Fig. 11.3. Model of part of the tetrahedral reflector under test in a 3·5 m solar simulation chamber

11.10. Polymers for reinforcing and prestressing concrete

Over the last 100 years, concrete has become a major structural material: it is relatively cheap; it can be cast into almost any shape; and, provided that a certain care is taken during the detailing stage in design and also on site, it is a durable material.

Its drawback is of course the very low tensile strength, typically about one tenth of the compressive strength. Thus, with the exception of certain mass concrete structures such as dams and arches, alternative provision must be made to carry the tensile stresses. Traditionally, two methods have been used, both of which use steel. To appreciate the problems associated with the potential use of polymers, it is necessary to look carefully at the way steel is used.

Reinforcing bars can be embedded in the concrete at places where tensile stresses are expected (see, for example, reference 11.4). Typically, a reinforced concrete beam in flexure will have about 1·5 per cent of its cross-sectional area replaced with steel near the tension face, with additional steel placed in other areas to carry shear stresses. Reinforcing steels have working stresses of about 250 MN/m^2 so the strain in the bars is about 0·00125. For normal concrete, with a strength of 30 MN/m^2 the working stress will be about 10 MN/m^2 and the strain, for long-term loads, will also be about 0·0012 (Fig. 11.4). The neutral axis of the beam is thus at about mid depth, and both materials are acting at their full working strength at strains (and hence curvatures) which do not cause problems.

As an alternative to reinforcing the concrete, it may be precompressed by the application of an external force.[11.5] This force puts the concrete into compression, which is then able to resist applied tensile stresses (Fig. 11.5). The force is generally applied by a jack and is reacted by a steel tendon passing through a duct in the concrete. It is the force that is important at the working load, not the tendon itself; the force puts the concrete into such a state of stress that the concrete is able to resist tensile loads. To ensure that some prestress remains after creep of the concrete has taken place, the tendon must have a large strain; with the fixed modulus of steel, this requires high stresses in the steel. It was the recognition of this by Freyssinet which allowed prestressed concrete to become a practical engineering material.

The use of steel as the tensile element in both reinforced and prestressed concrete raises problems of durability. Steel rusts; the high pH of concrete passivates the steel, but carbonation by atmospheric CO_2 may eventually reduce the alkalinity of concrete, which allows corrosion to take place if the concrete gets wet. The quality of the concrete can be improved, but to reduce the possibility of corrosion, the steel is embedded to a considerable depth in

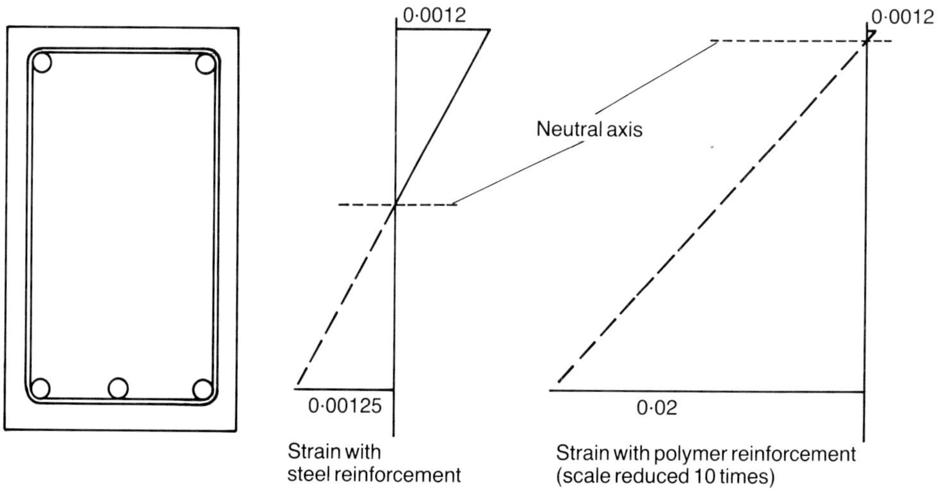

Fig. 11.4. Typical strains for concrete and various types of reinforcements

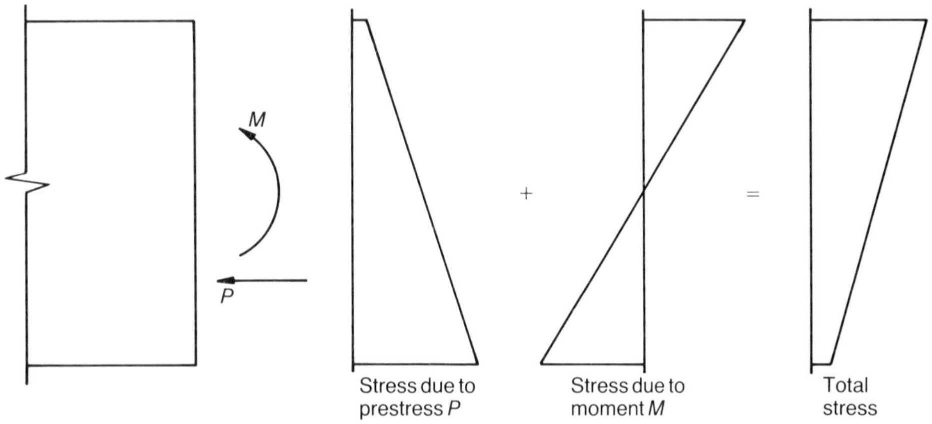

Fig. 11.5. Stresses in prestressed concrete

the concrete. This increases the cross-sectional area, thus increasing weight, and often also reduces the effective depth of the steel, thus reducing structural efficiency. A few attempts have been made to prestress bridges with external cables, more particularly in France[11.6] but much care has to be taken with the corrosion protection and this is not always successful.

Various attempts have been made to overcome corrosion problems, and still allow the use of steel. Galvanizing the steel is not a long-term solution, since the zinc coating is sacrificial and once it has corroded the steel is unprotected. Galvanizing also encourages hydrogen embrittlement in prestressing tendons. Stainless steel is used for both reinforcing[11.7] and prestressing, but is expensive and can suffer from stress corrosion in certain circumstances. Epoxy-coated reinforcing steel[11.8] is widely specified in North America, where bridge decks are not usually waterproofed, and epoxy-coated prestressing tendons[11.9] are also available. The bars are factory coated in epoxy, which then has to be made good on site if the coating is damaged.

Increasingly, cathodic protection is being considered as a method of inhibiting chloride-induced reinforcement corrosion.[11.10] Sacrificial anode systems may be appropriate for buried or submerged structures, but impressed current systems are generally necessary. Cathodic protection has been used extensively to protect bridge decks in North America, but there are relatively few examples of its full-scale application to other types of structure, and continuous monitoring is required to ensure that the system remains effective.

None of these systems fully meets the requirements of a non-corroding reinforcing bar or prestressing tendon; the potential new materials will be considered below.

11.10.1. Reinforcement

Various materials have been considered for use as non-corroding reinforcing bars. These include polyethylenes (PE),[11.11,11.12] polypropylenes (PP)[11.13] and glass fibres.[11.14] Polyesters[11.15] may also usefully be considered.

They all suffer from one primary drawback, however, which comes from the relatively low modulus of the materials. Fig. 11.6 shows the short-term stress–strain curves for typical examples of these materials, with high-yield reinforcing steel for comparison. Although the new materials have appreciable strengths, these strengths are only achieved at high strain levels. Thus, if concrete is reinforced with these materials, significant loads can only be carried once the concrete has undergone severe cracking. It may be argued that since the reinforcement does not corrode, such cracks do not matter, but they reflect a significant loss of stiffness in the structure, and hence a lack of serviceability.

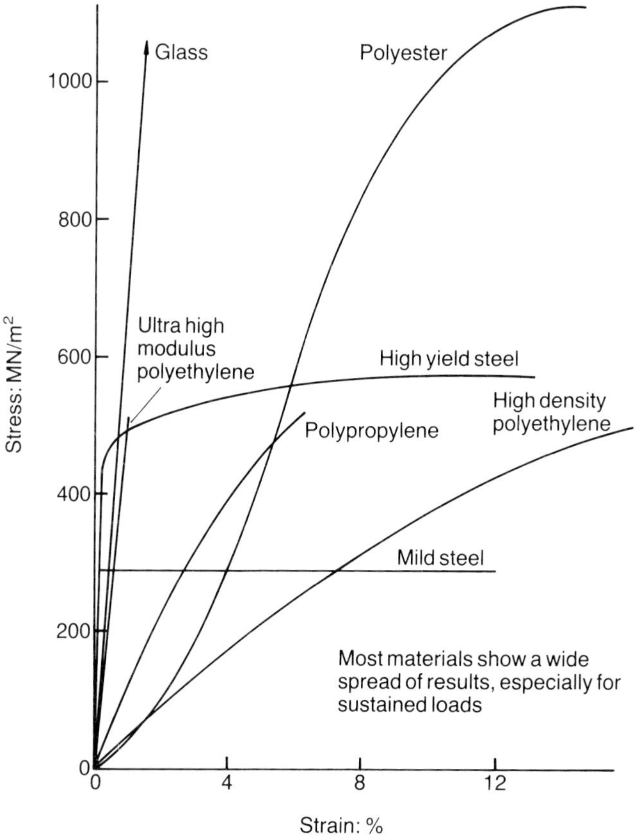

Fig. 11.6. Short-term stress–strain curves for actual and potential reinforcing materials (data taken from references 11.15, 11.19 and 11.24)

There are, however, additional problems. Glass fibres, after promising early uses, have been found to be susceptible to attack by the high alkalinity of concrete. Although alkali-resistant glasses are now available, their use is restricted to non-structural elements such as cladding and permanent formwork. These are widely covered elsewhere[11.16] and since glasses are not, strictly, polymers, they will not be discussed further here.

Most of the true polymers referred to above suffer significantly from creep and the effects of elevated temperature. Thus, sustained loads cannot be carried without significant increases in deflections, and temperatures must at all times be kept close to ambient. Polyethylene melts at 115°C and polypropylene at 175°C, but both lose stiffness significantly at much lower temperatures.

The combination of these effects means that, in the Author's opinion, there is no alternative to steel for reinforcing concrete for *structural* applications at present. Nevertheless, new materials are being developed all the time and the market for an alternative would be enormous, so it is perhaps worthwhile identifying the properties that such a material would need.

It must not corrode, either in the presence of water which might penetrate the concrete, or due to the alkalinity of the concrete. Steel fails the first test, if the passivating effect of the concrete is destroyed by carbonation, and glass fails the second.

The material must achieve a significant proportion of its tensile strength at strains in the range 0·0025–0·005. If the material cannot achieve this, either the deflections in the structure will be very high, or a significant proportion of the material's strength will be unused. Most existing polymers do not meet this criterion.

The strength must be at least 400 MN/m^2 otherwise the proportion of the cross-section taken up by the reinforcement becomes significant, and congestion of the reinforcement becomes a major problem.

The material should not creep significantly. A small amount of creep would be acceptable in most cases (say 10%–20% increase in strain for long-term loads), but creep factors of 3 or 4 for the reinforcement, when combined with factors of 2 or 3 for the concrete itself, are unacceptable.

The material should not soften at temperatures below 200°C, and ideally, should retain significant strength at higher temperatures.

The reinforcement must be capable of bonding to the concrete in order to achieve load transfer, and the bond mechanism itself must satisfy the same creep and temperature criteria as the material itself. This precludes the use of most resin formulations.

The material should be ductile. In reinforced concrete, the reinforcing material has the same strains locally as the surrounding concrete. Thus, there is no possibility of redistributing the high strains that occur in the flexural failure zone (as there is with unbonded prestressing tendons). Thus, to avoid brittle failure, higher partial safety factors would need to be employed for brittle reinforcement than are currently used for ductile steel.

Despite the fact that none of the existing materials is ideal, there is one material which seems to be gaining acceptance in a limited way for reinforcing concrete, primarily for nonstructural applications. This is polypropylene, either in the form of fibres, nets or bars.

Polypropylene is produced as an unoriented thermoplastic, with relatively low strength. It can be drawn into an elongated form, which increases both the strength and stiffness in the direction of the orientation. There is a wide variation both in the precise formulation of the material, and the final properties that are achieved by the drawing process. Four versions are described below which give an illustration of the properties that can be achieved.

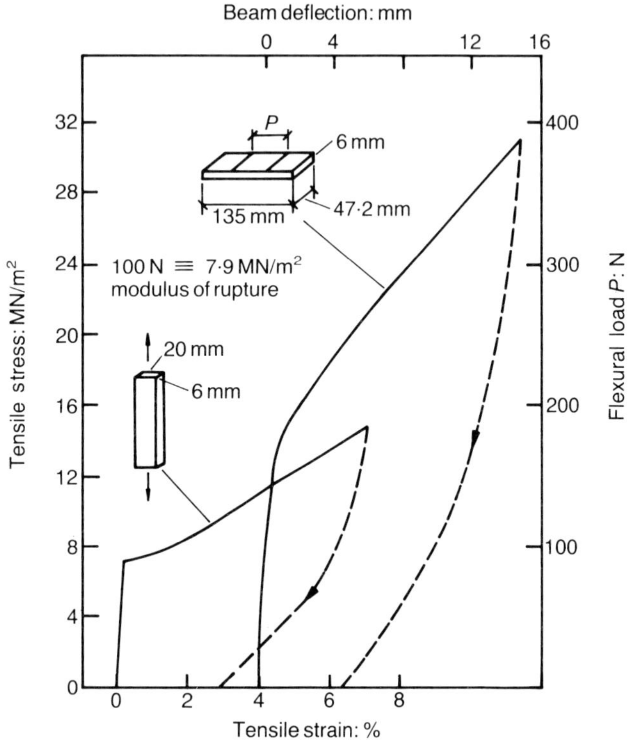

Fig. 11.7. Tensile stress–strain curve and load-deflection curve in flexure for a composite containing 5·7% by volume of fibrillated film networks parallel to direction of stress. 100 N (flexural load) = 7·9 MN/m^2 (flexural stress) (courtesy Royal Society)[11.20]

Fibermesh (which is a trade name of Fibermesh (UK) Ltd) is produced as fibrillated polypropylene, which is then cut into small pieces. The pieces are added to the concrete during mixing, and are broken down into their constituent fibres by the subsequent agitation. The result is a mix containing many individual fibres, each about 20 mm long, with a rectangular cross-section of about 0·5 mm by 0·06 mm. The fibres have an elastic modulus of about 3·5 GN/m^2 and a strength of about 600 MN/m^2. Typical mixes would contain about 0·1% of fibres by volume.[11.17]

The fibres are not primarily intended to contribute to the long-term strength of the concrete;[11.18] they should preferably be regarded as an admixture which serves to resist the short-term shrinkage strains that ocur during the first few hours of curing. However, significant improvements have been quoted for the impact resistance of hardened concrete containing polypropylene fibres.[11.19]

Netcem (a trade mark of Netcem Ltd) uses a material similar to Fibremesh, but the net is not chopped into pieces; the fibres are not separated, and much higher fibre fractions (typically 6% by volume) are used. The result is a cement-based thin sheet composite containing layers of continuous networks of fibrillated polypropylene film. A 6 mm thick sheet may contain between 40 and 60 individual layers of net, each about 0·08 mm thick. Tensile stress–strain curves and flexural load-deflection curves are shown in Fig. 11.7; ultimate strengths of more than 16 MN/m^2 in tension and 32 MN/m^2 in flexure are possible.[11.20] The mechanical properties are characterized by great toughness with closely spaced (1 to 5 mm) multiple cracks after matrix failure. The composite is marketed as flat or corrugated sheeting for roofing and cladding applications as an alternative to asbestos cement.

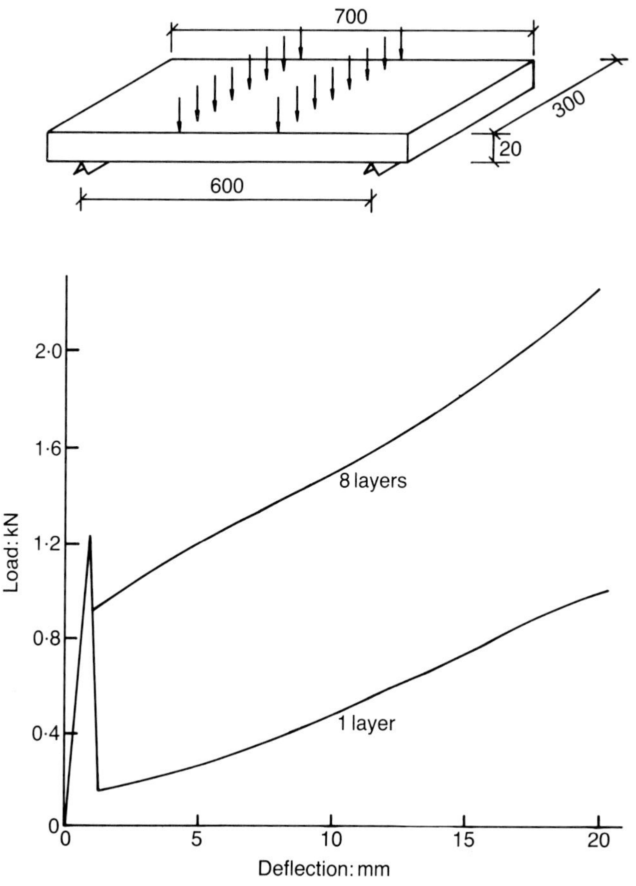

Fig. 11.8. Load-deflection response of slab reinforced with layers of Netlon[11.21]

Another form of polypropylene reinforcement, this time using much larger components, is the Netlon system (marketed by Netlon Ltd). The manufacture of this system (by the Tensar process) has been discussed in section 2.6.1 and its use for soil reinforcement has been described in section 10.3.2.2. It has also been proposed for use as a reinforcing material. When the polypropylene sheets have been punched into regular grids of holes and mechanically stretched it is possible, by varying the spacing of the holes and the degree of stretching, to produce nets with different properties in the longitudinal and transverse directions. Strips of orientated thermoplastic polymer may also be produced by stretching the material in one direction only.

The strips of polypropylene are not of uniform cross-section, being thickest where the strips intersect. Strengths of up to 500 MN/m^2 have been achieved in certain grids at the thinnest sections, where the amount of drawing has been largest, but 180 MN/m^2 is a more typical figure.

Attempts have been made to produce structural elements using these grids, but tests carried out only serve to show the limitations of the material. Swamy[11.21] has made flat slab elements 700 mm by 300 mm, which were reinforced with layers of Netlon grids. The slabs were 20 mm thick, and each layer of Netlon was 3 mm thick at the maximum point; layers were offset to minimize overall thickness. The slabs were then supported on two opposite sides 600 mm apart and loaded at the third points to produce uniaxial bending.

Typical load-deflection curves are shown in Fig. 11.8. The curves exhibit two distinct phases. On first loading, the concrete is initially uncracked, and is carrying significant tensile stresses. When the limiting tensile strain is reached, which will be between 100 and 200 microstrain, the concrete cracks. At this strain however, the Netlon is stressed only to 1 or 2 MN/m^2 so that a significant drop in load occurs. The load can then be increased again, with the Netlon providing the tensile resistance, but only at very large strains, which are reflected in large deflections. Two curves are illustrated; one shows the resistance with 1 layer of Netlon, the other with 8 layers of Netlon. The most heavily reinforced element eventually reaches a load higher than that of the initial uncracked peak, but at a deflection of about 1/30 of the span.

From this test it can be concluded that it is difficult to provide sufficient reinforcement by means of these grids to contribute significantly to the strength of the concrete. However, the material is extremely ductile, so it may be useful in situations where the ability to resist damage due to short-term impact loads is important.

Use has been made of polypropylene reinforcement using this system for floor slabs[11.22] in which the mesh serves to distribute cracking and prevents the slab breaking up into pieces. However, it is difficult to see how the material can serve any significant structural purpose, since with such a low modulus, integrity is lost before the polymer is carrying significant load.

Another potential use for the material is in repair. Where an additional layer is required on top of existing concrete there are difficulties in ensuring that the fresh concrete remains in place. Steel wire mesh is often used as an armature to support the fresh concrete. Although the mesh has no subsequent structural purpose, it can corrode and cause spalling of the repair concrete. Netlon has been used in this way for the repair of bridges in the West Midlands.[11.23]

The final example concerns an investigation of the use of polypropylene as bar reinforcement by Gowripalan.[11.24] Polypropylene reinforcement was drawn from billet to achieve strengths in tension of 550 MN/m^2 with a short-term modulus of about 12 GN/m^2. The bars were rolled through a mill to produce indentations on the sides of the bar, which enhanced the bond characteristics, and investigations were made of the properties of both the bars themselves, and of reinforced concrete beams and columns made using the reinforcement.

Perhaps the most significant tests carried out were the creep and recovery tests on polypropylene bars in tension, since these results will be reflected directly in the creep properties of concrete reinforced with the bars.

Tests carried out by subjecting the bars to tensile stresses of $20\,MN/m^2$ (i.e. about 4% of the strength of the bars) for 90 days gave a total strain of 4150 microstrain, equivalent to a creep modulus of $4{\cdot}8\,GN/m^2$; 88% of the total strain was recovered 90 days after unloading. At higher loads, much larger creep strains were observed. At stresses of about 50% of the strength, creep strains of about 3% were noted 3 days after loading. Other studies on polyethylene[11.25] and polypropylene[11.26] have observed critical stresses above which the material undergoes creep at constant rate until failure. The tests quoted above did not display this phenomenon, but it is not clear whether the tests were carried out for sufficient time for the effect to become apparent.

Additional tests were carried out by Gowripalan on beams, axially loaded columns and columns with eccentric loads. Broadly speaking, the results are as expected; in compression, the polypropylene has virtually no effect, since its modulus is very similar to that of the concrete it is displacing. In tension, the bars are effective for carrying load in the short term, but lose stiffness significantly when long-term flexural loads are applied to the structures.

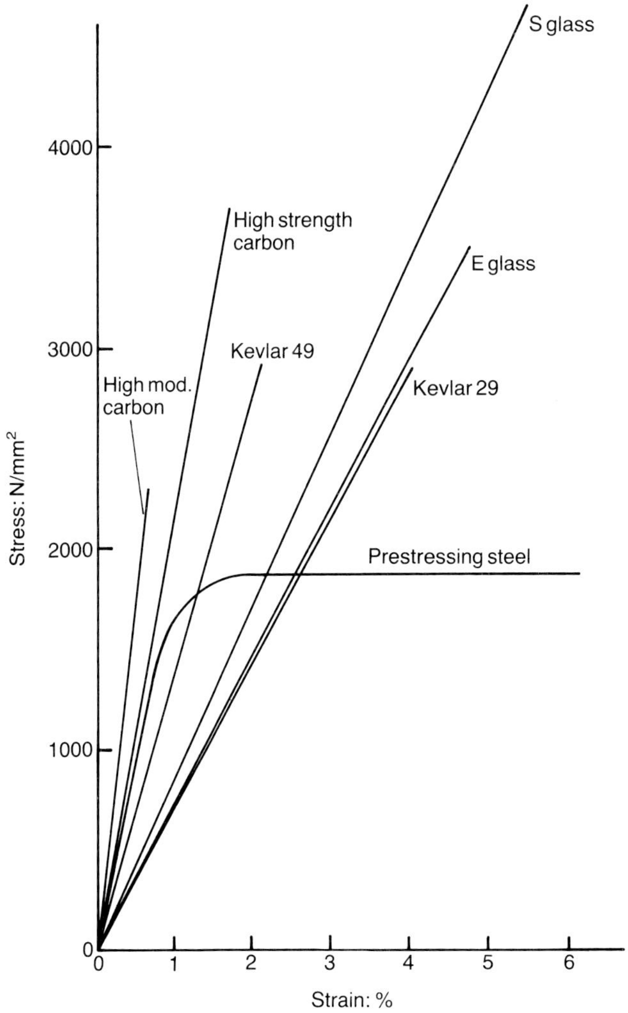

Fig. 11.9. Typical stress–strain curves for fibres that could be used for prestressing

The study concluded that the material is useful in areas where the loading is predominantly compressive, but where only occasional short-term or transient flexural loads are applied. This virtually precludes the use in beams, but may point to specialized use in columns subject to occasional transverse loads, especially when corrosion is a particular problem. Precast concrete piles were particularly identified, since these are subjected to tensile stresses during lifting, and during driving, both of which are known to be of short duration.

It may thus be concluded that the time for polymers as a general replacement for steel reinforcement has not yet arrived, but that there is a fortune to be made if the correct material is invented.

11.10.1.1. Prestressing tendons

The prestressing of concrete is a much more fruitful field of application for new materials. Prestressing is often viewed as a means of giving concrete some ability to resist tensile stresses: at the uncracked stage, that is valid. However, prestressing can also be regarded as a means of allowing, indeed requiring, the use of materials with a high strain capacity.

By taking out some of the stretch of the tendon during the prestressing exercise, the use of high strength but lower modulus materials becomes feasible. Concrete *prestressed* with cold drawn steel wire can be loaded so as to yield the steel without causing excessive deflections in the concrete, whereas concrete *reinforced* with such steel would undergo very large strains before the full strength of the steel was utilized. In a similar way, considerable strains can be applied to polymer tendons during the prestressing process, which thus allows them to be used economically for structural applications.

Prestressing tendons are subject to very high permanent stresses. Thus, resistance to creep becomes of paramount importance. Similarly, an ability to apply the prestressing force and to anchor the tendon must also be provided.

Three materials have been suggested as suitable materials for prestressing tendons: glass fibres, carbon fibres and aramids.[11.27] Typical stress–strain curves for bare fibres of these materials are shown in Fig. 11.9. Glass has the lowest modulus, but has the advantage of being cheap; carbon fibres are made in many grades: some have high strength and relatively low modulus, while others have lower strength and higher modulus. Aramids are available in a variety of forms which have similar strength but different modulus. The original aramids were developed by Du-Pont under the name of Kevlar; these fibres have been discussed in chapter 2. Similar, though not identical, materials are available under the names of Twaron (from Enka) and Technora (from Teijin).

Prestressing tendons utilizing both glass fibres and aramids are now commercially available, but the Author is unaware of any attempts to produce prestressing tendons from carbon fibres, which are the most expensive of the three. An advantage of carbon fibres over aramids in composite applications is the ability to resist compressive stresses as well as tensile stresses. These do not arise in prestressing tendons, so there is little advantage in the use of carbon fibres, and so they will not be considered further here.

The individual fibres must be aggregated to form tendons. Conventional laid ropes, which maintain their integrity by twisting together many fibres, are not suitable, since the individual fibres follow helical paths along the rope. These act like springs when stretched, and the axial stiffness of the rope is very low by comparison with that of the constituent fibres.[11.28] A parallel alignment of fibres is needed, which requires some method of holding the filaments together. This can be done by embedding them in a resin matrix, (as in the pultrusion technique, which is discussed in chapter 2) or by enclosing them within a sheath.

11.10.1.2. Anchorage

The ability of an element to carry significant tensile forces is only as good as the mechanism for getting the force into, or out of, the tensile member. If the range of methods used to load steel wires is considered, comparisons can be drawn between these and the methods used for fibres.

Prestressing tendons are anchored by external wedges, button heads, or bond.[11.29] Stay cables and wire ropes, on the other hand, which have more individual wires, are usually anchored by an internal spike, or by splaying out the wires and encapsulating the ends in a block of a low melting point metal (usually zinc).[11.30] Welding is not used because of the effect of the heat on the properties of the steel.

Which of these systems can work for elements made up from fibres? External wedges (Fig. 11.10(a)) develop high circumferential compressive fores between the outer wires in a strand, but have significantly less effect on

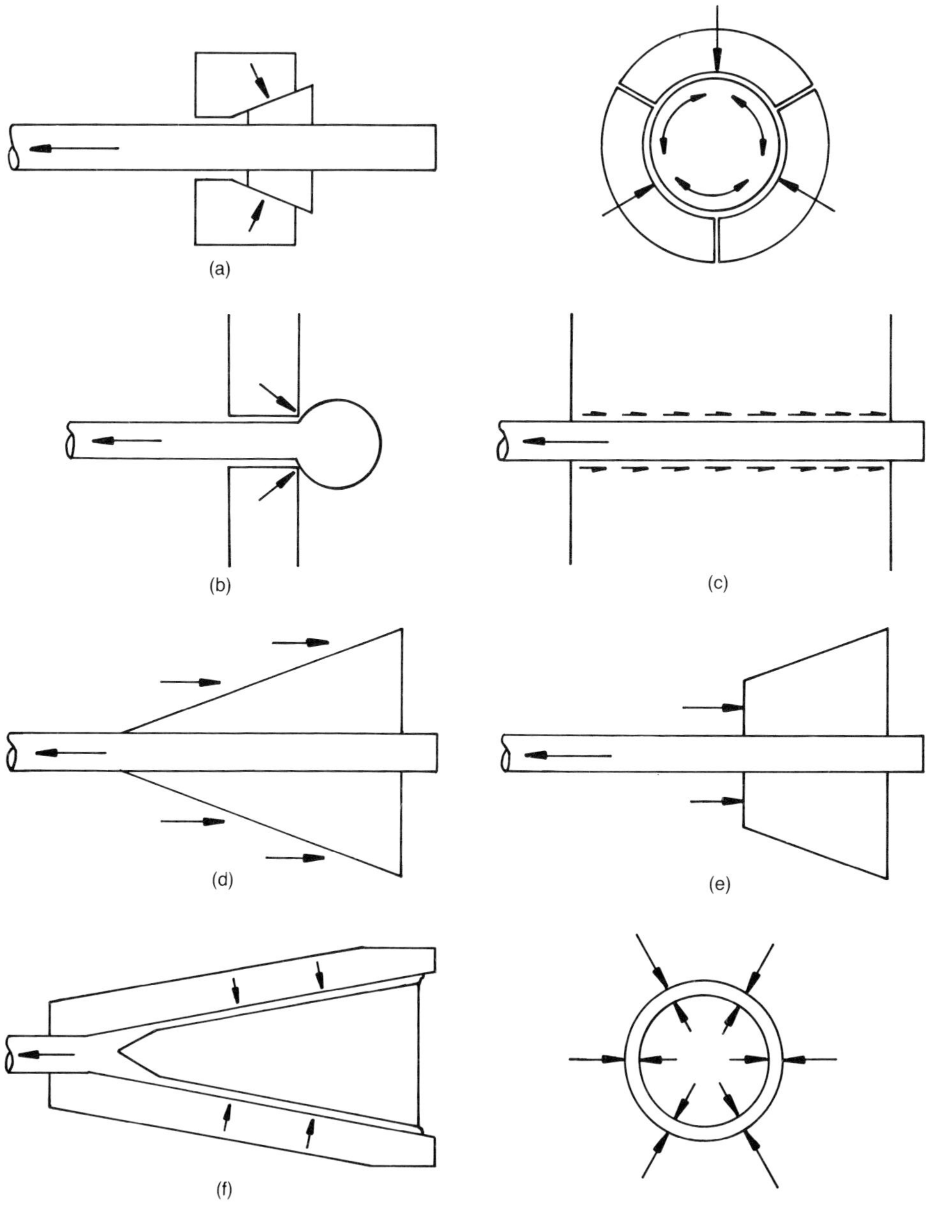

Fig. 11.10. Potential anchoring systems for composite prestressing strands: (a) external wedges; (b) button head; (c) bond; (d) cast cone; (e) truncated cone; (f) internal spike

the inner wires. With the seven wire steel strands usually used for prestressing tendons, this is not a major problem, but 19 wire strands, which have an extra layer of wires, are less popular for prestressing systems because of the difficulty of anchoring the central wire. Thus, the use of external wedges for systems containing millions of fibres is unlikely to be successful. Even in pultruded sections anchoring would be difficult because the anchoring loads of the inner fibres would have to be carried by high shear stresses in the matrix without the benefit of high radial compressive stresses.

Button-heads on steel wires (Fig. 11.10(b)), as used in the BBRV stressing system, rely on the deformability and the isotropic properties of steel. Modern oriented fibres have high axial strength, but little transverse strength, and they cannot be deformed, so a similar system cannot be envisaged for fibres.

Tension elements can be anchored by the use of bond when the element itself has cohesion (Fig. 11.10(c)) and when some distribution of anchorage along the length is acceptable. Pultruded sections can be anchored in this way when used as pretensioning tendons, but the bond mechanism has the properties of the resin matrix, not those of the fibres. In particular, the creep and thermal properties will be significantly worse than those of the fibres. Anchorage by bond is only possible if the fibres within the rope are themselves bonded together. Tests[11.31] have shown that negligible load transfer capacity is possible on parallel lay ropes which do not have a resin matrix.

An anchorage can be formed by splaying the fibres out into a mould, and then casting a resin cone around them. The cone can be anchored mechanically as required, usually by bearing on the side of the cone (Fig. 11.10(d)) or on the front face of a truncated cone (Fig. 11.10(e)). As with the bond technique, the integrity of the anchorage relies on the properties of the resin, and the shape of the cone is crucial in ensuring even load transfer to all fibres.

The best method for anchoring parallel fibre ropes is the internal wedge (or spike) (Fig. 11.10). The gripping force is provided by radial forces between the spike and the external body, so all the fibres are anchored. The length of the spike can be chosen to ensure that the transverse stresses are within the capacity of the fibres. Fig. 11.11 shows a typical termination for parallel lay aramid ropes. Once the load has been transmitted from the fibres to the terminal body, further connection to the structure can be made by fitting a variety of connections, such as clevis pins, anchorage plates, or whatever is required. The figure shows such a terminal modified for use as a prestressing tendon, in which the terminal body has two threaded regions. The inner thread is used for connection to a pull rod which is attached to the jack during stressing, while the outer thread is used to provide a connection for a permanent back nut, which also allows some adjustment to take account of slack. The anchorage is capable of achieving the full strength of the rope. Reasonable care must be taken to ensure that the spike is fitted centrally within the rope, but otherwise no special skills are needed. Anchorages for aramid ropes with capacities between 1 and 1500 t have been provided using this system.

Fig. 11.11. Longitudinal section through a termination of a 60 t type G Parafil rope modified for prestressing

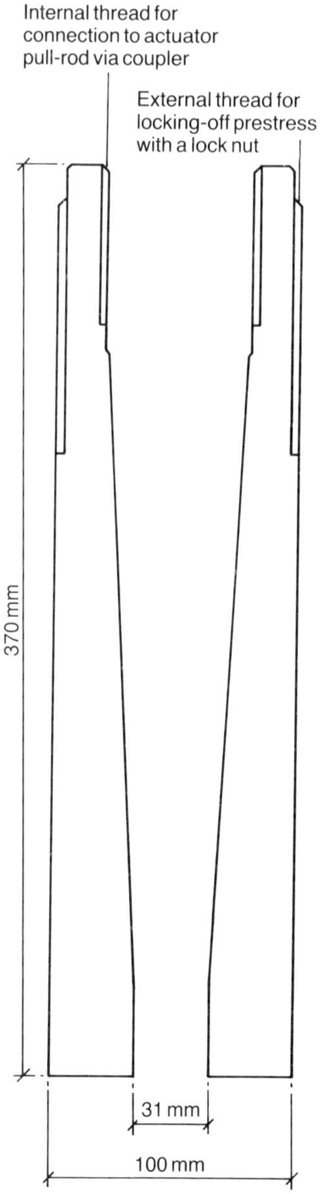

11.10.1.3. Polymer materials

Three materials are now commercially available for use as prestressing tendons. Polystal is a pultrusion of glass fibres contained within a resin matrix. Parafil is a parallel-lay rope which derives its strength from aramid yarns, while Arapree is a pultrusion of aramid fibres. The remainder of this chapter will concentrate on these three products.

Polystal. This is produced by Bayer AG in association with Strabag AG in West Germany; the product has been under development since 1978.[11.32] In essence, the tendon consists of bundles of bars or rods, each containing E type

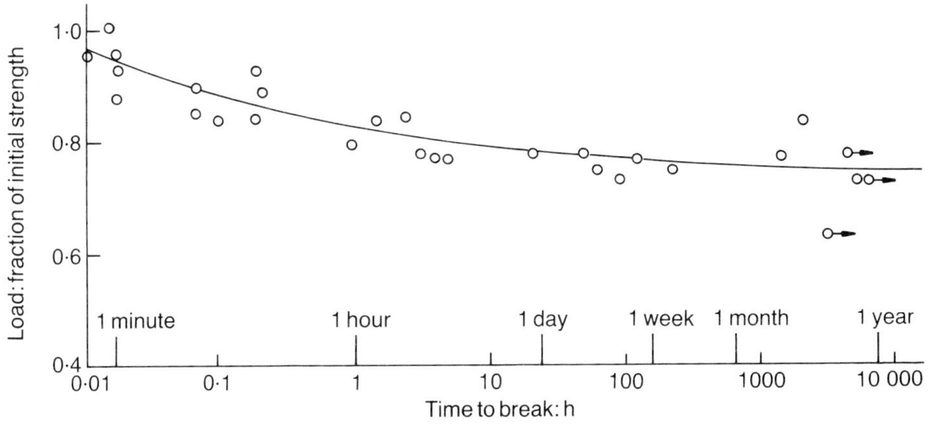

Fig. 11.12. Stress-rupture response of Polystal prestressing strand[11.34]

glass-fibre filaments in an unsaturated polyester resin matrix. A typical bar will be 7·5 mm in diameter, with the volume fraction of fibres being 68%; 19 of these bars will be grouped together to give a tendon with a working load capacity of 600 kN.

The anchorage of the tendons is either by a cast resin anchor, or by external wedges. Once stressed, the tendon is grouted in place by a resin-based mortar.

The Polystal bars have a modulus of elasticity of 51 GN/m^2 with a linear stress–strain curve to failure at 1520 MN/m^2. This corresponds to a strain at failure of 0·033, considerably higher than that of the other materials that are being considered.

Glass has long been known to suffer from stress-ageing, in which cracks develop at surface flaws and propagate through the thickness; indeed, early attempts to use glass-fibre tendons for prestressing[11.33] were not pursued for just this reason. Fig. 11.12, taken from reference 11.34, shows the time to break of Polystal, subject to loads of between 60% and 100% of the short-term breaking load. The published data appears to be limited to durations of about 10 000 h (about 1 year), and a certain amount of caution must be expressed about longer term extrapolations.

These bars are intended for use in grouted tendons, and the behaviour of beams prestressed in this way at the ultimate must be considered. Glass fibres are brittle, so to prevent sudden snapping of the tendon at high beam curvatures either the bond mechanism or the resin matrix must allow a significant degree of slip to occur at high load levels to accommodate the high strains. Alternatively, strains in the concrete at high loads must be kept small, so that the tendon is not approaching its ultimate load.

Considerable research and development have been undertaken in West Germany on the use of these strands for post-tensioning concrete bridges.[11.35] The first prestressed concrete bridge to be built using Polystal tendons was a small 7 m span footbridge in Dusseldorf which was completed in 1980 and was primarily used to demonstrate the feasibility of the prestressing system. The second bridge to be constructed using these materials was a continuous two span structure (die Brucke Ulenbergstrasse) of 21·3 and 25·6 m and was also built in Dusseldorf in 1986. The 15 m wide bridge consisted of a 1·44 m deep slab which was cast in situ with steel reinforcement. The slab was post-tensioned with 59 tendons, each made up from 19 glass-reinforced polymer rods of nominal 7·5 mm diameter anchored in a specially designed block and each tensioned to a working load of 600 kN.

Parafil. This is a rope produced by Linear Composites Ltd which has a core of parallel yarns contained within a polymer sheath. The core yarn can

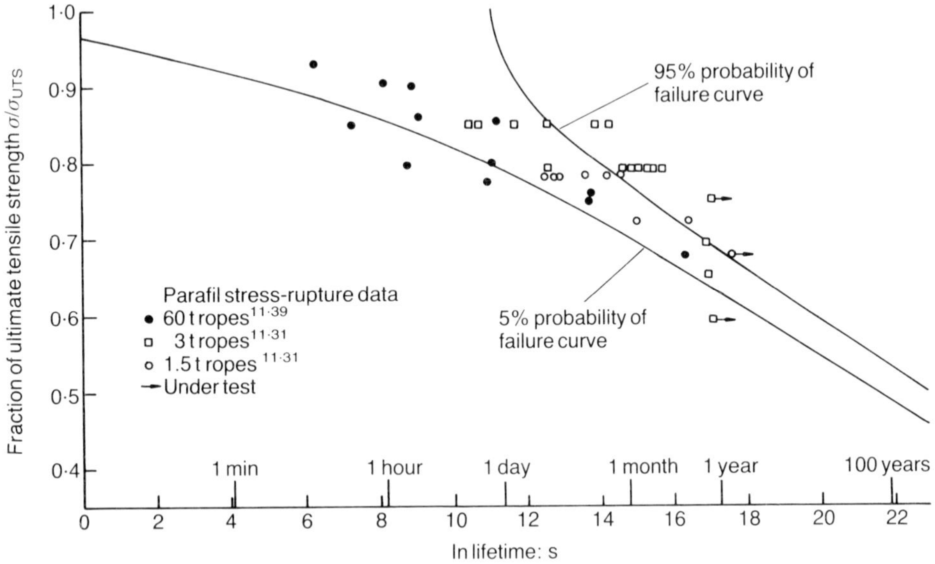

Fig. 11.13. Stress-rupture of type G Parafil ropes[11.31]

be of many different fibres, but the ones most frequently used are polyester (known as type A), Kevlar 29 (type F) and Kevlar 49 (type G). The lower modulus polyester ropes are used for guy ropes and some mooring applications, which is not applicable in this context.

The ropes are anchored by means of an internal spike which grips the fibres against the terminal body. This termination system has been developed after much testing, and can be used for ropes of all sizes (ropes have been made with breaking loads up to 1500 t). All the test results quoted here have been obtained with ropes anchored in this way. As discussed before, this system ensures that all fibres are anchored, and is not affected by heat or creep, which reduces the effectiveness of resin anchorages.

Further details of the properties and uses of Parafil ropes are given in references 11.36, 11.37, 11.38.

The most suitable rope for use as prestressing tendons is the type G rope, which contains Kevlar 49 as its core yarn. This is the stiffest of the available yarns and, although the losses caused by creep of the concrete would be reduced somewhat if the lower modulus versions were used, there are overall benefits from using the Kevlar 49 caused by the reduced jack extension and improved long-term properties of the fibres themselves. Studies[11.39] have shown that the total losses of force in a beam prestressed with type G Parafil are the same as those expected in beams prestressed with steel tendons.

The elastic modulus of Type G Parafil is about 120 GN/m^2 with a short-term ultimate strength of 1930 MN/m^2. As with Polystal, the ropes will creep to failure at high load levels; Fig. 11.13 shows this stress-rupture response. Extrapolations from short-term (up to 24 months) tests, combined with predictions based on the reaction rate theories of chemical processes, predict that a Parafil rope will sustain a load of 50% of the short-term strength for 100 years.[11.40] Applying a material factor of 1·5 to this load level gives a permanent sustainable working stress of about 650 MN/m^2. Work is currently underway to produce reliable cumulative damage laws which will allow the designer to take into account the reduction in stress that occurs in a prestressing tendon due to the creep of both the concrete and the tendon, thus allowing higher stresses to be used for the initial prestress.

Typical stress relaxation results are shown in Fig. 11.14. These are derived from tests carried out on 60 tendons that were stressed to various load levels

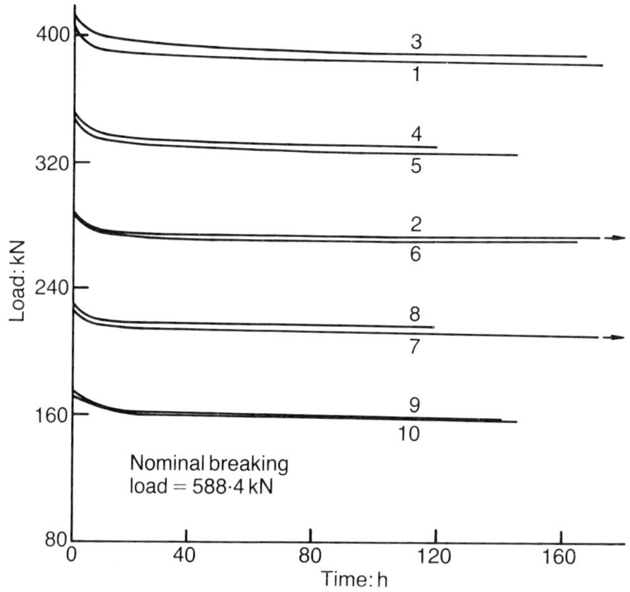

Fig. 11.14. Stress-relaxation response of 60t type G Parafil ropes from various initial loads[11.39]

against a steel reaction frame, and then locked into place mechanically.[11.39] Relaxation of load of about 8% of the nominal breaking load is predicted at high stress levels.

Test beams have been built using Parafil as the prestressing medium. The first, with an overall length of 5 m, was prestressed with a single 60 t rope, which passed through a straight duct on the centre line of the beam. The second was 8 m long, with two 60 t ropes placed outside the section. The tendons passed under deflectors at about the third points of the beam, so that the tendon was straight and near the bottom of the beam at mid-span. The anchorage was provided near the neutral axis of the beam in thickened end blocks. This beam was more representative of practical beams, since the resistance to corrosion means there is no need to place the tendons inside the concrete, thus saving weight and complexity. Fig. 11.15 shows the beam prior to testing, and Fig. 11.16 shows diagrammatically the stressing sequence that was adopted.

The beams were both loaded by two point loads applied near the centre of the beam, and both beams showed a ductile response to the applied load.

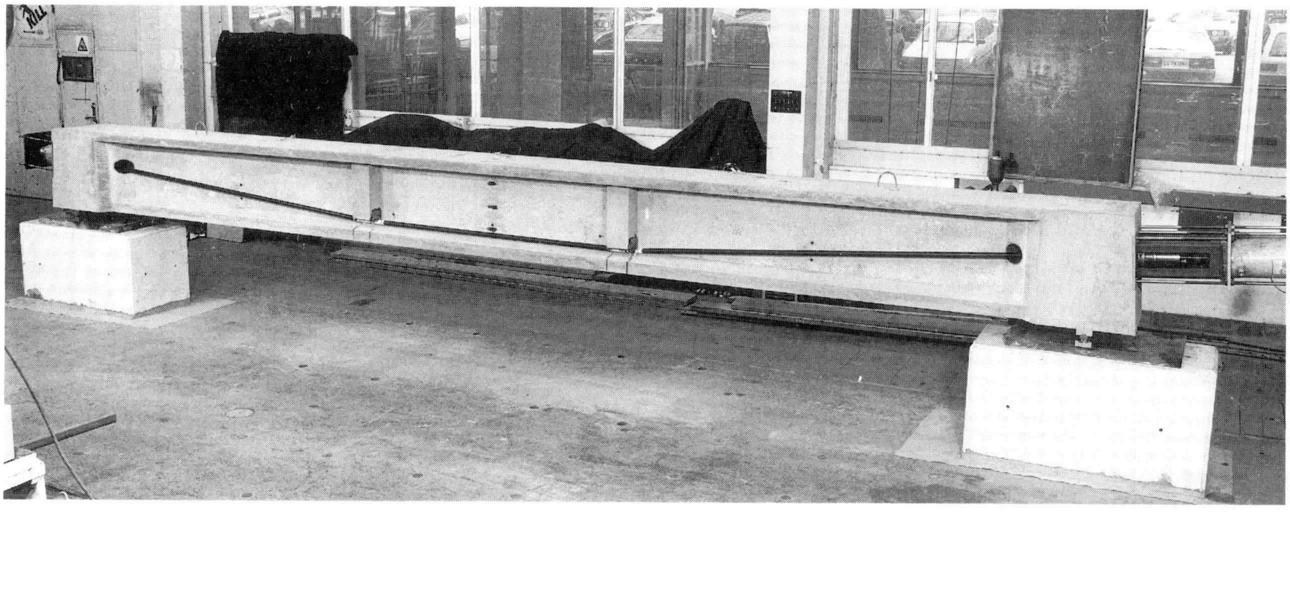

Fig. 11.15. 8 m beam externally prestressed with two 60 t Parafil ropes (each stressed to 30 t)

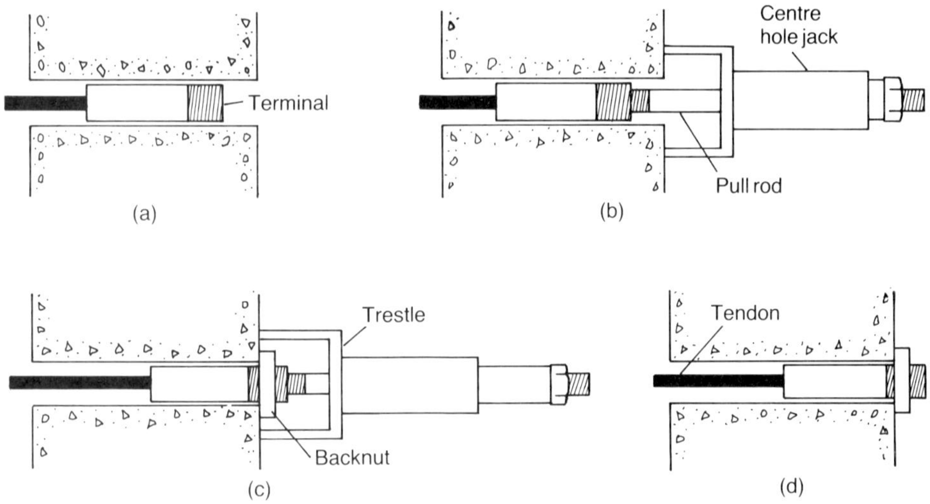

Fig. 11.16. Stressing sequence for Parafil prestressing tendons

When the tensile strength of the concrete is exceeded, cracks form, but the beam can carry significant increases in the load at reduced stiffness before final failure. Failure occurs by crushing of the concrete in the compression zone. Fig. 11.17 shows the load-deflection curve of one of the tests; the curve indicates that considerable energy is absorbed prior to failure, even though the two constituent materials are brittle.

One of the principal benefits of the use of Parafil tendons is that they can be used as external tendons, thus allowing significant weight savings to be achieved, since concrete does not have to be provided merely to give cover to steel tendons.

Parafil has been used for the repair of structures, and this is likely to be a field in which all noncorroding composite materials are useful. Many distressed structures can be reinstated by prestressing, which closes cracks and restores integrity. However, the absence of existing (or spare) ducts means that steel tendons are not suitable for use in these cases, so an alternative must be found. Three cooling towers at Thorpe Marsh Power Station were found to have

Fig. 11.17. Load-deflection curve for a 5 m beam prestressed with a single Parafil rope

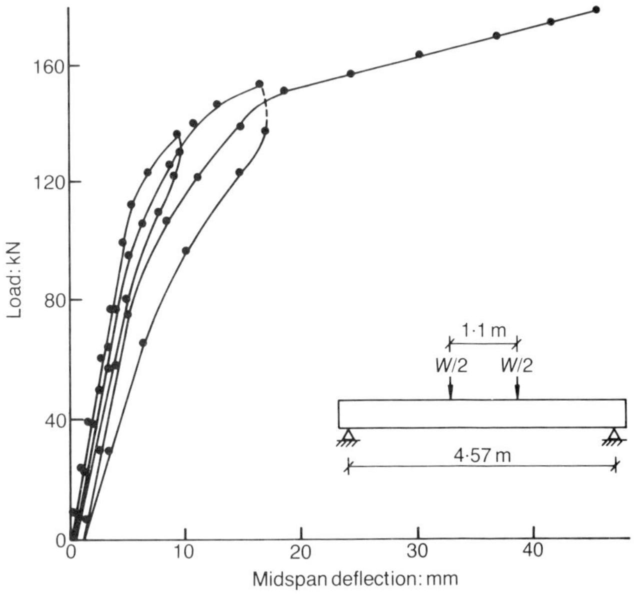

Fig. 11.18. Cooling towers at Thorpe Marsh Power Station repaired by external prestressing with Parafil ropes

large vertical cracks; these were resin-injected to fill the gaps, and then the towers were externally prestressed by circumferential Parafil cables, resting on stainless steel brackets bolted to the towers (Fig. 11.18). In this case, the low weight of the cables was also of benefit, since the whole prestressing system had to be assembled at the top of the towers by steeplejacks working from cradles.

Arapree. Arapree is a pretensioning tendon, produced by Enka, which is made in the form of pultrusions of the aramid fibre Twaron. The tendons rely on bond between the concrete and the pultrusion resin. Pretensioned planks have been made with these tendons[11.41] and the manufacture of railway sleepers has been proposed. The properties of Twaron are broadly similar to those of Kevlar, although higher relaxation figures have been reported for Arapree as compared with those for Parafil.[11.42]

11.10.2. Observations on the use of polymers for reinforcing and prestressing concrete

The materials discussed above offer the designer of concrete structures corrosion-free tendons, in the form of aramids or glass fibres, which are easier to handle on site because of greatly reduced weight. However, these tendons will have to compete with established steel tendon technology by demonstrating their improved long-term durability. Effective non-corroding reinforcement is possibly some way off, although the use of polypropylene to minimize early age shrinkage cracks and to provide an armature to support fresh concrete is now possible.

11.11. The future

The contents of this book have indicated that polymers have a considerable part to play in the construction engineering industries. The parts played by polymers cover a wide spectrum, from subterranean functions such as stabilization and reinforcement of soils, to surface structures involving composites, to extra-terrestrial systems involving the high technology composites; in all these functions the expertise of the civil/structural engineer must play an active role.

As new and improved manufacturing techniques for composites and new polymer materials are developed, even greater emphasis will be placed on the use of these materials for geosynthetics. In addition, as the architect and the structural engineer appreciate the advantages of composite materials in certain applications and the designer becomes more familiar with the analysis and design techniques for composite laminates, the use of composites in the construction industry cannot fail to increase.

In the design of composites it has been demonstrated that the material and structural designs must be undertaken simultaneously because they are both interdependent. This means that many interesting structural concepts which previously have been a figment of the engineer's imagination can be realized. Indeed some very exciting structural applications have been built in very recent times.

An ideal structural use for composites is the new "caretaker" bridge enclosure which was conceived by Maunsell, Consulting Engineers, London, as a membrane which could enclose the steel plate girders of composite or steel-decked bridges in order to provide access for inspection and maintenance, to protect the steelwork and bearings from corrosion and enhance the appearance of this type of bridge. In 1986, a detailed design of a GRP enclosure for the A19 Tees viaduct, the largest composite plate girder bridge in the British Isles, was undertaken. The design was carried out by Maunsell using the developed limit state design methods for composites (see chapter 4); the first panels were erected in 1988. This type of protective system is likely to be adopted for many bridge systems in the future.

The progress in the production and usage of high strength fibres in recent years have increased the potential for advanced structures and this has lead to the design, manufacture and evaluation of lightweight flexible tension members or long continuous lengths. If composite tension members are feasible the applications are enormous and include masts, cables to bridges, wind energy convertor stays, mooring lines for offshore platforms operating in deep water and long-span bridges where design requirements can only be obtained with materials of high specific strengths and stiffness.[11.43]

It has been claimed that suspension bridges with main spans of 4 km are possible with steel cables; but it is interesting to note that the Akashi bridge, with a main span of 1900 m, now under construction in Japan, has necessitated the development of special steels at the limit of existing technology;[11.44] 2 km seems a more practical limit for steel cables. Preliminary calculations have suggested that spans of over 8 km are possible using carbon-fibre composite cables. Richmond and Head[11.45] have suggested that maximum bridge spans have increased by a factor of 3·5 every 100 years since 1770; this would suggest a span of 5 km by 2070. Spans have risen in steps corresponding to the introduction of wrought iron in the 1800s, steel in the late 1800s and high strength steels in the 1930s.

For the future, man-made fibres and fibre-reinforced polymers provide the potential for the next step forward; however, so far innovations, which might have resulted in the use of these materials have not happened; it would seem that there is a time interval of some 30 to 40 years between the introduction of each new material and the engineer's acceptance and confidence in its use.

It is probable that new polymer composite materials for the construction industry would be manufactured from glass, carbon or aramid fibres in the primary phase. They can be bundled together in a protective sheath to form cables, embedded in polymer matrix to form bridge decks or beams and it has been shown in chapters 7 and 8 that the composite can meet most serviceability requirements such as fire retardance and resistance to ultraviolet radiation.

Composite strands made from glass, carbon or aramid fibres have substantially lower breaking strains than steel. However, the specific strength of the glass fibre strand and of the carbon and aramid fibre strand is about twice and four or five times that of steel respectively. Richmond and Head[11.45] have suggested that composites have immediate potential in the construction of bridge decks, with a weight saving, durability and aerodynamic form which could make the material cost-effective. One possibility for a deck system would be to use a sandwich construction in which the core would be of cellular units and the face materials would be steel. The cellular system would probably consist of glass-reinforced polyester composite I beams with or without foam infill.

The use of stress-analysis software packages has been mentioned in chapter 3. They invariably contain solutions to many strength-analysis problems met in the design of fibre-reinforced laminated composite structures. These will include failure criteria, plate vibration and buckling, analysis of bonded joints, stress concentrations, in addition to the calculation of basic stiffness and stresses including built-in thermal stresses. There are more than 600 finite element programs available to the structural engineer, but when the features are specifically addressed to the applications or structural analysis of composites the number of suitable software systems reduces to about ten. In future it will be necessary for all forward-looking engineers, who are engaged in analysis and design of composite structures and systems, to understand and to use the relevant stress-analysis packages.

It can be concluded that polymer composites and polymers do have a very large part to play in the construction industries and their contribution can only increase in the future. In addition, structural engineers need to develop expertise in the field of polymer composite associated with space, so that they can take advantage of the great potential waiting to be exploited.

References

11.1 Hollaway L. *Glass reinforced plastics in construction—engineering aspects.* Surrey University Press, Guildford, 1978.

11.2. Hollaway L. Pultrusion. In: *Developments in plastics technology.* **3**, Chapter 1, A. Wheelan and J.L. Craft (Eds). Elsevier, 1986.

11.3. Hollaway L. and Thorne A. Composite structural system for a large collapsible space antenna. *Composite Structures 4* **4**, chapter 1, I.M. Marshall, (Ed). Elsevier, London, 1987.

11.4. Park R. and Paulay T. *Reinforced concrete structures.* Wiley, New York, 1975.

11.5. Naaman A.E. *Prestressed concrete analysis and design fundamentals.* McGraw Hill, New York, 1982.

11.6. Virlogeux M. La precontraite exterieure. *Annales ITBTP*, 1983, No. 420, Serie Beton 219, 115–194.

11.7. Whiteley J.D. Selection of stainless steel for corrosion resistant applications. *Proc. Symp. Special steels and systems.* 59–71, Concrete Society, London, 1982.

11.8. Wills J. Epoxy coated reinforcement in bridge decks. SR667. Transport and Road Research Laboratory, Crowthorne, 1982.

11.9. Dorsten V. *et al.* Epoxy coated seven-wire strand for prestressed concrete. *PCI Jl*, 1984, **24**, No. 4, 45–51.

11.10. Wyatt B.S. and Irvine D.J. Cathodic protection of reinforced concrete. *Proc. UK Corrosion '86 Conf.* ICossST/CCEJV, Birmingham, 1986.

11.11. Capaccio G. and Ward I.M. Preparation of ultra-high modulus linear polyethylenes; effect of molecular weight and molecular weight distribution on drawing behaviour and mechanical properties. *Polymer*, 1974, **15**, 233–238.

11.12. Kamal M.M. *The reinforcement of concrete structures using high strength polyethylene.* PhD thesis, University of Leeds, 1983.

11.13. Naaman A.E. *et al.* Some developments in polypropylene fibers for concrete. *Proc. Symp. Fiber Reinforced Concrete.* ACI, Detroit, SP-81. 1984.
11.14. Allen H.G. *Glass-fibre reinforced cement—strength and stiffness.* Report No. 55, CIRIA, London, 1975.
11.15. Cook J.G. *Handbook of textile fibres* **2**, 5th edn. Merrow, Shildon, 1984.
11.16. Building Research Establishment. A study of the properties of Cem-Fil/OPC Composites, CP38/76. Building Research Establishment, Garston, 1976.
11.17. Dahl P.A. Plastic shrinkage and cracking of mortar and concrete containing Fibermesh. Report STF 65 A85039, SINTEF, Trondheim, 1985.
11.18. Correspondence in *Concrete*, November 1987 and subsequent months.
11.19. Hannant D.J. *Fibre cements and fibre concretes.* Wiley, Chichester, 1978.
11.20. Hannant D.J. and Zonsfeld J.J. Polyolefin fibrous networks in cement matrices for low cost sheeting. *Phil Trans. Roy. Soc. London.*, 1980, **A294**, 591–597.
11.21. Swamy R.N. *et al.* The behaviour of Tensar reinforced cement composites under static loads, *Proc. Symp. on Polymer Grid reinforcement.* 222–232, Institution of Civil Engineers, London, 1984.
11.22. Oliver T.L.H. and Morris M. Use of high tensile polymer grids in concrete. *Proc. 4th Int. Conf. on durability of building materials and components. Singapore*, 1987, 544–551.
11.23. Middleboe S. *New Civil Engineer*, 21 Feb. 1985, 4–5.
11.24. Gowripalan N. *Reinforcement of concrete elements with modified polymers.* PhD thesis, University of Leeds, 1987.
11.25. Wilding M.A. and Ward I.M. Creep and recovery of ultra-high modulus polyethylene. *Polymer*, 1981, **22**, 870–876.
11.26. Ward I.M. *Recent developments in the science and technology of ultra-high modulus polyolefins.* University of Leeds, 1983.
11.27. Mallet G.P. Prestressing steels and systems. In: *Developments in prestressed concrete*, **1**, F. Sawko (ed). Applied Science, London, 1978.
11.28. Ferer K.M. and Swenson R.C. *Aramid fibre for use as oceanographic strength members.* Report No. 8040. Naval Research Laboratory, Washington, 1976.
11.29. Andrew A.E. and Turner F.H. *Post-tensioning systems for concrete in the UK: 1940–1985.* Report 106. CIRIA, London, 1985.
11.30. British Standards Institution. *Sockets for wire ropes.* BS463. British Standards Institution, London, 1970.
11.31. Guimaraes G.B. *Parallel-lay aramid ropes for use in structural engineering.* PhD thesis, University of London, 1988.
11.32. Preis L. *Faserverbundwerkstoffe mit besonderer Zugfestigkeit, in Verbundwerkstoffe und Werkstoffeverbunde in der Kunststoffetechnik*, 127–143, VDI-Verlag, Dusseldorf, 1982.
11.33. Soames N.F. Resin-bonded glass-fibre tendons for prestressed concrete. *Mag. Concr. Res.*, 1963, **15**, No. 45, 151–158.
11.34. Rehm G. and Franke L. *Kunstharzgebundene Glasfaserstabe als Bewehrung im Betonbau.* DAfStb Heft 304. 1979.
11.35. Waaser E. and Wolff R. Ein neuer Werkstoff für Spannbeton. *Beton*, 1986, **36**, H7, 245–250.
11.36. Guimaraes G.B. Short term properties of Parafil. *Proc. Symp. Engineering Applications of Parafil Ropes.* Imperial College, London, 13–20, 1988.
11.37. Burgoyne C.J. Structural applications of Type G Parafil. *Proc. Symp. Engineering Applications of Parafil Ropes.* Imperial College, London, 39–48. 1988.
11.38. Burgoyne C.J. and Chambers J.J. Prestressing with Parafil tendons. *Concrete*, 1985, 12–15.
11.39. Chambers J.J. *Parallel-lay aramid ropes for use as tendons in prestressed concrete.* PhD thesis, University of London, 1986.
11.40. Chambers J.J. Long term properties of Parafil. *Proc. Symp. Engineering Applications of Parafil Ropes*, 21–28. 1988.
11.41. Gerritse A. and Schurhoff H.J. Prestressing with aramid tendons. *10th FIP Congr. New Delhi*, 1986.
11.42. Gerritse A. and Schurhoff H.J. Prestressed concrete structures with Arapree: Relaxation, *Proc. IABSE Symp.* Paris, 1987.

11.43. Meier U.J.S. Multiplication of the critical span of suspension bridges through the use of high performance composites. *Proc. 14th Reinforced Plastics Conference*, paper 40. British Plastics Federation, London, 1984.
11.44. *New Civ. Engr*, 1988, 4 Aug., 18–20.
11.45. Richmond B. and Head P. Alternative Materials in long span bridge structures. *Proc. 1st Int. Oleg Kerensky Memorial Conference, London*, 1988.

Glossary of terms used in composites

Accelerator: a highly active oxidizing material suspended in a liquid carrier in conjunction with an initiator to produce internal heat in a liquid resin to cure it.
Activator: see *Accelerator.*
Addition polymer: polymers formed into long-chain molecules by the chemical reaction of one or more types of monomer unit, each of which has a double bond prior to polymerization.
Air-inhibited resin: a resin in which surface cure is inhibited or stopped by the presence of air.
Catalyst: see *Initiator.*
Cavity: the space between matched moulds in which the laminate is formed; also used to describe the female mould half.
Condensation polymer: polymers formed by the chemical reaction of at least two monomer units with, in most cases, production of a by-product of low molecular weight.
Co-polymer: an addition polymer of at least two monomers.
Crazing: hair line cracks either within or in the surface of a laminate caused by cure generated stresses.
Critical length: the critical length of a fibre is the length which is required for the fibre stress to develop its maximum value when under a particular load condition.
Cross-linking: the setting up of chemical links between molecular chains.
Crystallite: the most rudimentary form of an embryonic crystal which can be identified as a certain species under the microscope.
Cure: the cross-linking or total polymerization of the molecules of a resin. Transformation of the liquid resin to the solid state.
Cure time: the time required for the liquid resin to reach a cured or fully polymerized state after the initiator has been added.
Degree of polymerization: an impression of the length of the average molecular chain in a polymer; it is assessed by an estimate of the average molecular weight and usually stated in terms of the number of repeating units in a chain.
Delamination: the separation of a laminate along the plane of its reinforcements.
Dimensional stability: the ability to retain constant shape and size.
Exotherm curve: a chart of temperature versus time made during the curing cycle of a resin mix. Peak exotherm is the point of maximum temperature recorded.
Exotherm heat: heat given off during a polymerization reaction by the chemical ingredients as they react and the resin cures.
Fibre: any material in an elongated form such that the ratio of its minimum length to its maximum average transverse dimension is 10 : 1, its maximum cross-sectional area $1{\cdot}975 \times 10^{-3}\,\text{mm}^2$ (corresponding to a circular cross-section of 0·25 mm diameter) and its transverse dimension is not greater than 0·25 mm.

(*a*) Filament: a continuous fibre.

(*b*) Yarn on tow: a number of filaments in a bundle which can be handled as a single unit. A tow is usually bigger than a yarn, having thousands of filaments, whereas a yarn usually has a few hundred filaments. A yarn may be spun and twined from staple fibre, but a tow is formed from constant filaments.

(*c*) Strand: a strand is a bundle of 204 filaments of glass fibre. The diameter of a filament is up to 1/400 mm.

(*d*) Chopped strand: these are made from continuous strands which are chopped into short lengths (usually 50 mm).

(*e*) Chopped strand mat: these are chopped strands which are held together by means of a size. The strands are completely randomly orientated and the mats are of uniform thickness. They are produced in various sizes, the most usual being 300 g/m^2 and 450 g/m^2.

(*f*) Woven rowings: these are continuous strands which may be unidirectionally or bidirectionally orientated.

(*g*) Woven cloth: this is a more refined product than above. It is usually a bidirectional reinforcement.

(*h*) Continuous fibres: fibres which are continuous throughout the whole length of the laminate, resulting in the load being applied directly to them; the stress throughout the length of the fibre is constant.

(*i*) Discontinuous fibre reinforced plastics: plastics whose reinforcing fibre have length-to-diameter (l/d) ratios (known as aspect ratios) varying between 100 and 5000. Whiskers reinforcement have l/d ratios between approximately 150 and 2500. Discontinuous glass fibre reinforced plastics which consist of premix thermosetting resins, chopped strand mat reinforced thermoset resins, and fibre reinforced thermoplastic resins have fibres with l/d ratios between 150 and 5000. The ultimate strength and modulus of short fibre reinforced composites can approach the values for continuous fibre composites providing that the short filaments can be aligned unidirectionally and that their length is much greater than the critical length (l_c) required for shear stress transfer.

Filler (*inert filler*)*:* an inorganic material added to resins to extend volume, improve properties, and reduce costs of the moulded part.
Gel: a partial cure of a resin; a semi-solid jelly-like state.
Gel coat: a protective surface coating of clear or pigmented resin.
Glass transition temperature: the temperature at which a sudden change in slope of various physical properties versus temperature curves occurs (commonly measured in terms of the standard heat distortion temperature). It very nearly approximates the temperature below which a polymer fails in a brittle manner and above which it behaves as a leathery or rubbery solid.
Gel time: the time required to change a flowable resin to a nonflowable gel condition.
Hardener: see *Initiator*.
Inhibitor: a material which retards polymerization of the monomer and increases the gel time of the resin mix.
Initiator: a material which produces free radicals to begin the cure of a resin. Frequently referred to as a catalyst.
Mould release: a material used to coat the mould to prevent the cured laminate from sticking to the mould.
Monomer: the monomer is the low molecular weight starting material from which the polymer is formed.
Orthophthalic resin: an unsaturated polyester resin of which phthalic anhydride is the starting point.
Plasticizers: materials deliberately added to polymers to reduce their stiffness.

Polymerization: the chemical reaction involved when, say, a liquid polyester resin sets to a solid. A comparatively simple chain molecule becomes a highly complex three-dimensional one.
Porosity: the formation of undesirable clusters of air bubbles in the surface of a laminate.
Preform: a process in which glass mats are formed over a perforated screen shaped like the mould in which the part will be made.
Prepreg: an intermediate product consisting of fibres of tows which have been coated with a matrix material such as resin. The fibres are aligned in the majority of cases to give a flat or tape. Usually the resin is not fully cured so that the aggregate remains flexible and the sheet can be built up in piles to form a composite.
Release agent: a lubricant, often wax, used to prevent the adhesion of the moulded part to the mould.
Shelf life: the length of time an uncatalyzed resin or resin mix remains workable while stored in a tightly sealed container.
Styrene monomer: a water-thin liquid monomer used to dissolve polyester resins and act as the cross-linking agent.
Thickener: a material added to a resin mix to chemically thicken the mix by raising its viscosity so that it will not flow as readily.
Thixotropy: the property of being a gel at rest and liquid upon agitation.
Under cut: negative or reverse draft on a mould.
Viscosity: a measure of the resistance to flow of a liquid.
Wet out: the ability of a resin mix to quickly saturate the reinforcement.